Environmental Issues and Waste Management Technologies in the Ceramic and Nuclear Industries IX

T0328414

Journal of the American Ceramic Society

www.ceramicjournal.org

With the highest impact factor of any ceramics-specific journal, the *Journal of the American Ceramic Society* is the world's leading source of published research in ceramics and related materials sciences.

Contents include ceramic processing science; electric and dielectic properties; mechanical, thermal and chemical properties; microstructure and phase equilibria; and much more.

Journal of the American Ceramic Society is abstracted/indexed in Chemical Abstracts, Ceramic Abstracts, Cambridge Scientific, ISI's Web of Science, Science Citation Index, Chemistry Citation Index, Materials Science Citation Index, Reaction Citation Index, Current Contents/ Physical, Chemical and Earth Sciences, Current Contents/Engineering, Computing and Technology, plus more.

View abstracts of all content from 1997 through the current issue at no charge at www.ceramicjournal.org. Subscribers receive full-text access to online content.

Published monthly in print and online. Annual subscription runs from January through December. ISSN 0002-7820

International Journal of Applied Ceramic Technology

www.ceramics.org/act

Launched in January 2004, *International Journal of Applied Ceramic Technology* is a must read for engineers, scientists,and companies using or exploring the use of engineered ceramics in product and commercial applications.

Led by an editorial board of experts from industry, government and universities, *International Journal of Applied Ceramic Technology* is a peer-reviewed publication that provides the latest information on fuel cells, nanotechnology, ceramic armor, thermal and environmental barrier coatings, functional materials, ceramic matrix composites, biomaterials, and other cutting-edge topics.

Go to www.ceramics.org/act to see the current issue's table of contents listing state-of-the-art coverage of important topics by internationally recognized leaders.

Published quarterly. Annual subscription runs from January through December. ISSN 1546-542X

American Ceramic Society Bulletin

www.ceramicbulletin.org

The *American Ceramic Society Bulletin*, is a must-read publication devoted to current and emerging developments in materials, manufacturing processes, instrumentation, equipment, and systems impacting the global ceramics and glass industries.

The *Bulletin* is written primarily for key specifiers of products and services: researchers, engineers, other technical personnel and corporate managers involved in the research, development and manufacture of ceramic and glass products. Membership in The American Ceramic Society includes a subscription to the *Bulletin*, including online access.

Published monthly in print and online, the December issue includes the annual *ceramicSOURCE* company directory and buyer's guide. ISSN 0002-7812

Ceramic Engineering and Science Proceedings (CESP)

www.ceramics.org/cesp

Practical and effective solutions for manufacturing and processing issues are offered by industry experts. CESP includes five issues per year: Glass Problems, Whitewares & Materials, Advanced Ceramics and Composites, Porcelain Enamel. Annual subscription runs from January to December. ISSN 0196-6219

ACerS-NIST Phase Equilibria Diagrams CD-ROM Database Version 3.0

www.ceramics.org/phasecd

The ACerS-NIST Phase Equilibria Diagrams CD-ROM Database Version 3.0 contains more than 19,000 diagrams previously published in 20 phase volumes produced as part of the ACerS-NIST Phase Equilibria Diagrams Program: Volumes I through XIII; Annuals 91, 92 and 93; High Tc Superconductors I & II; Zirconium & Zirconia Systems; and Electronic Ceramics I. The CD-ROM includes full commentaries and interactive capabilities.

Volume 155

Environmental Issues and Waste Management Technologies in the Ceramic and Nuclear Industries IX

Proceedings of the Science and Technology in Addressing
Environmental Issues in the Ceramic Industry and Ceramic
Science and Technology for the Nuclear Industry symposia at
The American Ceramic Society 105th Annual Meeting & Exposition held
April 27–30, 2003 in Nashville, Tennessee

Edited by

John D. Vienna
Pacific Northwest National Laboratory

Dane R. Spearing
Los Alamos National Laboratory

Published by
The American Ceramic Society
P.O. Box 6136
Westerville, Ohio 43086-6136
www.ceramics.org

Proceedings of the Science and Technology in Addressing Environmental Issues in the Ceramic Industry and Ceramic Science and Technology for the Nuclear Industry symposia at The American Ceramic Society 105th Annual Meeting & Exposition held April 27–30, 2003 in Nashville, Tennessee.

COVER PHOTO Backscattered SEM images of surface crust features dendritic crystals growing on external surface is courtesy of R. Evans, A. Quach, G. Xia, B.J. Zelinski, W.P. Ela, D.P. Birnie III, A.E. Sáez, H.D. Smith, and G.L. Smith and appears as figure 1, b in their paper "Microstructure of Emulsion-Based Polymeric Waste Forms for Encapsulating Low-Level, Radioactive and Toxic Metal Wastes," which begins on page 331.

For information on ordering titles published by The American Ceramic Society, or to request a publications catalog, please call 614-794-5890.

Printed in the United States of America.

4 3 2 1–07 06 05 04

ISSN 1042-1122
ISBN 1-57498-209-5

Contents

Ceramics for Waste or Nuclear Applications

Melter Processing and Process Monitoring

Waste Vitrification Programs

Glass Formulation and Property Models

Alternate Waste Forms and Processes

Preface

In 2003, The American Ceramic Society hosted several symposia focusing on main areas of ceramic manufacturing, product/process development, and science. Two key symposia foci on the use of ceramics in energy and environment restoration and on energy conservation and environmental issues in ceramics and manufacturing. Ceramics and glasses play a critical role in the nuclear industry. Nuclear fuels and waste forms for low-level and high-level radioactive, mixed, and hazardous wastes are primarily either ceramic or glass. Effective and responsible environmental stewardship is becoming increasingly more important in the world. In addition, with today's world of increasingly stringent environmental regulations and energy costs, it is critical to identify and adequately address environmental and energy consumption issues in the ceramic industry to ensure success.

We hope that through these symposia and subsequent proceedings, we are helping to foster continued scientific understanding, technological growth, and environmental stewardship within the field of ceramics. You hold in your hands the combined proceedings from two symposia, Ceramic Science and Technology for the Nuclear Industry and Science and Technology in Addressing Environmental Issues in the Ceramic Industry, presented at The American Ceramic Society 105th Annual Meeting & Exposition held April 27–30, 2003 in Nashville, Tennessee.

This proceeding represents the fifteenth volume published by The American Ceramic Society in the areas of waste management and environmental issues in relation to ceramics. Previous proceedings on nuclear waste management and environmental issues date from 1983 and include *Advances in Ceramics* volumes 8 and 20 and *Ceramic Transactions* volumes 9, 23, 39, 45, 61, 72, 87, 93, 107, 119, 132, and 143.

The editors gratefully acknowledge and thank Anne-Marie Choho, Connie Herman, Carol Jantzen, Gary Smith, SK Sundaram, and Lou Vance for help in organizing the symposia; the session chairs Connie Herman, Pavel Hrma, Jim Marra, Gary Smith, and Lou Vance for their contribution in keeping the presentations running smoothly; and most importantly, the authors and reviewers, without whom a high quality proceedings volume would not be possible. Lastly, the editors thank Teresa Schott, Debbie Vienna, and the team at The American Ceramic Society: Jackie Davis, Greg Geiger, Chris Schnitzer, and Marilyn Stoltz. Their support and contributions were instrumental in the symposia organization and publication of this volume.

John D. Vienna

Dane R. Spearing

Ceramics for Waste or Nuclear Applications

URANIUM VALENCES IN PEROVSKITE, CaTiO$_3$

E. R. Vance, M. L. Carter, Z. Zhang, K. S. Finnie, S. J. Thomson and B. D. Begg,
Materials and Engineering Science, Australian Nuclear Science and Technology
Organisation, Menai, NSW 2234, Australia

ABSTRACT

Attempts to produce trivalent U substituted into the Ca site of perovskite by
hydrogen reduction and association with Ti metal were not succcessful. The
tetravalent U solid solubility in the Ca site of perovskite is enhanced from ~ 0.05
formula units (f.u.) on firing in an Ar atmosphere (Al charge compensators or Ca
vacancies) to ~ 0.2 f.u. under reducing conditions. This behaviour may reflect
enhanced U solubility with Ti^{3+} charge compensation, but is still not understood in
detail. From diffuse reflectance spectroscopy of samples sintered in argon, optical
absorption bands due to U^{4+} over the 4000-12000 cm^{-1} range increase in intensity
with U content faster than linearly for U concentrations in excess of ~ 0.01 f.u.,
and new bands also appear. This behaviour is ascribed to the formation of more
distorted sites at higher U concentrations. However we cannot rule out the
occurrence of U^{5+} complexes, although the substitution of U^{5+} into the Ca or Ti
sites of perovskite could not be positively demonstrated.

INTRODUCTION

Perovskite, CaTiO$_3$, is an important constituent of synroc, a titanate-based
ceramic designed for the immobilization of high-level nuclear waste from nuclear
fuel reprocessing [1]. Perovskite incorporates the Sr and some of the rare earth
fission products as well as some of the actinides in the Ca sites.

In a previous attempt to incorporate U^{3+} in the Ca site of perovskite, we
fabricated [2] a single-phase sample of Ca$_{0.9}$U$_{0.1}$Ti$_{0.9}$Al$_{0.1}$O$_3$ stoichiometry in a
3.5%H$_2$/N$_2$ atmosphere at 1400°C. In the simplest interpretation of this result, 0.1
formula units (f.u.) of trivalent U in the Ca site would have been compensated by
the Al in the Ti (all Ti being assumed in this case to be Ti^{4+}) site. However X-ray
near edge spectroscopy (XANES) in the vicinity of the L$_{III}$ edge near 17.16 keV
indicated little difference (~ 0.2 eV at most) in the edge energies between the
perovskite sample and tetravalent U in brannerite, UTi$_2$O$_6$ [2]. We concluded
that the U was actually in the tetravalent form and that a further 0.1 f.u. of Ti^{3+}
(and the 0.1 f.u. of Al) was compensating the U^{4+}. We interpreted the further

observation [2] that upon reheating the sample in argon at 1400°C, the amount of U in the perovskite was reduced to ~ 0.05 f.u., because of the 0.1 f.u. of Ti^{3+} being converted to Ti^{4+}, thus leaving only the 0.1 f.u. of Al as a charge compensator for the U^{4+}. Pyrochlore-structured $CaUTi_2O_7$ also formed after the argon treatment. It was further concluded [2] that the solid solubility limit of U^{4+} in reduced perovskite was at least 0.1 f.u.

However, Hanajiri et al. [3] reported that the solubility of U^{4+} in the Ca site of $CaTiO_3$ using Al compensation in the Ti site is only 0.05-0.075 f.u. and we have now revisited the issue of U valence states and their solid solubilities in perovskite.

EXPERIMENTAL

Perovskite samples were prepared by the standard alkoxide route[4] in which aqueous solutions of Ca, Cr, Ga, In, uranyl, and Al nitrates were mixed with an ethanolic solution of Ti isopropoxide. After stir-drying and calcination, the powders were wet-milled, dried, cold-pressed, and sintered at 1400°C.

Scanning electron microscopy was carried out with a JEOL 6300 instrument operated at 15 keV, and fitted with a NORAN Instruments Voyager IV X-ray microanalysis System (EDS) which utilised a comprehensive set of standards for quantitative work, giving a high degree of accuracy [5]. Powder X-ray diffractometry was carried out with a Siemens D500 instrument, using Co Kα radiation.

Diffuse reflectance spectra of the sintered pellet samples were collected at ambient temperature using a Cary 500 spectrophotometer equipped with a Labsphere Biconical Accessory, which utilises a beam spot size ~ 1 mm diameter. Spectra were converted into reflectance units by ratioing them to the spectrum of a Labsphere certified standard (Spectralon), and then transforming them into Kubelka-Munk units, $F(R)=(1-R)^2/2R$ [6]. At low chromophore concentrations and similar packing densities, $F(R)$ is expected to be proportional to the concentration of the chromophore. Bandfitting was conducted using Bio-Rad Win-IR software. The purpose of such measurements was to detect electronic absorption bands, tens or hundreds of wavenumbers in width, due to different valence states of U, notably U^{4+} and U^{5+}. Electronic absorptions due to U^{4+} would be expected right across the infrared to ultraviolet sections of the spectrum, but those from U^{5+} would be observed only in the near infrared (see for example [7]).

Polished $(Ca_{0.9}U_{0.1})(Ti_{0.9}Al_{0.1})O_3$ and $(Ca_{0.97}U_{0.03})(Ti_{0.94}Al_{0.06})O_3$ surfaces were ultrasonically cleaned in acetone for 10 min before XPS measurements in ultrahigh vacuum with a Kratos XSAM 800 system. The Mg anode of the X-ray source ($K_α$: 1253.6 eV) was operated at 180 W, and the spectrometer pass energy was set at 20 eV for regional scans. The analysis area was 4x6 mm^2, much larger than the average grain size. The thickness of the probed surface layer was approximately 5 nm. The binding energies were calibrated by fixing the C 1s peak (due to adventitious carbon) at 285.0 eV. Because XPS has the ability to discriminate between different oxidation states and chemical environments, it has been used widely since the 1970s to determine the surface oxidation state of U in many uranium compounds [8]. The binding energy of the U $4f_{7/2}$ peak is directly related to the oxidation state of uranium. As the oxidation state of uranium increased from U^{4+} (UO_2) to U^{6+} (UO_3), the binding energy of the U $4f_{7/2}$ peak was observed to increase by 1.7-1.8 eV [9,10].

X-ray absorption near edge spectroscopy (XANES) was performed at the Stanford Synchrotron Radiation Laboratory (SSRL) on powdered samples in fluorescence mode over the U L_{III}-edge at 80K.

RESULTS AND DISCUSSION

U^{3+} in perovskite?

U^{3+} is a known valence state in halides and organics but not in oxides. But in studies of titanate oxides containing Np and Pu, it was found [11] to be much easier to form trivalent actinides in the Ca site of perovskite than was the case in zirconolite ($CaZrTi_2O_7$) and in the actinide oxides themselves. Hence it appeared that perovskite could be likely to allow U^{3+} formation. To maximize the chances of reduction of the U, the samples, having $Ca_{(1-x)}U_xTiO_3$ stoichiometries (x = 0.01 or 0.03) were mixed with 5 wt% of with Ti metal powder and then fired for 16 h in 3.5% H_2/N_2 atmosphere. However XANES (see Fig. 1) showed the U in the samples was tetravalent.

U^{4+} in perovskite

Samples formed in H_2/N_2. Having found that the solid solubility of U^{4+} in perovskite had not been exceeded in the $Ca_{0.9}U_{0.1}Ti_{0.9}Al_{0.1}O_3$ sample (see above), we made a further perovskite sample of $Ca_{0.7}U_{0.3}Ti_{0.7}Al_{0.3}O_3$ stoichiometry. This upon firing at 1400°C in 3.5% H_2/N_2 yielded a perovskite phase containing ~ 0.2 f.u. of both U and Al, i.e. of $Ca_{0.8}U_{0.2}Ti_{0.8}Al_{0.2}O_3$ stoichiometry, plus minor UO_2 and Al_2O_3. The outside regions of the samples contained more U in solid solution in the perovskite. Given that the U was present as U^{4+}, the solid solution limit of 0.2 f.y. is a factor of 3 or 4 higher than that given by Hanajiri et al. [3].

To explain the discrepancy, all the Ti in the perovskite could conceivably be in the form of Ti^{4+}; in previous work on the stability of zirconolite in 3.5%H_2/N_2 at 1400°C [12] we have shown that $CaTiO_3$ (essentially all Ti as Ti^{4+}) is a stable decomposition product. If all the Ti in the present sample was tetravalent, charge compensation for U^{4+} to inhabit the Ca site would need to take place by cation vacancies. However cation vacancies are relatively high-energy defects and Ti^{3+} can readily be formed in reducing conditions if the charge balancing conditions are appropriate. Hence the ~ 0.2 f.u. of U^{4+} found in solid solution in the $Ca_{0.7}U_{0.3}Ti_{0.7}Al_{0.3}O_3$ sample is very likely to be charge compensated by 0.4 f.u. of Ti^{3+}.

The results for U then suggested that the solid solution limit of U^{4+} in perovskite depends on the charge compensating ions present in the Ti site. Alternative charge compensating M^{3+} ions which could be accommodated in the

Ti^{4+} sites are Cr, Ga or In. However Ga and In would form Ga^+ and In metal respectively in reducing atmospheres, so argon atmospheres were utilised to try to put Ga and In in their trivalent states.

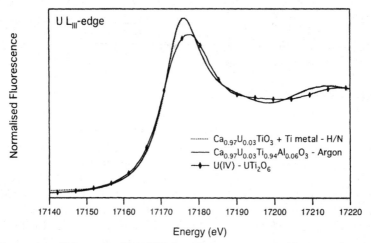

Figure 1. U L_{III}-edge XANES data from two perovskite samples along with a tetravalent UTi_2O_6 standard.

Samples fired in argon atmospheres

We first studied U^{4+} solubility in $Ca_{(1-x)}U_xTi_{(1-2x)}M_{2x}O_3$ preparations in which two trivalent M ions in a Ti site acted as charge compensators for each U^{4+} ion in a Ca site. In samples with M = Al and x = 0.1 and 0.2, SEM showed that the perovskite contained only about 0.05 f.u. of U, charge compensated by 0.1 f.u. of Al ions. This result agreed with that of Hanajiri et al. [3]. Residual UO_2, Al_2O_3 and pyrochlore-structured material having a stoichiometry fairly close to $CaUTi_2O_7$ were also observed (the actual pyrochlore observed had a composition of approximately $Ca_{0.9}U_{0.9}Ti_{2.2}O_7$). With M = Cr, Ga or In, we conducted preliminary tests with x = 0.05. The samples were fired for 16 h at 1400°C. For M = Cr, the limiting solid solubility of U was found to be ~0.04 f.u. The U solid solubility in the Ga samples was lower, ~ 0.02 f.u., and the In samples retained very little In, with evidence of metallic In the interiors of the samples-hence there was little solid solubility of U.

We also studied the effect of using Ca vacancies as charge compensators for tetravalent U. Samples of $Ca_{(1-2x)}U_x\square_xTiO_3$ stoichiometeries ($\square$ = vacancy) were fired in argon at 1400°C; for x = 0.2 the limiting U^{4+} solubility in perovskite was found as ~ 0.05 f.u., with the overall stoichiometry being approximately $Ca_{0.9}U_{0.05}\square_{0.05}TiO_3$ as above, with UO_2, brannerite (UTi_2O_6) and TiO_2 (rutile) being observed as minor phases.

To check on U valences in perovskite explicitly by XANES, we prepared an Al-compensated perovskite with only 0.03 f.u. of U^{4+} being present, by firing in argon at 1400°C. The overall stoichiometry was $Ca_{0.97}U_{0.03}Ti_{0.94}Al_{0.06}O_3$. Fig. 1 compares the U L_{III} edges from the reduced sample of the $Ca_{0.9}U_{0.1}Ti_{0.9}Al_{0.1}O_3$ stoichiometry (see above) and the Ar-treated sample and it is seen that the spectra are identical, and close to the edge of U^{4+} in brannerite, indicating that U was tetravalent in both perovskites.

The U 4f XPS spectra of the reduced $Ca_{0.9}U_{0.1}Ti_{0.9}Al_{0.1}O_3$ sample [2] and $Ca_{0.97}U_{0.03}Ti_{0.94}Al_{0.06}O_3$ are similar. The U $4f_{7/2}$ peak is composed of a main "bulk" peak and a shoulder on the higher binding energy side. The presence of this shoulder is believed to be the result of atmospheric oxidation at the surface. In order to separate the "bulk" component from the "surface" component, the U $4f_{7/2}$ region was fitted with two Gaussian peaks after linear background subtraction. The result of curve-fitting is shown in Figure 2. The "bulk" and "surface" peaks are located at 379.7 and 381.4 eV, respectively, for both samples. These peak positions are consistent with the presence of a "bulk" U^{4+} peak and a "surface" U^{6+} peak [8-10]. Note that the U^{6+} component is around 9% and 18% for $Ca_{0.9}U_{0.1}Ti_{0.9}Al_{0.1}O_3$ and $Ca_{0.97}U_{0.03}Ti_{0.94}Al_{0.06}O_3$ respectively, indicating these U-containing perovskite surfaces are relatively stable towards atmospheric oxidation. U is clearly tetravalent in both samples, in agreement with the XANES results (see above and [2]).

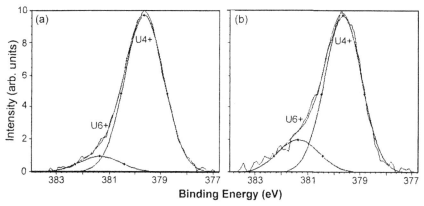

Figure 2:   XPS results for $Ca_{0.9}U_{0.1}Ti_{0.9}Al_{0.1}O_3$ sintered in H_2/N_2(LH side) and $Ca_{0.97}U_{0.03}Ti_{0.94}Al_{0.06}O_3$ (RH side).

SEM showed that traces of Ca uranates were present, but evidently the quantities of such materials were too low to affect the U signal from the perovskite. The U content of the perovskite itself agreed with the starting stoichiometry.

DRS of samples doped with 0.001, 0.003, 0.01 and 0.03 f.u. of U substituted into the Ca site ($Ca_{1-x}U_xTi_{1-2x}Al_{2x}O_3$) were studied and the results are shown in Fig. 3. The spectra are multiplied by appropriate factors to give constant intensities of the bands if the bands were proportional to U concentration. The first impression is that the intensities of the absorption bands at ~ 4500 (doublet), ~ 6000, 7500, 8500, 8900 and 9500 cm^{-1}, due to electronic transitions of U are highly non-linear in U concentration. Moreover in the sample containing 0.03 f.u. of U, there appear to be significant "extra" bands at ~ 4100, 7200 and 10700 cm^{-1}. In addition, there was a considerable change in the visual appearances of the samples; those having x = 0.001 and 0.003 were pale brown, whereas at x ~ 0.01, the samples became quite dark in colour. All these features have the hallmarks of significant changes of site symmetry, with supralinearity of intensity reflecting increasing lack of inversion symmetry for electric dipole transitions.

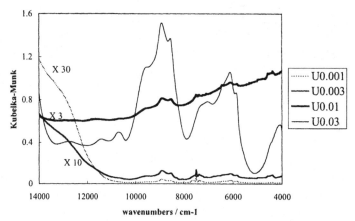

Figure 3. Near infrared diffuse reflection spectra from $Ca_{1-x}U_xTi_{2-2x}Al_{2x}O_3$ perovskites

Targeting U^{5+}

Finally, we prepared samples of (a) $Ca_{(1-x)}U_xTi_{(1-3x)}Al_{3x}O_3$ stoichiometry in which U on the Ca site was targeted as U^{5+}, compensated by three Al ions replacing Ti ions, and (b) $CaTi_{(1-2x)}U_xAl_xO_3$ stoichiometry in which U on the Ti site was targeted as U^{5+}. These were fired in air and argon, noting that we have obtained evidence for U^{5+} in both air- and argon-fired brannerite and zirconolite containing appropriate charge compensators [13,14].

In the samples with U notionally substituted in the Ca site, we found in argon-fired samples that 0.03-0.05 f.u. of U were incorporated, plus Al in amounts approximately double the U content. In air-fired samples only ~ 0.025 f.u. of U ions were incorporated, plus 0.05 f.u. of Al ions. These result correspond to the U valence being +4. The fact that smaller amounts of U^{4+} were deduced to be present than in argon-fired samples may reflect the essential metastability of U^{4+} in air-fired $CaTiO_3$ preparations. The formation of $CaUO_4$ was also observed and this reaction would have competed with the possible entry of U into the perovskite lattice. Note that in previous work on air-fired perovskite doped with 0.02 f.u. of U [6], we had deduced from XRD that U^{4+} ions can occupy the Ca site in perovskite if twice as many Fe^{3+} ions enter the Ti site, in an air atmosphere.

While U^{4+} is ~ 10% smaller in ionic radius than Ca, U^{5+} is ~ 20% smaller [15] and so would not be expected to have significant solubility in the Ca site. In the samples with U notionally substituted in the Ti site (see (b) above), no evidence from SEM was obtained that U was actually substituted in this way; the size disparity between U^{5+} and Ti^{4+} is also about 20% [15].

CONCLUSIONS

The solubility of U^{4+} in the Ca site of perovskite appears to depend on the charge compensation mechanism employed. With trivalent Ti ions substituted in the Ti^{4+} site under reducing firing conditions, the solid solubility of U^{4+} could be as much as 0.2 f.u., but when compensated by Al ions in the Ti site or one Ca vacancy per U^{4+} ion the solid solubility was found to be ~ 0.05 f.u. DRS spectra of U (probably all as U^{4+}) in perovskite showed complex behaviour as the doping exceeded ~ 0.01 f.u. No evidence of U^{3+} or U^{5+} was observed.

ACKNOWLEDGEMENTS

This work was mainly supported by the Environmental Science Management Project, operating out of the Office of Basic Energy Sciences. The XANES experiments were conducted at the Stanford Synchrotron Research Laboratory, a national user facility operated by Stanford University on behalf of the U.S. Department of Energy, Office of Basic Energy Sciences.

REFERENCES

[1] A. E. Ringwood, S. E. Kesson, N. G. Ware, W. Hibberson and A. Major, "Geological immobilisation of nuclear reactor wastes", *Nature* (London) 278, 219-23 (1979).

[2] E. R. Vance, M. L. Carter, B. D. Begg, R. A. Day and S. H. F. Leung, "Solid Solubilities of Pu, U, Hf and Gd in Candidate Ceramic Phases for Actinide Waste Immobilisation", in Scientific Basis for Nuclear Waste Management", Eds. R. W. Smith and D. Shoesmith, Materials Research Society, Pittsburgh, PA, USA, pp. 431-6 (2000).

[3] Y. Hanajiri, H. Yokoi, T. Matsui, Y. Arita, T. Nagasaki and H. Shigematsu, "Phase equilibrias of $CaTiO_3$ doped with Ce, Nd and U", *J. Nucl. Mater.*, 247, 285-8 (1997).

[4] A. E. Ringwood, S. E. Kesson, K. D. Reeve, D. M. Levins and E. J. Ramm, "Synroc", in Radioactive Waste Forms for the Future, Eds. W. Lutze and R. C. Ewing, Elsevier, Netherlands, pp. 233-334 (1988).

[5] E. R. Vance, R. A. Day, Z. Zhang, B.D. Begg, C. J. Ball and M. G. Blackford, Charge Compensation for Gd in $CaTiO_3$", *J. Solid State Chem.,* 124, 77-82 (1996).

[6] G. Kortum, "Reflectnce Spectroscopy", Springer-Verlag, Berlin, 1969.

[7] K. S. Finnie, Z. Zhang, E. R. Vance and M. L. Carter, "Examination of U valence states in the brannerite structure by near-infrared diffuse reflectance and X-ray photoelectron spectroscopies", *J. Nucl. Mater.,* 317, 46-53 (2003).

[8] B. W. Veal and D. J. Lam, "Photoemission Spectra", pp 176-210 in Gmelin Handbook of Inorganic Chemistry, Suppl. Vol. A5, Springer, Berlin, 1982.

[9] D. Chadwick, "Uranium 4f Binding Energies Studied by X-ray Photoelectron Spectroscopy" *Chem. Phys. Lett.,* 21, 291-4 (1973).

[10] G. C. Allen, J. A. Crofts, M. T. Curtis, P. M. Tucker, D. Chadwick and P. J. Hampson, "X-ray Photoelectron Spectroscopy of Some Uranium Oxide Phases" *J. Chem. Soc., Dalton Trans.,* 1296-1301 (1974).

[11] B. D. Begg, E. R. Vance and S. D. Conradson, "The Incorporation of Plutonium and Neptunium in Zirconolite and Perovskite", *Journal of Alloys and Compounds,* 271-3, 221-6 (1998).

[12] B. D. Begg, E. R. Vance, B. A. Hunter and J. V. Hanna, "The structural effects of reduction on Zirconolite", *J. Mater. Res.,* 13, 3181-90 (1998)

[13] E. R. Vance, G. R. Lumpkin, M. L. Carter, D. J. Cassidy, C. J. Ball and B. D. Begg, "The Incorporation of U in Zirconolite ($CaZrTi_2O_7$)", *J. Amer. Ceram. Soc.,* 85 [12], 1853-9 (2002).

[14] E. R. Vance, J. N. Watson, M L. Carter, R. A. Day and B. D. Begg, "Crystal Chemistry and Stabilization in Air of Brannerite, UTi_2O_6", *J. Amer. Ceram. Soc.,* 84 [1], 141-4 (2001).

[15] R. D. Shannon, "Revised Ionic Radii and Systematic Studies of Interatomic Distances in Halides and Chalcogenides", *Acta Cryst.,* A32, 751-67 (1976).

IRON-SUBSTITUTED BARIUM HOLLANDITE CERAMICS FOR CESIUM IMMOBILIZATION

F. Bart, G. Leturcq and H. Rabiller
CEA Valrho Marcoule
Nuclear Energy Division – Waste Confinement and Vitrification
DEN – DIEC – SCDV
BP 17171, 30207 Bagnols-sur-Cèze France

ABSTRACT

French legislation passed in December 1991 has spurred research on enhanced separation for conditioning of long-lived radionuclides. With regard to incorporation capacity and high chemical durability, barium hollandite ($BaAl_2Ti_6O_{16}$) ceramics suit cesium immobilization. However, Cs incorporation in hollandite requires the reduction of a part of the titanium and implies specific reducing conditions for the synthesis process. To avoid this constraint, a new hollandite composition was studied, in which Fe^{III} was substituted for Ti^{III}.

A five wt% Cs_2O-doped Fe-substituted hollandite was synthesized on a laboratory scale (100g), using an alcoxide route. The hollandite crystallisation was achieved after a 5 hours calcination at 1000°C, in air atmosphere. After milling and cold pressing steps, ceramic pellets were sintered in air at 1250°C for 15 hours. The resulting materials were dense single phase pellets, and were characterized by SEM, XRD and leaching tests.

Leaching experiments were carried out in order to study the chemical durability of this specific hollandite. Initial leaching rates were measured at 100°C using a Soxlhet apparatus.

INTRODUCTION

A French law passed in 1991, specified one major area of research on long-lived radioactive waste management: develop processes capable of enhanced separation of long-lived radionuclides, and investigate conditioning processes and long-term interim surface or subsurface storage options. Cesium is probably one of the most difficult long-lived elements to confine durably because of its high thermal power and mobility in aqueous media. A containment matrix suitable for radioactive cesium must present the following characteristics: high long term resistance to aqueous alteration; good thermal stability; good chemical stability when subjected to cesium decay to barium.

Although borosilicate glasses are currently used to condition fission product solutions, the prospect of separate conditioning of radionuclides raises the possibility of using other types of materials, notably new glasses[1] or ceramic materials with greater chemical durability.

Hollandite is a mineral with the general formula $A_2B_8O_{16}$. The twenty million year old natural analog, priderite $(K,Ba)_x(Mg,Fe,Ti)_8O_{16}$, weas found in silica poor rocks in Australia, in the Kimberley region. The existence of natural iron-containing hollandite minerals, such as henrymeyerite $BaFe^{2+}Ti_7O_{16}$ and redledgeite $BaCr^{3+}_2Ti_6O_{16},H_2O$, indicates that iron and chromium can enter into these structures[2,3]. Moreover, hollandite with the formula $(Ba,Cs)Al_2Ti_6O_{16}$ is one of the major phases of Synroc, a matrix dedicated to the immobilization of the PUREX fission products solutions[4,5].

We studied a set of iron-containing hollandite ceramics, in the $Ba_xCs_y(Al,Fe)^{3+}_{2x+y}Ti^{4+}_{8-2x-y}O_{16}$ compositional range, for x varying between 0.5 and 1. The aim of this work was to develop a single phase, dense hollandite ceramic using a natural air sintering process[6]. The ceramics were designed to contain 5 wt% of cesium oxide included in the structure of the crystals, without any secondary phase. Furthermore, we investigated the leaching resistance of the ceramic pellets by measuring their initial leaching rate in pure water, at 100°C, using Soxlhet tests.

After a description of the experimental protocols used to fabricate the materials, the second part of the paper presents the microstructure of the ceramics, and its evolution depending on the composition. The last part of this work concerns the leaching behavior of the iron containing hollandite ceramics.

MATERIAL SYNTHESIS

Sample composition

The composition of the materials was chosen following several criteria.

First, the composition was chosen near the $(Ba,Cs)Al_2Ti_6O_{16}$ composition, based on the fact that the chemical durability of hollandite ceramics is well established in this compositional range[7].

According to previous work, the presence of trivalent titanium in the crystalline structure is necessary to ensure the incorporation of cesium in the A site of the $A_2B_8O_{16}$ type structure [5]. This criterion implies a close control of the atmosphere for calcination and sintering steps, in order to prevent the soluble undesirable phase appearing under oxidizing conditions ($Cs_2Al_2Ti_2O_8$). For ceramic fabrication, reducing conditions were generated by the addition of metallic titanium in the mixture after calcination. In natural sintering conditions, we observed, in the past, that the presence of metallic particles induced a high heterogeneity in the microstructure of the ceramics. Moreover, the ceramics were constituted of two phases, rutile + hollandite, and this could heve been a disadvantage considering the loading factor of the final waste form. To overcome this difficulty, we replaced trivalent titanium by trivalent iron in the mixture, the trivalent oxidation state of iron being stable under oxidizing conditions. This substitution was expected to have 2 important effects. Concerning the sintering process, it was supposed to allow us to work in oxidizing atmosphere through out the process. Concerning the material, it should have allowed us to obtain a single phase material.

Finally, the last point was related to the cesium loading factor. It had to be high enough to limit the final waste volume. However, careful attention had to be paid to the high thermal energy which would be generated by the short-lived isotopes of cesium. Based on these two opposing points, 5 wt% of cesium oxide were introduced in the preparations[*].

Table 1 gives the molar and weight composition of the 5 samples under study. In the $Ba_xCs_y(Al,Fe)^{3+}_{2x+y}Ti^{4+}_{8-2x-y}O_{16}$ compositional range, we studied the stability of hollandite as a function of the barium stoichiometry x varying between 0.5 and 1. The cesium proportion y was fixed at about 0.28 atoms/unit formula. The trivalent ratio Al/Fe was equal to 1.8 on average for the 5 samples.

Table 1. Theoretical composition of the samples (H = Hollandite, R=Rutile)

Sample	Theoretical Formula	BaO Wt%	Cs$_2$O Wt%	Al$_2$O$_3$ Wt%	Fe$_2$O$_3$ Wt%	TiO$_2$ Wt%
H1	$Ba_{0.5}Cs_{0.26}(Fe_{0.46}Al_{0.8})Ti_{6.74}O_{16}$	10.51	5.02	5.59	5.04	73.83
H2	$Ba_{0.66}Cs_{0.26}(Fe_{0.57}Al_{1.01})Ti_{6.42}O_{16}$	15.38	4.97	6.88	6.06	68.58
H3	$Ba_{0.75}Cs_{0.27}(Fe_{0.6}Al_{1.17})Ti_{6.23}O_{16}$	15.17	5.02	7.87	6.32	65.63
H4	$Ba_{0.85}Cs_{0.27}(Fe_{0.71}Al_{1.264})Ti_{6.026}O_{16}$	16.89	5.00	8.35	7.35	62.40
H5	$Ba_1Cs_{0.28}(Fe_{0.82}Al_{1.46})Ti_{5.72}O_{16}$	19.42	5.00	9.43	8.29	57.87

Sample preparation

Precursor synthesis following an alcoxide route

The alcoxide route was used for the preparation of the ceramic precursor (100 g batches). A mixture of aluminum secbutoxide and titanium isopropoxide was dissolved in ethanol. An aqueous solution of cesium, barium and ferrous nitrates was then added to the alcoxide preparation. The chemical reaction

---

[*] Some calculations were made to evaluate the internal temperatures of a typical waste container, as a function of the Cs loading of the material. The conclusion of the thermal modeling leads us to limit the Cs loading factor to a value close to 5 wt%, whatever may be the waste form (glass, glass-ceramics or ceramics).

between organometallic and aqueous reagents led to the hydrolysis of the mixture and the formation of a precursor, which was supposed to have the calculated stoechiometric proportion (Table 1). The residual solvent present in the precursor was then removed in a rotating evaporator at about 70°C. Until this step, the precursor was amorphous.

Calcination of the precursor

The denitration of this mixture, and its crystallization into the hollandite structure was obtained for the next step, consisting in a calcination, which was carried out in a muffle furnace in air. According to the differential thermal analysis realized on the dried H1 precursor (see figure 1), there was a clear exothermic reaction, corresponding to the peak appearing around 850°C[**]. The temperature of the calcination was consequently varied from 750°C to 1000°C, for duration varying from 2 to 10 hours, in order to obtain the complete transformation of the powder into hollandite phase during this step, without any cesium losses. It is important to underline that the samples were further analyzed by X-ray fluorescence[***]: the cesium oxide content was the same for the whole set of materials (4,3 wt% in Cs_2O versus 5wt% initially introduced to the composition). Figure 2 shows the influence of the temperature and duration of the calcination on the crystallinity of the material, obtained XRD experiments on ground calcined powders. The calcination at 1000°C for 2 hours resulted in the entire crystallization of the precursor into the hollandite phase.

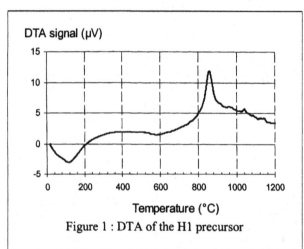

Figure 1 : DTA of the H1 precursor

[**] The experiment was done on 4 different samples coming from the same precursor, in order to compensate to chemical heterogeneity which id expected at this first step of the synthesis.

[***] This XRF analysis was performed in the COGEMA Marcoule laboratory, after a dilution of the ceramic powder in a lithium borate glass at 900°C, which is supposed not to involve supplementary cesium losses.

Milling

The calcined powder, mixed with deionized water, was then ground in a planetary ball milling, using zirconia balls and bowls. In this step, the average grain size was lower than 1 micrometer, as shown by granulometric distribution measurements (results obtained on the H5 sample). The corresponding surface area was about 25 m^2/g, as measured by the nitrogen chemisorption experiments (BET).

Sintering

After drying in an oven, the powders were cold pressed between 100 and 120 MPa. The 5 mm thick, 30 mm diameter pellets were finally naturally sintered, in a muffle furnace under uncontrolled air conditions. The sintering temperature was determined from dilatometry measurements carried out on the H5 sample. The temperature of 1250°C, corresponding to the maximum sintering rate, was used for the whole set of samples. The sintering process was also optimized, in order to reach a good densification of the pellets. After a sintering step of 15 hours at 1250°C, the residual porosity of the materials was considered to suit further characterization.

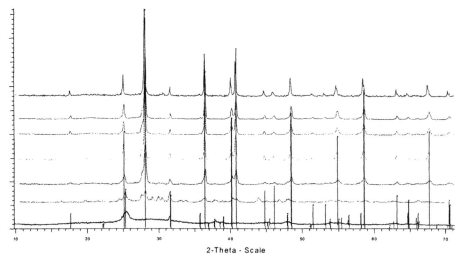

2-Theta - Scale

Figure 1 : X-Ray diffractogramms of the H1 materials showing the influence of the temperature and duration of the calcination on the crystallinity. The vertical blue lines are related to the Hollandite (ref JCPDS 33-0133), whereas the red lines correspond to anatase (ref JCPDS 21-1272) and the pink lines to barium iron oxide phase (ref JCPDS 23-1023). From the bottom to the top : 750°C for 2 hours, 850°C for 2 hours, 850°C for 20 hours, 1000°C for 2 hours, 1000°C for 5 hours, 1000°C for 10 hours. The upper diffractogramm corresponds to a sintered pellet (750°C for 2 hours, ground, cold pressed and natural air-sintered for 96 hours at 1200°C).

MICROSTRUCTURE OF THE SINTERED CERAMICS

The sintered materials were characterized with SEM and XRD, in order to study their microstructure with regardds to the compositional range (see Table 1). Note that we compared the sintered samples after a calcination at 750°C for 3 hours, followed by ball milling and natural sintering at 1200°C for about 96 hours. This process slightly differs from the previous optimized scheme, but it was nevertheless convenient for comparing the resulting phase assemblage discussion.

Whereas H1, H2, H3 and H4 showed the same two phases, i.e. hollandite plus TiO_2, only H5 presented hollandite as a single phase. Figure 3 shows the microstructure of H1, H2 and H5 samples[****]. These Scanning Electron Microscope images, in the backscattered electron detection mode, show that, in the $Ba_xCs_y(Al,Fe)^{3+}_{2x+y}Ti^{4+}_{8-2x-y}O_{16}$ compositionnal range, hollandite stability is very dependent on the Barium stoechiometry x, (when y is fixed at about 0.28 atoms/unit formula and for the trivalent ratio Al/Fe equal to 1.8 in average).

The evolution of the microstructure of the sintered ceramics with regards to their composition showed that the titanium oxide proportion was higher when barium stoechiometry x was lower. The anatase proportion clearly increased when x diminished from 1 (no anatase detected) to 0.5 (same range for the quantities of anatase and hollandite, see figure 3).

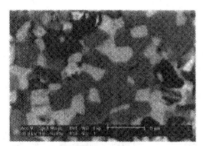

Figure 3a : SEM image of H1
$Ba_{0.5}Cs_{0.26}(Fe_{0.46}Al_{0.8})Ti_{6.74}O_{16}$
Dark crystals = TiO_2
Light crystals = Hollandite

Figure 3b : SEM image of H2
$Ba_{0.66}Cs_{0.26}(Fe_{0.57}Al_{1.01})Ti_{6.42}O_{16}$
Dark crystals = TiO_2
Light crystals = Hollandite

[****] At this point, the samples were once again analyzed by X-ray fluorescence: the cesium oxide content was the same in the whole set of materials, equal to 4.3 wt% in Cs_2O versus 5wt% initially introduced in the composition. This percentage did not change between the calcination step and the sintering step.

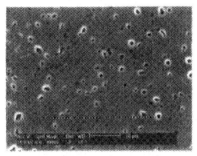

Figure 3c : SEM images (BSE mode) of the sample H5 $Ba_{0.5}Cs_{0.26}(Fe_{0.46}Al_{0.8})Ti_{6.74}O_{16}$
Hollandite is the only phase detected in the sample (Dark zones = porosity)

Considering that no other phase was present in the samples, it was possible to calculate the stoechiometry of the resulting hollandite crystals, by subtracting a given quantity of TiO_2 from the theoretical compositions. The result of this rough calculation, given in table 2, did not demonstrate, but was coherent with the fact that the more stable hollandite was obtained when the Ba stoechiometry was equal to 1. Furthermore, since the measured 10% relative cesium losses did not depend on the sample composition, these results show that the hollandite crystals should have included in their structure up to 0.4 moles of Cs per unit formula. These hypotheses could be tested in the future, via to microprobe analyses.

Table 2. Calculated composition of the hollandite present in the samples

Sample	Estimated molar fraction of TiO_2	Recalculated possible sample composition
H1	0.6 (estimated)	$0.4\ Ba_1Cs_{0.5}(Fe_{0.9}Al_{1.6})Ti_{5.4}O_{16} + 0.6\ TiO_2$
H2	0.4 (estimated)	$0.6\ Ba_1Cs_{0.4}(Fe_{0.8}Al_{1.5})Ti_{5.7}O_{16} + 0.6\ TiO_2$
H3	0.3 (estimated)	$0.7\ Ba_1Cs_{0.35}(Fe_{0.8}Al_{1.5})Ti5.7O_{16} + 0.6\ TiO_2$
H4	0.2 (estimated)	$0.2\ Ba_1Cs_{0.32}(Fe_{0.8}Al_{1.5})Ti_{5.7}O_{16} + 0.2\ TiO_2$
H5	**0 (measured)**	**Only $Ba_1Cs_{0.28}(Fe_{0.82}Al_{1.46})Ti_{5.72}O_{16}$**

The following characterization concerns the chemical durability properties of the materials. The results that will be given are exclusively related to the single phase H5 ceramics.

LEACHING BEHAVIOR

Test Description

The initial alteration rate r_0 in pure water at 100°C was determined by Soxhlet-mode testing in which a 22 mm diameter ceramic pellet sample was subjected to a continuously renewed pure water flow at 100°C for 56 days, with leachate samples taken at regular intervals for ICP-MS analysis to determine the normalized mass loss***** at various intervals between 1 and 56 days. The apparatus that we currently use for this experiments was made of stainless steel, thus preventing from obtaining information about iron normalized losses.

Cesium is the more leachable element in Cs-doped hollandites, compared to Ba, and especially to Ti and Al[7], it can thus be used as a suitable tracer of the alteration to the ceramics.

Figure 4 shows the evolution of the normalized cesium losses for a 56 day leaching test for 3 samples having the same H5 composition (square, losanges and triangles) and for a titanium 1.5 wt% Cs_2O doped composition (dots).

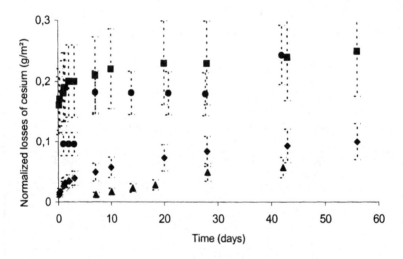

Figure 4 : Evolution of the normalized cesium losses for a 56 days leaching test, for 4 hollandite samples :
- ■ polished sample H5
- ◆ and ▲ sintered unpolished samples H5 (duplicate test)
- ● iron-free 1.5 wt% Cs_2O doped composition

***** The normalized mass loss $NL(i)$ for element i represents the quantity of glass per unit area $(g \cdot m^{-2})$ that must be dissolved to obtain the concentration of element i measured in solution.

The leaching results showed that there was an initial release of cesium at the beginning of the test, which lead to a rate of 2.10^{-2} g.m^{-2}.d^{-1} if calculated between 0 and 1 day. The alteration rate in pure water at 100°C was equal to 6.10^{-4} g.m^{-2}.d^{-1} if calculated between 25 and 56 days. The average rate was equal to 2.10^{-3} g.m^{-2}.d^{-1} if the total mass loss of the pellet was considered.

This graph also showed that the polishing of the pellets induced an increase in the mass losses. This may be due to the presence of very small particles created at the surface for this operation. It may also indicate a mechanical weakness that was created by the polishing. A fragile, possibly amorphous, thin layer may then be present at the surface of the pellet, and this layer could be more leachable. These hypotheses could be confirmed using grazing incidence X-Ray diffraction.

However, the chemical durability of iron hollandites, loaded with 5 wt% of cesium oxide was very good, and compared to that of Synroc hollandite: initial leaching rates in pure water at 100°C were small (in the range of 10^{-3} g.m^{-2}.d^{-1}), and they rapidly decreased with time.

CONCLUSION

Iron substituted barium hollandite compositions, suitable for containment of radioactive cesium, were studied to assess the microstructure with regards to the composition, in the $Ba_xCs_y(Al,Fe)^{3+}_{2x+y}Ti^{4+}_{8-2x-y}O_{16}$ chemical composition range. The ceramic materials are all well suited in terms of their cesium loading capacity. The value of 1 for x allowed us to get a single phase hollandite dense material. The initial alteration rates in pure water at 100°C are in the range of 10^{-3} g.m^{-2}.d^{-1}, for tests conducted for 56 days in Soxlhet apparatus.

This work demonstrates the suitability of iron-substituted barium hollandite ceramics for cesium conditioning, from the process and material points of view. The perspectives of this work concern long term behavior of the waste form, related to leaching and irradiation stability.

REFERENCES

[1] F. Bart, S. Sounilhac, J.L. Dussossoy, A. Bonnetier and C. Fillet, "Development of new vitreous matrices for conditioning cesium" pp. 353-360 in American Ceramic Society Conference Proceedings 1999, CERAMIC TRANSACTIONS, vol 119, 2001.

[2] R.H. Mitchell et al, "Henrymeyerite, a new hollandite-type Ba-Fe titanate from the Kovdor complex, Russia", *The Canadian Mineralogist*, **38** 617-626 (2000).

[3] Dmitriyeva et al, "Crystal chemistry of natural Ba-(Ti,V,Cr,Fe, Mg,Al)-hollandite", *Dokl. Acad. Sci. USSR, Earth Sci. Sect.*, **326** 158-162 (1992).

[4] A.E. Ringwood and S.E. Kesson, "Synroc" p. 233-334 in *Radioactive Waste Forms for the Future*, edited by W. Lutze et R. Ewing, 1988.

[5] Kesson S.E. and White T.J., "[Ba$_x$Cs$_y$][(Ti,Al)$^{3+}_{2x+y}$Ti$^{4+}_{8-2x-y}$]O$_{16}$ Synroc-types hollandites Part I : Phase chemistry", *Proceeding of the Royal Society of London* pp73-101 **A 405** (1986), and "[Ba$_x$Cs$_y$][(Ti,Al)$^{3+}_{2x+y}$Ti$^{4+}_{8-2x-y}$]O$_{16}$ Synroc-types hollandites Part II : Structural Chemistry", *Proceeding of the Royal Society of London* pp 295-319 **A 408** (1986).

[6]Leturcq G., Bart F. Compte A., Brevet intitulé Céramique de structure hollandite incorporant du césium utilisable pour un éventuel conditionnement de césium radioactif et ses procédés de synthèse N°E.N. 01/15972 (Décembre 2001).

[7] Carter M.L. and Vance E .R., "Leaching studies of Synroc-types hollandites" in *PacRim2 CD*, edited by P. Walls symposium **16** (1998)

HOLLANDITE-RICH TITANATE CERAMICS PREPARED BY MELTING IN AIR

M.L. Carter[1,2], E.R. Vance[1] and H. Li[1]
[1]Australian Nuclear Science and
Technology Organisation, New
Illawarra Rd, Lucas Heights, NSW
2234, Australia
[2]School of Chemistry, The University
of Sydney, Sydney, NSW 2006,
Australia

ABSTRACT

Leach resistant titanate hollandite-rich ceramic melts incorporating Cs have been prepared in air by melting, in spite of the previously observed high Cs leach rates of (hollandite-bearing) synroc-C melted in air. The absence of Mo in the samples facilitated the incorporation of the Cs into the hollandite phase. Detailed analysis of the samples by electron microscopy is presented on materials rich in Fe- and Mn- substituted titanate hollandites which were fabricated by melting in air. These samples had normalised Cs leachate concentrations of < 0.3 g/l in the PCT-B test. This paper also reports the solid solubility limits of Cs in single-phase hollandite $Ba_xCs_y(M^{3+})_{2x+y}Ti_{8-2x-y}O_{16}$ and $Ba_xCs_y(M^{2+})_{x+y/2}Ti_{8-x-y/2}O_{16}$ ceramic formulations where M = Mn or Fe.

INTRODUCTION

Immobilisation of high-level radioactive wastes in stable matrices for long term storage or geological disposal is an important step in the closed nuclear fuel cycle. In recent years there has been renewed interest in the development of ceramic waste forms for separated radionuclei, notably in Europe and Japan. Of particular interest are the cesium isotopes ^{137}Cs and more particularly [135]Cs which has been identified in performance assessment studies as a major contributor to long-term releases from repositories. These are among the most difficult radioisotopes to immobilise because of their volatility at high temperatures and tendency to form water-soluble compounds.

The hollandite group of minerals have the general formula $A_xM_yB_{8-y}O_{16}$. The M and B cations are surrounded by octahedral configurations of oxygens. Each of these $(M,B)O_6$ octahedra share two edges to form paired-chains running parallel to the c-axis. These chains are corner-linked to neighbouring paired-chains to form a 3-dimensional framework with tunnels running parallel to the c-axis. The large A cations are located in these tunnels. The synroc-type hollandite $(Ba_xCs_yM^{3+}{}_{2x+y}Ti_{8-2x-y}O_{16}$ where trivalent M = Al in air and Al and Ti in reducing conditions) is well known for its ability to incorporate Cs when produced by hot pressing[1,2].

Attempts to melt (hollandite-bearing) synroc-C under air atmospheres resulted in the formation of all the synroc titanate phases, including hollandite[3,4]. However leachable Cs molybdate phases were also formed due to the presence of Mo^{6+}. Day et al. [5] found that the Cs did not enter the hollandite phase in melted synroc-C and in order to incorporate the Cs in the hollandite the synroc-C melts had to be processed under reducing atmospheres. The reducing atmosphere had two effects on the hollandite and the Cs in the melt system. Firstly the Mo was reduced to metal and no longer was available to combine with the Cs to form soluble phases, and secondly the hollandite contained Ti^{3+} in the M site.

In the current work we investigate alternate formulations for hollandite in hollandite-rich ceramics produced by the melt route in air. The strategy was to replace the Al in the hollandite with cations of larger ionic radii to expand the tunnels in the structure and thereby facilitate the entry of Cs. The elements chosen for this purpose were Ga, Fe and Mn.

EXPERIMENTAL

Adding 7.5wt% Cs_2O to synroc-B (the precursor of synroc-C), giving rise to a hollandite-rich sample in which the B site atom was Al, the formulation was then further adjusted to contain more hollandite (resulting in Hollandite-rich Al). The strategy used to formulate the Hollandite-rich Ga, Fe and Mn samples was to replace the Al_2O_3 in the Hollandite-rich Al samples with the same wt% of the other oxides (Ga, Fe and Mn) assuming they were in the 3+ oxidation state.

Samples (see Table 1) for melting were produced by the alkoxide-route[6]. This method involves mixing the correct molar quantities of aluminum sec-butoxide (where necessary), titanium (IV) isopropoxide and zirconium tertbutoxide in ethanol with nitrates of other components dissolved in water, while continuously stirring. The mixture was then heated to ~110°C to drive off the alcohol and water. The dry product was then calcined in air for two hours at 750°C. The samples were placed in Pt crucibles and melted in air at 1450-1550°C.

Nominally pure hollandite samples for sintering were produced using the above method up to the calcination stage, then wet ball milled in cyclohexane and sintered in air or argon at 1300°C for 20 hours. The compositions of the sintered samples were $Ba_xCs_y(M^{3+})_{2x+y}Ti_{8-2x-y}O_{16}$ with M = Fe or Mn and x + y = 1.2 for the air sintered samples, and $Ba_xCs_y(M^{2+})_{x+y/2}Ti_{8-x-y/2}O_{16}$ with M = Fe or Mn and x + y = 1.2 for the argon sintered samples, x and y were varied from 0 to 1.2. In the sintered samples x + y = 1.2 was chosen but the exact Ba + Cs occupancy is unknown. However previous workers [7,8] have shown the titanate hollandites to have (Ba, Cs) occupancies in the range $1.03 < x + y < 1.51$. All sintered samples were made with an additional 0.5 moles of TiO_2 (nominal). This was because the occupancy of (Ba + Cs) was not known in advance (see above) and the rutile would act as a buffer.

In addition three sintered samples, made by the above method, were subsequently melted at temperatures between 1550 and 1650°C. The nominal compositions of these samples were $Ba_{0.9}Cs_{0.3}Fe_{2.1}Ti_{5.9}O_{16}$ + 0.5 TiO_2, $BaCs_{0.1}Ga_{2.1}Ti_{5.9}O_{16}$ + 0.5 TiO_2 and $BaCs_{0.1}Al_{2.1}Ti_{5.9}O_{16}$ + 0.5 TiO_2.

A JEOL JSM6400 scanning electron microscope (SEM) equipped with a Noran Voyager energy-dispersive spectroscopy system (EDS) was operated at 15 keV for microstructural work.

Leach testing was carried out using the Product Consistency Test (PCT-B)[9].

Table 1: Compositions of samples for melt experiments

Sample	Composition	wt% oxide	Sample	Composition	wt% oxide
Synroc-B + 7.5wt% Cs_2O	Al_2O_3	5.10	Hollandite- rich Fe(b)	Fe_2O_3	10.32
	BaO	5.18		BaO	5.83
	CaO	10.18		CaO	7.34
	ZrO_2	6.11		ZrO_2	4.61
	TiO_2	66.05		TiO_2	64.40
	Cs_2O	7.50		Cs_2O	7.50
Hollandite- rich Al	Al_2O_3	6.74	Hollandite- rich Fe/K	Fe_2O_3	10.30
	BaO	5.08		K_2O	1.8
	CaO	9.97		CaO	7.30
	ZrO_2	5.98		ZrO_2	4.60
	TiO_2	64.72		TiO_2	68.4
	Cs_2O	7.50		Cs_2O	7.3
Hollandite- rich Ga	Ga_2O_3	6.74	Hollandite- rich Mn	Mn_2O_3	6.74
	BaO	5.08		BaO	5.08
	CaO	9.97		CaO	9.97
	ZrO_2	5.98		ZrO_2	5.98
	TiO_2	64.72		TiO_2	64.72
	Cs_2O	7.50		Cs_2O	7.50
Hollandite- rich Fe	Fe_2O_3	6.74			
	BaO	5.08			
	CaO	9.97			
	ZrO_2	5.98			
	TiO_2	64.72			
	Cs_2O	7.50			

RESULTS AND DISCUSSION

Sintered Samples of Nominal Hollandite Composition

Before studying the melted samples the intrinsic behaviour of Cs in titanate hollandites was explored by studying samples of nominal hollandite stoichiometries after sintering.

The Cs solubility limit in Al-substituted Ba hollandite was reported as $Cs_{0.25}Ba_{0.67}Al_{1.6}Ti_{6.4}O_{16}$ [10] but Vance and Agrawal [11] reported titanate hollandite with x up to 0.45 in the formula $Ba_{1-x}Cs_{2x}Al_2Ti_yO_{2y+4}$ (y = 5 or 6). Thorogood et al. [12] found single phase Al-containing hollandites of $Ba_xCs_y(M^{3+})_{2x+y}Ti_{8-2x-y}O_{16}$ stoichiometry to have 0.5 f.u. of Cs ions on the A site. At higher Cs substitutions, 90% of the Cs inventory was found on the A site of hollandite, but other Cs containing phases, mainly $Cs_2Al_2Ti_2O_8$, were also formed. The maximum Cs solid solubility in the hollandite corresponded to y = 0.9. They also established that single-phase Ga -containing hollandites could be made with up to 91% of the barium ions replaced by Cs ions, and in fully Cs- substituted Ga-containing material (nominally $Cs_{1.5}Ga_{1.5}Ti_{6.5}O_{16}$) hollandite of the nominal stoichiometry was formed along with other Cs-and Ga-rich phases. No literature values could be found for Cs solubility limits in Fe- and Mn- containing hollandites.

SEM investigation of the Fe-hollandites sintered in air (Fe^{3+}) and argon (Fe^{2+}) found the samples to contain major hollandite, rutile and minor amounts of Fe titanate. All the Cs had entered the hollandite even in the fully Cs substituted samples, indicating that the Fe was mainly in the targeted oxidation state because the samples displayed the nominal stoichiometries. The presence of a small amount (< 1%) of Fe titanate in some of the samples could be due to slight Cs volatilisation losses at the relatively high sintering temperature.

SEM investigation of the Mn-hollandites sintered in air (Mn designed to be 3+) showed the samples to contain major hollandite, Mn titanate and minor amounts of rutile. The EDS analysis of these samples however indicated that the Mn in the hollandite was not in the 3+ state but in the 2+ state. The argon-sintered samples targeting the Mn as 2+ resulted in all the samples containing hollandite and only minor rutile across the full range of Cs substitutions, with no additional Mn titanate forming. The hollandite stoichiometries were consistent with all the Mn being divalent in both air and argon atmospheres.

Melting Experiments

Single Phase Hollandites

The viability of melting near single phase hollandites (TiO_2 secondary phase) was investigated. SEM of the melted (1550°C) and previously sintered Fe-containing hollandite showed it to consist of major hollandite containing Cs, rutile and Cs titanate. Both the Ga and Al containing samples did not melt until the temperature was 1650°C. SEM investigation showed the samples to contain major Cs-free hollandite (by EDS), with minor amounts of rutile and $Cs_2Al_2Ti_2O_8$ in the Al-containing sample and $Cs_2Ga_2Ti_2O_8$ in the Ga-containing ones.

Synroc-B + 7.5wt% Cs_2O and Hollandite-rich Al

As discussed earlier, the presence of Mo was possibly one mechanism which prevented Cs from entering hollandite when synroc-C was melted in air. Therefore it was interesting to see whether the removal of Mo would allow the Cs to be incorporated in the hollandite in a mixed titanate ceramic on melting in air.

[Another possibility was the necessity of having Ti^{3+} in the hollandite to retain the Cs (see above).]

The Synroc-B containing 7.5wt% Cs_2O was melted in air at 1550°C. The SEM investigation found the sample to contain hollandite (~50%) plus zirconolite, perovskite, rutile and $Cs_2Al_2Ti_2O_8$ (see figure 1). EDS analysis indicated that the Cs would enter the hollandite under oxidising conditions but the presence of a small amount of highly leachable $Cs_2Al_2Ti_2O_8$, containing <1% of the Cs inventory, rendered the sample unsuitable as a wasteform.

Attempts were made to adjust the formulation by reducing the Cs_2O content, and/or increasing amount of hollandite in the sample. The Hollandite-rich Al sample was reformulated to contain more hollandite (~60%) by the addition of Al_2O_3 to avoid the formation of $Cs_2Al_2Ti_2O_8$ but all samples melted in air contained this undesirable phase at approximately the same levels as the Synroc-B + 7.5 wt% Cs_2O sample. It was concluded that the Mo in the synroc-C air- melted samples of earlier workers [3,4,5] and not the absence of Ti^{3+} stops the Cs entering

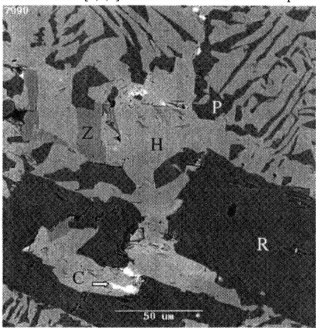

Figure 1: SEM backscattered electron image of Synroc-B containing 7.5wt% Cs_2O by melting in air at 1550°C. H = hollandite, Z = zirconolite, P= perovskite, R = rutile and C = $Cs_2Al_2Ti_2O_8$

the hollandite. From this result and the result obtained above on the air-melted previously sintered single phase Al-hollandite, it would appear that the addition of the additional titanate phases not only reduces the melting temperature but also aids in the incorporation of the Cs in the hollandite structure.

Hollandite-rich Ga

The Hollandite-rich Ga sample was melted in air at 1550°C. The SEM investigation found the sample to contain major hollandite containing the majority of the Cs and rutile plus minor zirconolite, perovskite, $Cs_2Ga_2Ti_2O_8$ and Cs titanate (see figure 2). The Ga-containing sample behaved in a similar manner to the Synroc-B plus 7.5 wt% Cs_2O when attempts were made to design melts to prevent $Ga_2Al_2Ti_2O_8$ and Cs titanate from forming, insofar as all resultant melt samples contained some $Cs_2Ga_2Ti_2O_8$. As $Cs_2Ga_2Ti_2O_8$ is an analogue of $Cs_2Al_2Ti_2O_8$ and very reactive with water, samples containing this phase were regarded as unsuitable as wasteforms.

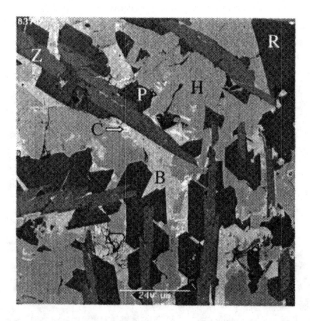

Figure 2: SEM backscattered electron image of Hollandite-rich Ga. H = hollandite, Z = zirconolite, P= perovskite, R = rutile, C = $Ga_2Al_2Ti_2O_8$ and B = Cs titanate

Hollandite-rich Fe, Fe(b) and Fe/K

The Hollandite-rich Fe and Fe(b) samples were melted in air at 1500°C and the Hollandite-rich Fe/K sample in air at 1450°C. SEM of the Hollandite-rich Fe sample found it to contain Cs-bearing hollandite, zirconolite, perovskite, rutile, and Cs titanate (see figure 3a). The sample was redesigned (Hollandite-rich Fe(b)) to try to remove the Cs titanate phase, by adding more Fe_2O_3 and BaO. The effect of this was to increase the proportion of hollandite in the sample. Thus

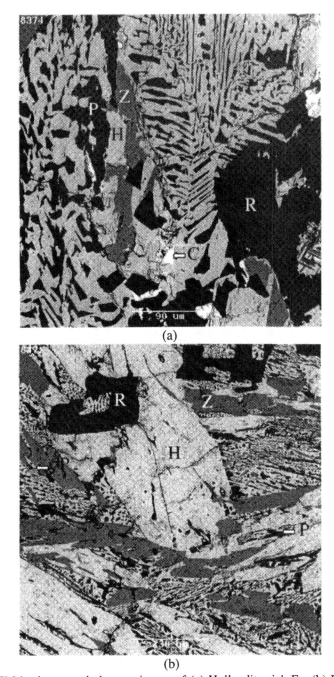

(a)

(b)

Figure 3: SEM backscattered electron image of (a) Hollandite-rich Fe, (b) Hollandite-rich Fe(b). H = hollandite, Z = zirconolite, P= perovskite, R = rutile and C = Cs titanate

the sample contained more hollandite and less of the other titanate phases. The SEM investigation of Hollandite-rich Fe(b) (see figure 3b) found the sample to contain major Cs-bearing hollandite, zirconolite, perovskite and rutile. This sample was deemed to have a composition suitable for a wasteform because no leachable phases were present. The Hollandite-rich Fe/K sample was designed in an attempt to lower the melting temperature. The SEM investigation of Hollandite-rich Fe/K sample found the sample to contain only major Cs-bearing hollandite, zirconolite, perovskite and rutile so again this composition should be suitable for a wasteform.

Hollandite-rich Mn

The Hollandite-rich Mn sample was melted in air at 1550°C. The SEM investigation found it to contain major Cs-bearing hollandite, zirconolite, perovskite, and rutile (see figure 4). Yet again the absence of easily leachable Cs-rich phases suggests it would also be potentially suitable as a wasteform.

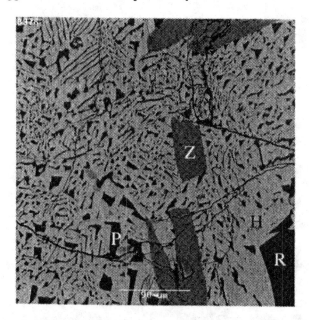

Figure 4: SEM backscattered electron image of Hollandite-rich Mn. H = hollandite, Z = zirconolite, P= perovskite and R = rutile

PCT-B Leach Testing

PCT-B leach testing was carried out on the air melted hollandite-rich Fe and Mn samples. Table 2 lists the results for the normalised Cs leachate concentrations. The leach results for the Hollandite-rich Fe sample were appreciably higher than those for the other samples (normalised Cs leachate concentration of 2.39 g/l), reflecting the presence of the Cs titanate in this sample. (see Figure 3a). By designing this phase out of the sample, to form Hollandite-rich

Fe(b) (see Figure 3b), the normalised Cs leachate concentration is reduced to 0.12 g/l. The leach results for the other two samples were comparatively low. Note that the normalised PCT-B leachate concentrations for the reference EA glass[9] for Na, Li and B are 13-16 g/l, two orders of magnitude above our lowest concentrations. The reason for the variation in Cs leachate concentration between the Titanate Fe(b), Fe/K and Mn samples is unknown, but were only a factor of 5 different between highest and lowest.

Table 2: PCT–B leach test results showing normalised Cs leachate concentrations.

Sample	normalised Cs leachate concentrations, g/l
Hollandite-rich Fe	2.39*
Hollandite-rich Fe(b)	0.12
Hollandite-rich Fe/K	0.62
Hollandite-rich Mn	0.27
Reference EA glass	13-16 (for Na, Li and B)

* Sample contained small amount of Cs titanate (see figure 3a)

CONCLUSIONS

Leach resistant titanate hollandite-rich ceramic melts incorporating ~7.5 wt% Cs_2O were prepared when the hollandite in the sample contained Fe or Mn. The normalised PCT-B Cs leachate concentrations for the melted samples are two orders of magnitude better than the Na, B and Li concentrations for reference EA glass. The absence of Mo in the samples studied facilitated the incorporation of the Cs into the hollandite phase in mixed titanate formulations. Cs could be fully substituted in the single-phase hollandite $Ba_xCs_y(M^{3+})_{2x+y}Ti_{8-2x-y}O_{16}$ and $Ba_xCs_y(M^{2+})_{x+y/2}Ti_{8-x-y/2}O_{16}$ where M = Mn or Fe in sintered samples.

References

[1] S.E. Kesson, 'The immobilization of Cesium in Synroc Hollandite', *Radioactive Waste Management Nuclear Fuel Cycle*, **2**, 53-71 (1983)

[2] M. L. Carter, E. R. Vance, D. R. G. Mitchell , J. V. Hanna, Z. Zhang, E .Loi, 'Fabrication, Characterisation, and Leach Testing of Hollandite $(Ba,Cs)(Al,Ti)_2Ti_6O_{16}$', *Journal of Materials Research*, **17** (10), 2578-89 (2002)

[3] R. A. Day, 'Report to CEA on Phase Composition and microstructure of Cold Crucible melted Synroc-C', ANSTO report to Commissariat à l'Energie Atomique (CEA), December 1995

[4] A. V. Kudrin, B.S. Nikonovans and S.V. Stefanovsky, 'Chemical Durability Study of Synroc-C ceramics produced by Through-Melting Method', in Scientific Basis for Nuclear Waste Management XX, edited by W.J. Gray and I.R. Triay, Materials Research Society, Pittsburgh, PA, USA, pp 417-423, (1997)

[5] R. A. Day, M. La Robina, S. Moricca, T. Eddowes, N. Blagojevic, and M.L. Carter, ' Influence of Melting Atmosphere on Synroc-C Microstructure and Phase

Composition', presented at HLW and Pu Immobilization meeting, Saclay, France, April 22-23, 1999

[6] A.E. Ringwood, S.E. Kesson, K.D. Reeve, D.M. Levins and E.J. Ramm, 'Synroc', in *Radioactive Waste Forms for the Future*, edited by W. Lutze and R.C. Ewing, (North-Holland, New York,) p. 233-334 (1988).

[7] R.W. Cheary, 'A Structural Analysis of Potassium, Rubidium and Caesium Substitution in Barium Hollandite', *Acta. Cryst.* **B43,** 28-34 (1987)

[8] S.E. Kesson and T.J. White,'$[Ba_xCs_y](Ti, Al)^{3+}_{2x+y}Ti_{8-2x-y}O_{16}$ Synroc-type Hollandites I. Phase Chemistry', *Proceedings of the Royal Society (London)* **A408,** 73 -101 (1986).

[9] PCT is based on the ASTM Designation: C 1285-97 [Determining Chemical Durability of Nuclear Hazardous and Mixed Wastes]

[10] R. W. Cheary and J. Kwiatkowska, 'An X-ray Structural Analysis of Cesium Substitution in the Barium Hollandite Phase of Synroc', *Journal of Nuclear Materials,* **125,** 236-43 (1984)

[11] E.R. Vance and D. K. Agrawal, 'Incorporation of Radionuclides in Crystalline Titanates', *Nuclear and Chemical Waste Management*, **3,** 229-34 (1982)

[12] G.J. Thorogood, M.L. Carter, J. V Hanna and E.R. Vance , 'Synthesis and Characterisation of Ba-Cs Titanate Hollandites', presented at Austceram 2002, Perth,WA, Australia, 29 September-5 October 2002

HYPERFINE INTERACTION STUDY OF SHORT RANGE ORDER IN ZIRCON

H. Jaeger and K. Pletzke
Department of Physics
Miami University
Oxford, OH 45056, USA

J. M. Hanchar
Department of Earth and Environmental Sciences
The George Washington University
Washington, DC 20006, USA

ABSTRACT

In recent years zircon has been studied as model substance for the immobilization of actinide-bearing radioactive waste. Metamict zircon recrystallizes during heating above 800°C, a process that has been linked to a displacive phase transition. We used perturbed angular correlation spectroscopy to determine the short range order of zircon near a *Zr* lattice site. Measurements of the electric field gradient (EFG) at the zircon lattice site were performed between room temperature and 1100°C. Both commercially obtained zircon, as well as a natural specimen from the *Mud Tank* region in central Australia were studied. All specimens show a narrow distribution of EFGs and have asymmetry values close to, but distinctly different from zero. While the quadrupole interaction frequency for all specimens decreases linearly with increasing temperature, the quadrupole frequency for the *Mud Tank* zircon specimen shows a change in slope near 800°C. This change in slope is thought to be due to a subtle rearrangement of the unit cell first observed in a synthetic zircon crystal. It is not known why the commercial zircon does not exhibit this behavior.

INTRODUCTION

Zircon (ZrSiO4) is one of the oldest minerals found on earth, and in the solar system, and serves as the primary commercial resource for the production of zirconium metal and zirconia-based advanced ceramics. In addition zircon is widely used in the ceramic, foundry, and refractory industry. More recently zircon and zircon-based materials have been studied a great deal as model substance for the immobilization of thorium, uranium and plutonium-bearing radioactive waste materials [1]. Zircon is a common accessory mineral in igneous, metamorphic, and sedimentary rocks [2]. Zircon crystals typically have about 1 at% *Hf* impurities but have been found to contain up to several weight percent [3]. Moreover, *U* and *Th* substitute quite easily on the *Zr* lattice site and preclude the incorporation of Pb, a fact that makes zircon one of the most useful minerals for radioactive *U-Th-Pb* geochronology [4,5,6]. α-decay of *U* and *Th*, and, to a lesser degree, spontaneous fission of *U* give rise to radiation damage. Thus most zircons are found with partial or completely amorphous crystal structure commonly referred to as a *metamict* state. Metamict zircon differs in a number of physical and chemical properties from crystalline zircon [7,8,9]. Metamict zircon recrystallizes after annealing at temperatures above 800°C [10], a process that has been studied in detail over the years, as it allows characterization of the structural states during the recovery process and, in a way, represents the inverse of the damage process. It has been suggested that the transformation from metamict to crystalline zircon is related to a displacive phase transition that was found to occur in a synthetic zircon crystal near 800°C [11]. The transition is initiated by a sudden increase of the *Si-O* bond length, which then affects the *Zr-O* coordination and results in a significant change of the thermal expansion of the unit cell above 800°C. The purpose of this work is to study the short-range order of crystalline zircon near a *Zr* lattice site using time-differential perturbed angular correlation spectroscopy (PAC) with the goal to find evidence for a change in the *Zr-O* coordination that occurs during the displacive phase transition seen in synthetic zircon. PAC is a nuclear probe technique that measures the interaction of radioactive probe nuclei with electric and magnetic fields due to the surrounding charges and spins. This technique is particularly suitable here because a *Hf* isotope is employed as probe nucleus, and we have used it previously in the study of annealing behavior of metamict zircon [12,13,14]. In this paper we report the results of PAC experiments performed with a commercially obtained zircon (*Aldrich*), as well as a zircon specimen from the *Mud Tank* carbonatite in central Australia. The latter is unique in that it contains very little impurities; in particular the concentration of radioactive impurities is very small so that *Mud Tank* zircon is practically free of radiation damage. The commercially obtained zircon has a *U* & *Th* concentration sufficient to cause moderate radiation damage, however, the material was annealed as part of the preparation process to achieve a highly crystalline form.

EXPERIMENTAL DETAILS AND PROCEDURE

Perturbed angular correlation spectroscopy (PAC) is a nuclear probe technique that measures the hyperfine interaction of a nuclear moment with extra-nuclear fields. Probe nuclei introduced into the material under study decay by successive emission of two angularly correlated γ-rays. If the nuclei interact with electric or magnetic fields due to charges and spins surrounding the probe nuclei the angular correlation is altered and described by a perturbation function $G_2(t)$. In our experiments the quadrupole moment of the probe nuclei interacts with the electric field gradient (EFG) due to surrounding charge distributions, and the perturbation function is of the form

$$G_2(t) = s_{20}(\eta) + \sum_{i=1}^{3} s_{2i}(\eta)\cos(g_i(\eta)v_Q t)\exp(-g_i(\eta)\delta t) \qquad (1)$$

where v_Q is the quadrupole frequency

$$v_Q = \frac{eQV_{zz}}{h} \qquad (2)$$

that expresses the interaction of the probe nuclei's known quadrupole moment, Q, with the magnitude of the EFG, V_{zz}, due to the surrounding charge distribution. The asymmetry parameter

$$\eta = \frac{V_{xx} - V_{yy}}{V_{zz}} \qquad (3)$$

may vary between 0 and 1 and is 0 for axially symmetric crystal structures, such as zircon; in this case the frequencies $\omega_i = g_i(\eta)v_Q$ in the perturbation function have the characteristic ratio $\omega_1 : \omega_2 : \omega_3 = 1 : 2 : 3$. The exponential damping parameter δ describes the width of a quadrupole frequency distribution about v_Q due to crystal defects and impurities. The coefficients s_{2i} and g_i are independent of v_Q and depend only weakly on η. The PAC spectrometer used in this work records delayed coincidences of γ-ray pairs with four BaF_2 detectors arranged in planar 90° geometry and a standard slow-fast logic [15]. Taking appropriate ratios of the delayed coincidence spectra results in the experimental spectra ratio $R(t)$ which is the product of the perturbation function $G_2(t)$ with a known effective anisotropy A_2. Computer fitting $R(t)$ to equation (1) allows determination of the EFG parameters v_Q, η, and δ.

In any PAC experiment the first questions that arise are about which probe to employ and how to introduce the probe nuclei into the material of interest. In our case a natural choice of probe is ^{181}Ta, a decay product of Hf, the most abundant impurity in zircon. ^{181}Ta is produced *in-situ* by irradiation of a zircon sample with thermal neutrons. This turns natural ^{180}Hf into radioactive ^{181}Hf which decays to the PAC probe nucleus ^{181}Ta with a half-life of 42.5 days. Since ^{181}Ta is a decay product of *Hf*, the PAC probe resides on a *Zr* lattice site. For our experiments zircon specimens were ground to a fine powder in an agate mortar, sealed in an evacuated fused silica capsule, and irradiated for two hours in a thermal neutron flux at *Ohio State University's Nuclear Reactor Laboratory*. Typically this produced 30 µCi of ^{181}Hf activity. In addition a number of other isotopes are produced, but in most cases these are short-lived and decay within a few days and do not interfere with the measurements. A sample retains adequate activity to be used for three months. Typical spectra accumulation times were 20 – 50 h. During experiments the sample was located in a small furnace in the center of the detector arrangement, and the sample temperature was controlled to ±1°C. Series of measurements were performed between room temperature and 1100°C.

RESULTS AND DISCUSSION

Crystalline zircon has a tetragonal unit cell (space group I4₁/amd). Zr atoms are 8-fold coordinated by oxygen forming ZrO_8 triangular dodecahedra, which are separated by isolated SiO_4 tetrahedra forming chains along the crystallographic c-axis [16]. The Zr-sites have axial symmetry, and the EFG of crystalline zircon is expected to have zero asymmetry ($\eta = 0$). Typical PAC spectra of *Aldrich* and *Mud Tank* zircon are shown in Figures 1 and 2, respectively. The solid lines in the *R(t)* graphs are results of least-squares computer fits to Eq. 1. The Fourier transforms *A(ω)* show three frequencies characteristic of an electric quadrupole interaction with nearly axially symmetric probe environment. The spectra for the two zircon-specimens are similar, however, the *Aldrich* zircon spectra show somewhat greater damping and a slightly greater asymmetry parameter that

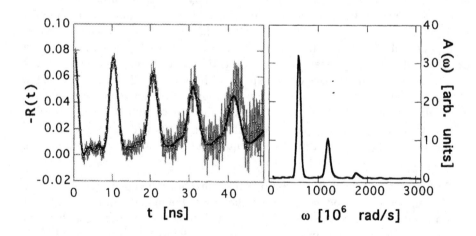

Fig. 1: The experimental perturbation function R(t) (left) and its Fourier transform (right) for *Aldrich* zircon.

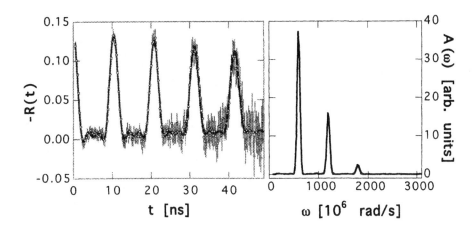

Fig. 2: The experimental perturbation function R(t) (left) and its Fourier transform (right) for *Mud Tank* zircon.

manifests itself in a less regular pattern of the perturbation function $R(t)$ and broader lines in $A(\omega)$. Figure 3 shows the temperature dependence of the EFG parameters of the *Aldrich* zircon specimen. The quadrupole frequency v_Q decreases linearly with increasing temperature from 675 MHz at room temperature to 637 MHz at 1000°C with a slope of -39 ± 2 kHz/°C. In the same temperature range the lattice parameters of a synthetic zircon crystal were reported to increase about 0.5 to 0.7 % [11]; since V_{zz} in Eq. 2 scales like r^{-3} the measured decrease of v_Q is somewhat greater than that expected due to the increase of the lattice parameters. The asymmetry parameter η and the EFG distribution δ are pretty much constant over the investigated temperature range, with average values of $\eta_{av} = 0.086 \pm 0.002$ and $\delta_{av} = 21.4 \pm 0.1$ MHz. It is interesting to note that the average η is significantly different from zero, a value that would be expected for a tetragonal structure such as zircon. It turns out that most commercial zircons and naturally occurring zircons that were annealed show small asymmetries of order 0.1 due to impurities, lattice imperfections, and residual radiation damage from α-emitting impurities such as *U* and *Th* [12].

Figure 4 shows the temperature dependence of the EFG parameters of the *Mud Tank* zircon specimen. As with the *Aldrich* zircon sample η and δ do not vary significantly with temperature; the averages over the investigated temperature range are $\eta_{av} = 0.041 \pm 0.007$ and $\delta_{av} = 5.8 \pm 0.3$ MHz, significantly smaller than the corresponding values of the *Aldrich* zircon. Chemical analyses

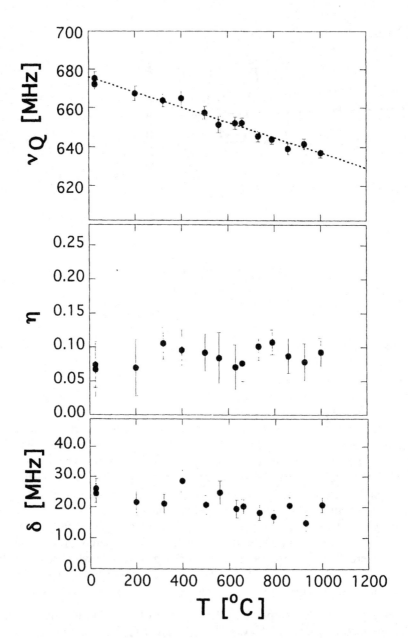

Fig. 3: EFG parameters for the *Aldrich* zircon specimen. The dotted line is a
guide to the eye.

show that *Mud Tank* zircon is very pure; besides the about 1 mol% *Hf* impurities it has fewer than 20 ppm *U* and *Th* and a total of 10 ppm rare earth elements [17, 18]. In particular, the very low concentrations of radioactive *U* and *Th* is the reason that zircon from the *Mud Tank* carbonatite shows no radiation damage and is not metamict as many other zircons are. In contrast, the *Aldrich* zircon sample has, besides 1 mol% Hf, about 210 ppm of *U* and 120 ppm of *Th* impurities, as well as *Al, Ti, Fe*, and *P* on the order of a few hundred ppm each [19]. Unit cell parameters of *Mud Tank* zircon determined by X-ray diffractometry are practically identical with those of synthetically grown zircon. Because of the low impurity content in the *Mud Tank* zircon, all PAC probe nuclei experience very nearly the same hyperfine interaction, which in turn results in a very narrow distribution of quadrupole frequencies v_Q, and an asymmetry parameter value very near zero, the theoretical value for the zirconium site in tetragonal zircon. It is conspicuous that the error bars of the *Mud Tank* η *vs. T* data are significantly bigger than those of the *Aldrich* data. The asymmetry η is calculated from the frequency ratio ω_2/ω_1. This ratio becomes very steep near $\eta = 0$, and propagation of error magnifies the errors of $\omega_i = g_i(\eta)v_Q$. This effect grows increasingly notable for $\eta < 0.1$. Attempts of fitting the spectra with an η fixed to zero consistently returns fits of lesser quality, therefore we conclude that η is significantly different from zero and the large error bars are an unfortunate artifact of the analysis. The temperature dependence of v_Q for the *Mud Tank* zircon differs somewhat from that of the *Aldrich* specimen. The v_Q values for *Mud Tank* and *Aldrich* zircon agree well from room temperature to about 800°C; the slope of the dotted line below 800°C in Figure 4 is -35 ± 4 kHz/°C, equal to the slope of the dotted line in Figure 3. At higher temperatures v_Q of *Mud Tank* zircon decreases faster with increasing temperature; the slope of the dotted line above 800°C in Fig. 4 is -70 ± 12 kHz/°C, nearly twice the value below 800°C. The change in the slope of the v_Q *vs. T* data indicates a change in the environment of a *Zr* lattice site. The oxygen atoms in the ZrO_8 dodecahedron can be divided in two groups, with four *O* are at a distance of 2.13 Å (group I) and four at 2.27 Å (group II) [16]; the group-I *O*-atoms share a corner with an adjacent SiO_4 tetrahedron, while the group-II *O*-atoms share an edge with a SiO_4 tetrahedron. As the *Si-O* bond lengths begin to increase at 800°C the *O*-atoms in the *Zr*-coordination are forced to rearrange as well. Because of the distinct coupling between the ZrO_8 and SiO_4 groups (edge-sharing *vs.* corner-sharing), this results in a different thermal behavior of the *Zr-O* bond lengths involving group-I and group-II oxygen atoms. While the *Zr-O$_{II}$* bond lengths increase steadily with increasing temperature, the *Zr-O$_I$* bond lengths remain nearly constant between 800 and 1000°C. Since the EFG at the *Zr* lattice site is determined to a large part by the location of the charge distribution around the *O*-atoms, the EFG will reflect any change in the *Zr-O* coordination and show a different thermal behavior then before the transition. Our v_Q *vs. T* data in Fig. 4 suggest a change in the EFG magnitude beginning near 800°C consistent with the one described by *Mursic et al.* [11]. Naturally the

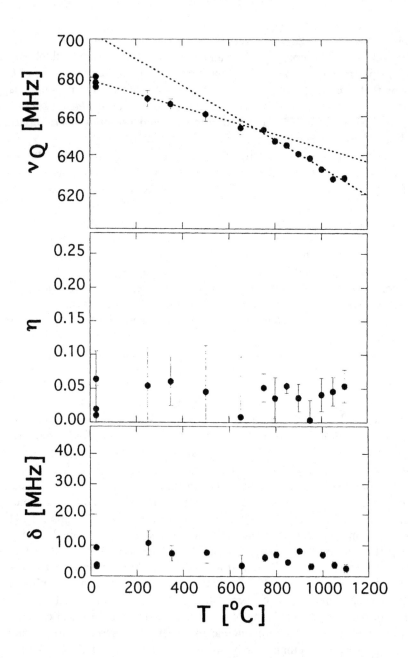

Fig. 4: EFG parameters for the *Mud Tank* zircon specimen. The dotted lines are a guide to the eye.

question arises why evidence for such a transition is not seen in the v_Q vs. T data of the *Aldrich* zircon (Fig. 3). We do not have a clear answer for this. The primary difference between the *Mud Tank* and *Aldrich* zircons is the different impurity concentration that results in a factor two lower asymmetry and a factor four lower quadrupole frequency distribution for the *Mud Tank* zircon. It is conceivable that the higher impurity concentration in the *Aldrich* zircon suppresses the subtle change in the bond length of the $Zr\text{-}O_I$. More experiments need to be performed to test this hypothesis.

SUMMARY AND CONCLUSIONS

We performed PAC experiments on crystalline zircon specimens with the goal to find evidence for a subtle rearrangement in the $Zr\text{-}O$ coordination. PAC spectroscopy is well suited for this study as it employs a nuclear probe derived from *Hf*, the primary impurity of zircon. Therefore PAC views the structure of zircon on a nanoscopic level from the vantage point of a *Zr* lattice site. Spectra for both the commercially obtained zircon and the zircon from the *Mud Tank* carbonatite in Australia show well-defined perturbation functions and spectra characteristic for a nearly axially symmetric probe environment. Because of the higher impurity concentration in the *Aldrich* zircon the values for asymmetry η and quadrupole frequency distribution δ are significantly higher than those in the *Mud Tank* zircon. The quadrupole frequencies v_Q for both samples are identical for temperatures below 800°C. Above 800°C the *Mud Tank* zircon v_Q data decreases faster with temperature than the corresponding *Aldrich* data. This change in slope of the v_Q vs. T data is interpreted as evidence of a subtle rearrangement of the $Zr\text{-}O$ coordination. It is consistent with the description given by *Mursic et al.*, who observed such a rearrangement in a synthetic zircon crystal. It is not known why evidence for this transition is not seen in the *Aldrich* zircon data, but we believe that the distinctly higher impurity concentration in the *Aldrich* zircon is at least in part responsible.

ACKNOWLEDGMENTS

The authors wish to acknowledge the *DOE Reactor Sharing Program* for supporting the neutron irradiations performed at *Ohio State University's Nuclear Reactor Laboratory*.

REFERENCES

[1]   R. C. Ewing, W. Lutze, and W. J. Weber, J. Mater. Res. **10** (1995) 243.

[2]   P. W. O. Hoskin and U. Schaltegger, in Zircon. Reviews in Mineralogy. J. M. Hanchar and P. W. O. Hoskin (eds), Mineralogical Society of America, Washington, D.C. (in press).

[3]   R. J. Finch and J. M. Hanchar, in <u>Zircon. Reviews in Mineralogy</u>. J. M. Hanchar and P. W. O. Hoskin (eds), Mineralogical Society of America, Washington, D.C. (in press).

[4]   L. Heaman and R. R. Parrish, in <u>Applications of Radiogenic Isotope Systems to Problems in Geology</u>, Mineralogical Association of Canada, short course handbook, **19** (1991).

[5]   R. R. Parrish and S. R. Noble, in <u>Zircon. Reviews in Mineralogy</u>. J. M. Hanchar and P. W. O. Hoskin (eds), Mineralogical Society of America, Washington, D.C. (in press).

[6]   E. Roth and B. Poty, editors, *Nuclear Methods of Dating*, Kluwer Academic Publishers, Dordrecht (1989).

[7]   H. D. Holland and D. Gottfried, Acta Cryst. **8** (1955) 291.

[8]   H. Ozkan, J. Appl. Phys. **47** (1976) 4772.

[9]   R. C. Ewing, R. F. Haaker, and W. Lutze, in <u>Scientific Basis for Radioactive Waste Management V</u>, W. Lutze, Editor, Materials Research Society Proceedings, North Holland, New York (1983).

[10] W. J. Weber, R. C. Ewing, and L. Wang, J. Mater. Res. **9** (1994) 688.

[11] Z. Mursic, T. Vogt, and F. Frey, Acta Cryst. **B48** (1992) 584.

[12] H. Jaeger, M. P. Rambo, and R. E. Klueg, Hyperfine Interactions, **136/137** (2001) 515.

[13] H. Jaeger, L. J. Abu-Raddad, and D. J. Wick, Appl. Radiat. Isotopes **48** (1997) 1083.

[14] H. Jaeger, M. P. Rambo, R. E. Klueg, Ceramic Transactions, Vol. 107, G. T. Chandler and X. Feng, Editors, American Ceramic Society, Westerville, Ohio, 2000.

[15] H. Jaeger and L. J. Abu-Raddad, Rev. Sci. Instrum., **66** (1995) 3069.

[16] K. Robinson, G. V. Gibbs, and P. H. Ribbe, Amer. Mineral. **56** (1971) 782.

[17] J. C. Jain, C. R. Neal, and J. M. Hanchar, Geostandards Newsletter, **25** (2001) 229.

[18] E. A. Belousova, W. L. Griffin, S. Y. O'Reilly, and N. J. Fisher, Contrib. Mineral. Petrol. **143** (2002) 602.

[19] H. Jaeger, M. P. Rambo, M. Uhrmacher, and K.-P. Lieb, Recent Res. Devel. Phys., **2** (2001) 81.

SCALE-UP OF LITHIUM ALUMINATE PELLET MANUFACTURING WITH A FLOWABLE POWDER

G. Hollenberg and L. Bagaassen
Pacific Northwest National Laboratory
Richland, Washington 99352

D. Tonn
Carpenter Advanced Ceramics
Auburn, California 95602

R. Kurosky
Praxair
Woodinville, Washington 98072

W. Carty
Alfred University
Alfred, New York 14802

ABSTRACT

Thin-walled, high-density lithium aluminate pellets are challenging to manufacture for nuclear reactor applications. The key to scale-up of production was the development of flowable, high density, lithium aluminate powder that permitted (1) automated isostatic pressing, (2) low compaction during pressing, (3) low shrinkage during firing, (4) elimination of chlorine-containing fumed alumina and (5) near-net shape forming. A triple spray drying process was developed that included: (I) a unique-feedstock blend cycle, (II) a post-calcination grinding cycle, and (III) a high- pH final cycle with high solids loading slurry that was spray dried into flowable high-density spheres with large, uniform diameters. Today, pellet manufacturing at a rate of more than 400,000 per year is possible.

INTRODUCTION

γ-LiAlO$_2$ has been considered a stable refractory material for production of tritium for several decades through the capture of either fusion or fission neutrons.[1-2] Recently, tritium producing burnable absorber rods (TPBARs) were being manufactured to produce tritium in commercial light water reactors (CLWR).[3] Unlike earlier pellets and fusion components, the CLWR pellets are thin-walled, high-density tubes. As in most ceramic manufacturing, the scale up to production was limited by the ability to synthesize a consistent, flowable powder. The transformation of a historical laboratory based powder manufacturing to one that supports present production is presented in this paper.

Prior to $LiAlO_2$ being adopted as a fusion and CLWR tritium producing target material, a flurry of powder development work had been completed in relation to its use in molten carbonate fuel cells.[4] In the 1960's and 1970's, the concept of reacting alumina with lithium carbonate to form different types of $LiAlO_2$ was repeatedly investigated by a variety of researchers.

In 1981, Arons et al. developed a powder synthesis process specifically aimed at sintering an $LiAlO_2$ pellet.[5] A high surface area, fumed-alumina (derived from aluminum chloride) was pretreated to remove the residual chlorine by a long heat treatment in moist air. After spray drying a slurry of the dechlorinated alumina and lithium carbonate, the resulting powder was calcined at 700°C so that the α-$LiAlO_2$, rather than the high temperature phase, γ-$LiAlO_2$, would be formed.[6] During sintering of the pellet, Arons took advantage of the 25% volume expansion during the transformation from α-$LiAlO_2$ to γ-$LiAlO_2$ as a densification enhancement mechanism since the densities of these phases are 3.40 gm/cm^3 and 2.55 gm/cm^3, respectively. Arons also observed that the high vapor pressure of lithium above $LiAlO_2$ would result in excessive lithium loss if higher temperatures were used to achieve densification.

Numerous researchers in the past two decades have attempted to prepare $LiAlO_2$ powder via an aqueous, sol-gel type of process with little success.[7] Aqueous solutions of Li and Al precipitate $LiAl_2(OH)_7 \cdot nH_2O$.[8] Calcination of this phase resulted in a powder depleted in lithium (i.e., Li/Al ratio of 0.5). Researchers have attempted to inject lithium via solutions, lithium rich phases, etc., into the sol-gel derived lithium-depleted calcine, but this is at best a poorly controlled process. Vollath,[7] on the other hand, proposed and demonstrated that Aron achieved success with their aqueous spray drying approach because $LiAl_2(OH)_7 \cdot nH_2O$ and LiOH crystals were trapped within each of the spray dried agglomerates.

PREVIOUS ANNULAR PELLET MANUFACTURING
Prior to the development work described herein, 2500 TPBAR pellets were manufactured using a powder synthesis process basically as described by Arons.[5] TPBAR pellets are nominally short thin-walled tubes: 50 mm long and 8 mm in diameter with only a 1mm wall. The pellets are sealed into 12 ft long stainless steel cladding prior to being inserted into the CLWR reactor core. Their specification includes requirements to be greater than 94% TD, and less than 7 μm grain size with low impurities, especially neutron absorbing isotopes and halides. The most important TPBAR pellet characteristic, however, is the precisely determined axial distribution of ^6Li (^6Li/in) within the pellets since this affects the accurate calculation and the maintenance of nuclear criticality. ^6Li enrichment over the natural level (7%) was necessary for all TPBAR designs.

Several commercial grades of α-LiAlO$_2$ for molten-carbonate fuel cells were available in 1997 but none possessed sufficiently high surface area to obtain the required sintered densities without using temperatures that would vaporize excessive amounts of lithium. Also, none of these materials had the required ^{6}Li enrichments. One of Arons' coworkers, R. Poeppel, synthesized the powder for that 1997 campaign utilizing multiple 1 kg batches. The powder was then hand-packed into molds consisting of a hardened steel, central mandrel and an outer rubber tube with end plugs at top and bottom. Because of the low density and poor flowability of the powder, workers had to tamp the powder into the relatively thin gap between the mandrel and the outer tube. A set of individual molds was loaded into a large isostatic press and, after pressing, the mold was manually disassembled and the green pellet was removed.

The entire process in 1997 was slow, manually intensive and costly because:.
1. Dechlorination of the fumed alumina took approximately 8 days and had to be monitored continually. The chlorine was highly corrosive and would require special controls in production equipment.
2. Three craftsmen could handcraft one isostatically pressed green pellet at a rate of one per minute through an intensive effort with significant potential for rejected green pellets. A zinc sterate release agent was added to the mandrel surface for every pressing. The job was repetitive and expectations for worker turnover were high.
3. Grinding losses were approximately 400%, which required excessive amount of powder to be produced with its enriched lithium going to controlled waste disposal. Because of the powder's poor flowability and its high compaction ratio (>4) the die was designed with a large "elephant's foot" of wasted mass at each end to provide uniform density within the actual final pellet.
4. Inefficient spray drier operation was caused by the low solids loading of the slurries (<100 g/l) used in order to have an acceptably low viscosity.
5. The small batches of powder in ball milling (<200 grams) and calcining (1.5 kg) obviously needed to be scaled up during production. Because production levels are expected to rise to almost 300,000 pellets per year, considerable scale up of the powder synthesis was required.

With these manufacturing issues as background, a powder development program was undertaken that worked to (1) eliminate de-chlorination, (2) switch to automatic isostatic presses, (3) achieve near net-shape forming, and (4) allow efficient spray drying of large batches of powder. This article presents the development work used to generate the triple spray dried process (Figure 1) with its chemical blending, milling, and conditioning cycle. In the chemical blending cycle, alumina and lithium feedstocks are blended in a slurry and spray dried to form a homogeneous mixture prior to calcination. During the milling cycle, the powder is milled to reduce the rigid agglomeration produced during calcination.

During the conditioning cycle, binders are added to a high solids loading slurry to achieve a flowable powder during spray drying. In the development program, a key revelation was that low solids loadings during the milling cycle were necessary to achieve effective milling while high solids loadings during the conditioning cycle were necessary to achieve flowable powders during spray drying.

DEVELOPMENT OF THE BLENDING CYCLE

The first task was to select a better alumina feedstock than the chloride bearing, fumed alumina (F-1) from the matrix in Table I. Although a lower chloride, fumed alumina (F-2) became available at that time, it was never made available in quantities that would support production. Although a sulfate-derived colloidal alumina possessed a higher surface area than any of the other alumina feedstocks, initial experiments found that it formed a gel when mixed with lithium hydroxides and presented uniformity issues in production of a high quality powder. Alkoxide-derived powders provided much better uniformity and sinterability. Among these powders the highest surface area powder was favored for producing a more reactive, final lithium aluminate powder that would sinter at lower temperatures.

A variety of lithium compounds were investigated as feedstocks in the chemical blend cycle as shown in Table II. The traditional lithium carbonate was found to be less than optimum, since slurries from it appeared to become unstable over a time period of a few hours. A significant effort was made to use lithium acetate, however acetate slurries were also unstable and formed sintered products that exhibited discontinuous grain growth with enclosed porosity at the grain boundaries. Processing lithium nitrate was attempted until violent releases of energy were observed during synthesis of the feedstock. Lithium peroxide did not produce the alpha phase that Arons found to be necessary for sintering.

Table I. Comparison of Alumina Feedstocks

Alumia	Basis	Surface Area m²/g	Impurities
F-1	AlCl₃	91	<0.6% Cl
F-2	AlCl₃	58	<0.03Cl
C	Sulfate	120	NA
A-0.5	Alkoxide	8	<0.05% Total
A-0.2	Alkoxide	42	<0.05% Total
A-100	**Alkoxide**	**71**	**<0.05% Total**

Triple Spray Dried LiALO₂ Powder

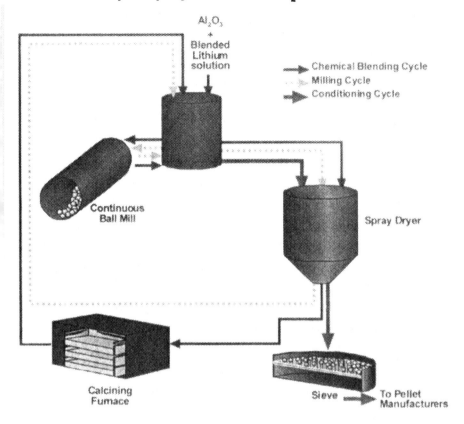

Figure 1. Schematic diagram of Triple Spray Dried Process for synthesis of LiAlO₂. The first cycle is the chemical blend cycle which is followed by the milling cycle and then finally the conditioning cycle.

Reflecting on the instability of the lithium carbonate and lithium acetate slurries, it was concluded that the slow leaching of lithium from these salts made them unstable with respect to pH. It was found that saturating a slurry of lithium carbonate with lithium hydroxide improved slurry stability. In addition, [6]Li-enriched lithium hydroxide could be obtained directly as a feedstock without additional processing. Mixing [6]Li-enriched lithium hydroxide with quantities of natural lithium carbonate and natural LiOH presented a convenient way to tailor the [6]Li/[7]Li ratio to the desired level. Consequently a blend of 70% Li₂CO₃ and 30%LiOH was adopted for the initial chemical blending cycle.

Table II. Comparison of Lithium FeedStocks

Feedstock	Disadvantages	Advantages
Li_2CO_3	Slurry Instability	Historic Baseline
$LiC_2H_3O_2$	Excessive Grain Growth Slurry Instability	
$LiNO_3$	Dangerous	
Li_2O_2	No α Phase	
70%Li_2CO_3 + 30%LiOH		Stable Slurry Direct use of enriched LiOH

CALCINATION

Arons proposed the advantages of using α-$LiAlO_2$ for higher density sintered products.[5] Figure 2 compares the sintering of a 100% α-$LiAlO_2$ and 100% γ-$LiAlO_2$ powders as measured by dilatometry. The sintering of 100% gamma powder is consistent with other ceramics. After slight thermal expansion up to 900°C, the dilatometer recorded the sintering of γ-$LiAlO_2$ up to almost 1200°C where it had sintered by 14% linear (42% volumetric). In the case of α-$LiAlO_2$, there is a similar thermal expansion up to 850°C where the transformation to the lower density γ-$LiAlO_2$ initiates. As the transformation starts there is a small expansion, as expected, but then the transformation initiates relatively rapid sintering at only 950°C. The combination of sintering and the transformation result in a integrated dimensional change of only 8% linear (24% volumetric) shrinkage. With the additional volumetric change from alpha to gamma of 25% (due solely to density differences) the overall densification in this case was greater (59%) than the simple sintering of γ-$LiAlO_2$. Figure 2 confirms Arons' original proposal that the alpha phase assists in the sintering of lithium aluminate.[5]

Figure 3 presents the effect of calcination temperature on alpha content and powder surface area for the traditional fumed alumina and the alkoxide-derived alumina feedstocks. The calcination temperature is an important parameter in developing a consistent powder production process. The temperature must be high enough to ensure complete reaction of the lithium carbonate and alumina to form lithium aluminate but low enough to prevent the transformation from alpha to gamma, nominally occurring at approximately 700°C, and the resultant loss of reactivity from surface area reduction. Although lithium aluminate from the traditional fumed-alumina starts with a higher surface area, the surface area drops rapidly at higher calcination temperatures. The amount of alpha lithium aluminate also drops sharply with increasing temperature from powder produced from fumed alumina. Both effects are a potential concern for a production process since slight changes in the calcination temperature could result in final powders with highly variable sintering properties. However, lithium aluminate powder synthesized from the high-surface-area, alkoxide-based powder did not transform to the gamma phase when calcined at temperatures as high as 800°C and maintained a constant surface area over a broad range of sintering temperatures. The high-surface area alkoxide powder was therefore adopted for production because of its robust calcination behavior and its low Cl concentration.

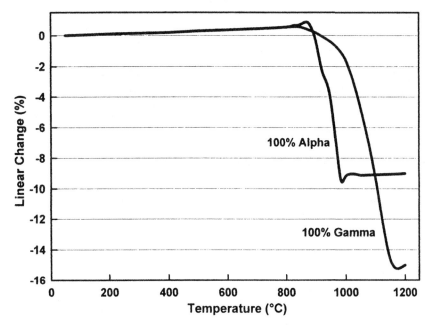

Figure 2. Effect of alpha content on the sintering of $LiAlO_2$ measured using dilatometry.

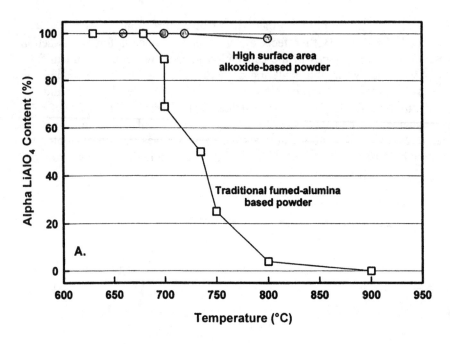

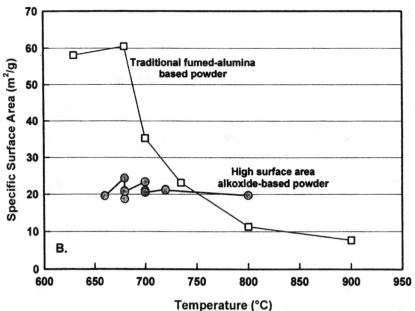

Figure 3. Effect of calcination temperature on (A) α-LiAlO$_2$ content and (B) the specific surface area for alkoxide-based and traditional fumed-alumina based powders.

MILLING

After calcining, the powders produced from the alkoxide alumina and the $Li_2CO_3/LiOH$ feedstocks possessed high surface area and sintered well, but would not flow in automated pressing equipment. Many unsuccessful attempts were made at producing a flowable powder through the use of these feedstocks. Flowability was monitored by the Hall flow rate, which consisted of measuring the time required for 25 grams of powder to flow through a standard funnel.[8] Table III lists the effects of extended post calcination milling of a powder synthesized from intermediate surface area, alkoxide alumina-based powder slurry. As with the other powders, milling for only 60 minutes was insufficient to permit large granule formation during the subsequent spray drying process, e.g., only 21 μm D_{50} granulate was obtained. This small granule size led to essentially no flow in the Hall flow meter and a high compaction ratio—the ratio of the volume of powder in an uncompacted state to its volume after compaction at 10,000 psi. A high compaction ratio indicates that greater powder volume and hence a larger die is required to make green lithium aluminate tubes. High compaction ratios are also inconsistent with the operation of automated isostatic presses. Compaction ratios of up to 4.7 were observed with powders synthesized directly from the high surface area, alkoxide alumina feedstock.

After milling for 300 minutes, the final spray dried powder from the conditioning cycle had a D_{50} granule size of over 50 μm, the sintering shrinkage increased, the compaction ratio reduced and, most importantly, a Hall flow rate of 25 g/180 seconds was achieved. The extensive milling introduced some zirconia impurity from the media after 300 minutes, indicating that processing time needed to be optimized achieve both flowable powder characteristics and low impurity levels. Fortunately, zirconia is a relatively benign impurity in lithium aluminate for this application. In order to further enhance the flowability of this conditioned powder, the size distribution of these powders was narrowed by sieving. The – 400 mesh (<38μm) granules were removed from the product further enhancing powder flowability.

Table III. The effect of milling time on granulate properties for powder synthesized from intermediate surface area alkoxide-derived alumina.

Milling time (min)	D_{50} Granules (μm)	Shrinkage (%)	Compaction Ratio	Hall Flow Rate (g/sec)
60	21.6	11.7	2.9	25/∞
300	51.9	13.2	2.5	25/180

POWDER CONDITIONING

Powder conditioning had historically been aimed at the addition of binder and other additives in preparation for pressing. In 1997 elvanol and glycerin were used as binder and plasticizer additives during the conditioning step. In the Triple Spray Dried Process, 1% polyvinyl alcohol and 1% polyethylene glycol (8000 molecular weight) were added as binder and plasticizer. These additives were selected because they (1) provided adequate pressing and green ware characteristics, (2) were stable in the caustic conditioning slurry and (3) were more stable during high temperature spray drying.

In addition to extensive milling, this study found that pH control of the final conditioning slurry to achieve high solids loading was an important part of producing a highly flowable powder. Spray drying produces nearly spherical granules, which flow well if they are large, high in density, rigid (strong), and have a narrow size distribution. These types of granules are commonly achieved by spray drying a high solids loading slurry since there is less shrinkage due to water loss as the droplets dry into spheres.

It is common in the alumina industry to add acid in order to reduce the pH and generate a slurry that possesses a high solids loading at a reasonable viscosity.[9] The ζ-potential trends of lithium aluminate (Figure 4) were consistent with other powders and the isoelectric point (i.e.p.) was approximately pH 6.8, or nearly neutral. Normally in electrostatic stabilization, a pH is selected that is as substantially removed from the i.e.p. to create either highly positively (below the i.e.p.) or highly negatively (above the i.e.p.) surface potentials thus producing a suspension in a dispersed state with high solids loading but low viscosity. With a nearly neutral i.e.p., either acid or base side processing is possible.

As acid or base was added to a $LiAlO_2$ suspension (ignoring the contribution of solids loading), as shown in Figure 5, the viscosity was reduced substantially to less than 1000 mPa·s. In the low pH case, however, even though the initial viscosity reduction appears greater than the high pH case, the leaching of lithium resulted in pH instability with time, causing the pH to increase, and therefore increasing the suspension viscosity. Hence addition of lithium hydroxide and high pH slurries were adopted for synthesis of lithium aluminate slurries with high solids loadings.

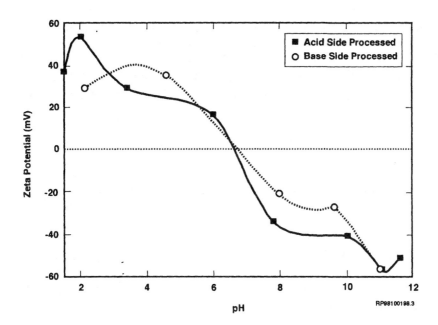

Figure 4. Zeta-potential of lithium aluminate powder

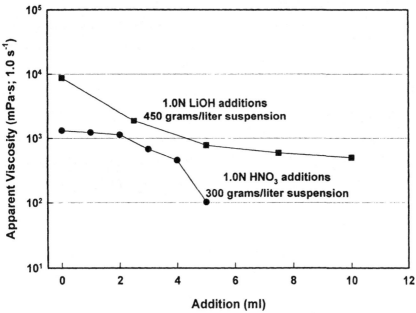

Figure 5. Effect of acid and base additions to the viscosity of lithium aluminate slurries

The additional lithium hydroxide added for pH adjustment at the later stages of manufacturing was accommodated by a sub-stochiometric formulation of $LiAlO_2$ in the original batching of the chemical blending cycle. Significant care and skill was employed to adjust spray dryer parameters (e.g. atomizer speed, feed rate, etc.) to maximize the size of droplet formation during the last spray drying cycle to generate a flowable powder.

DISCUSSION

The success in obtaining a flowable lithium aluminate powder is attributed to the Triple Spray Dried Process. In Figure 5, a granule from the 1997 campaign and one from the present triple spray dried process are compared. In Table IV, the parameters for the two powders are compared. The earlier granules were fragile, had a diameter of less than 10 μm and were at a bulk density of less than 4%. The bulk powder was fluffy, easily dispersed in air, and very difficult to compact into a die. In contrast, the Triple Spray Dried Powder has a granule size greater than 50 μm with a density of 18%. A Hall flow rate of 25 gms in 125 seconds and a compaction ratio of 2.6, means that the Triple Spray Dried Process provides a powder that can be used in automated isostatic presses.

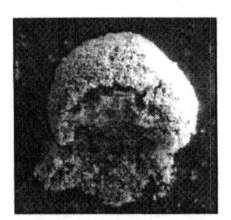 

Figure 5. Comparison of a granule (10 μm) from the 1997 production campaign with one of the higher-density, larger granule produced from the Triple Spray Dried Process.

Table IV. Comparison of powder from the Triple Spray Dried Powder
and that made in 1997 Campaign.

Characteristic	1997 Process	Triple Spray Dried Process
Hall Flow Rate (gm/sec)	25/∞	25/125
Granule Size (D_{50}, μm)	<10	63
Bulk Density (% TD)	<4	18% TD
Compaction Ratio	>4	2.6
Sintered Density (% TD)	95	98

It is worthwhile to consider the chemical/physical changes that take place
throughout the Triple Spray Dried Process in order to understand its value.

Chemical Cycle Slurry. In the chemical blending slurry, Vollath's work
suggests that the lithium carbonate, lithium hydroxide, and alumina slurry form
lithium dialuminate (or alumina coated with lithium dialuminate) plus lithium
carbonate particles suspended in the lithium hydroxide solution. When spray
dried, a mixture of individual particles of lithium dialuminate, lithium carbonate,
and lithium hydroxide is formed.

Calcination. During heating, lithium hydroxide melts at approximately
470°C and bonds lithium dialuminate particles together (via capillary forces) and
subsequently reacts them. At higher temperatures, the lithium carbonate begins to
react with the adjacent lithium dialuminate to form α-$LiAlO_2$. The carbonate and
moisture are lost rapidly in the open saggers but the α-$LiAlO_2$ particles partially
sinter to form a rigid skeletal structure from the low-density agglomerates.

Milling Cycle. During ball milling after calcination, the rigid skeletal
structure of the as-calcined α-$LiAlO_2$ powder is destroyed producing submicron,
high density agglomerates. The second spray drying does little to this structure
except to remove the excess water in preparation for creating a highly loaded
$LiAlO_2$ suspension.

Conditioning Spray Drying. The high solids loading suspension in the
conditioning cycle is formed from the submicron, high density agglomerates via
electrostatic stabilization using lithium hydroxide. This high density slurry is
significantly different than the previous slurry that possessed agglomerates with a
low density skeletal structure that prevented the generation of a highly loaded
suspension. The high solids loading, which was possible after milling, permits
spray droplets to contain a higher concentration of lithium aluminate (with some
lithium dialuminate). Therefore, when dried, the high density of particles in the
slurry droplets leads to the high density large diameter granules, producing the
flowable powder necessary for an automated pressing operation.

CONCLUSIONS

A method for synthesizing a flowable $LiAlO_2$ powder has been developed. This powder facilitated the use of an automated dry bag isostatic press and near net shape forming. The Triple Spray Dried Process used to synthesize the flowable powder was conceived on the basis of producing smooth, rigid, ~50μm granules, and possessed high bulk density and a narrow particle size distribution. Conditioning of the powder to achieve these characteristics required a post calcination grinding step to break up low density agglomerates, a high solids loading slurry made under high pH conditions, and a post-spraying sieving to narrow the distribution of granule sizes. Pellet pressing has improved to 12 pellets/minute made by one worker instead of 1 pellet per minute made by 3 workers. In addition, grinding losses have been reduced from 400% to 60%.

REFERENCES

[1] D. Guggi, H. Ihle, A Neubert, and R. Wolfle, "Tritium Release from $LiAlO_2$, Its Thermal Decomposition and Phase Relationship of γ-$LiAlO_2$-$LiAl_5O_8$," *Radiation Effects and Tritium Technology for Fusion Reactors*, CONF-750989, p416-431, ORNL Oak Ridge TN, 1976

[2] W. Gurwell, Fabrication of Ceramic Target Cores by the Bulk-Sinter-Vipac Process for PT-NR-57. BNWL-CC-429, Battelle, Pacific Northwest Laboratory, Jan, 1966.

[3] Final Environmental Impact Statement for Production of Tritium in a Commercial Light Water Reactor, DOE/EIS-0288, March 1999, USDOE, Washington, D.C.

[4] K. Kinoshita, J. W. Sim, and J. P. Ackerman, "Preparation and Characterization of Lithium Aluminate," *Mat. Res. Bull.*, **13** pp 445-455 (1978).

[5] R. M. Arons, R. B. Poeppel, M. Tetenbaum, and C. E. Johnson, "Preparation, Characterization, and Chemistry of Solid Ceramic Breeder Materials," *J. Nucl. Mat.*, **103 & 104**, 573-578 (1981).

[6] L. Cook and E. Plante, "Phase Diagram of System Li_2O-Al_2O_3," in *Fabrication and Properties of Lithium Ceramics III*, eds. I. Hastings and G. Hollenberg, *Ceramic Trans.* **27**, 193-222, (1992).

[7] D. Vollath and H. Wedemeyer, "Techniques for Synthesizing Lithium Silicates and Lithium Aluminates," in *Fabrication and Properties of Lithium Ceramics*, edited by I Hastings and G. Hollenberg, *Adv. in Ceramics*, **25,** 93-103, (1989).

[8] "Standard Test Method for Volumetric Flow Rate of Metal Powders using Arnold Meter and Hall Funnel," ASTM Standard B855-94, *Annual Book of ASTM Standards* **02.05**, (1994).

[9] W. H. Gizen, *Alumina as a Ceramic Material*, American Ceramic Society, Westerville, OH, 1970.

Melter Processing and Process Monitoring

LABORATORY MEASUREMENTS OF GLASS MELTING RATE

Dennis F. Bickford,
Savannah River Technology Center,
Westinghouse Savannah River Co.,
Aiken SC 29803

ABSTRACT

The Department of Energy has operated High Level Waste immobilization glass melters at the Savannah River Site and the West Valley Demonstration Project, and is constructing High and Low Level Waste melting facilities at the Hanford Site. Glass production rates are a critical parameter in both construction and operational costs, since they drive the scale of related equipment and determine the facility operational lifetime required to vitrify current inventories. During the initial stages of waste glass process development large-scale melters were available and were operated to demonstrate design construction details. At Savannah River small pilot scale melters and melt rate testing furnaces have replaced this experimental function for glass batch development and demonstration of evolutionary enhancements to melter operation. The smaller scale operations permit more economical testing, have lower fixed costs, and permit rapid testing with smaller quantities of experimental feed. For example, matrices of tests can be conducted for statistical process modeling that would not be practical with larger systems. This paper discusses the philosophy of simulating production melting in the laboratory with ties to engineering studies of glass production, and some examples of critical parameters that have been tested.

THE WASTE GLASS MELTING PROCESS

Aside from the requirements to accept melter feed, deliver molten glass, allow glass/gas separation, and containment of the reactions, the principal purpose of a melter is to deliver thermal energy to a element of the melter feed. As the supplier enthalpy is absorbed, the feed undergoes a series of physical and chemical changes that result in the molten glass product. The sequence of physical and chemical changes for typical DOE waste glasses is discussed in [1] and has been updated for proposed Hanford feeds in [2].

Separation of Effects

In any technical process, there are usually several process control parameters or variables that have strong influence on both the quality and rate of production of the product. The production of DOE waste glass is no exception. Waste compositions, melter design details and production rate requirements are different between the DOE HLW sites, and waste can vary in composition inside of individual tanks. The most unifying influences in this arena are the Waste Acceptance Product Specifications, and the limitations of the materials of construction used in melters.

Steady state equations

By performing a heat balance on the cold cap, Equation 1 below follows. Here, the sum of the convective heat from the glass pool, Qcon, and the net radiant heat absorbed by the cold cap, -Q1, is used to evaporate the feed water, melt the glass, and raise its temperature to the operating temperature.

$$Qcon-Q1 = SFR*cpw*(100-25) + SFR*\Delta Hevap + MR*Hbatch \qquad [1]$$

Where,

$$Hbatch = (\int_{20°C}^{1150°C} c_p dT + \int_{20°C}^{1150°C} \Delta H_r dT) = cpg*(Tsoft-25) + \Delta Hmelting + cpg*(1150-554),$$

$\Delta Hmelting$ is estimated to be 120 cal/gm for endothermic reaction heat and −80 cal/gm for exothermic reaction, which includes 20 cal/gm for silica melting, SFR is the water (or steam) feed rate, $\Delta Hevap$ is the heat of evaporation of water, cpw and cpg are the specific heats of water and glass, respectively. The average value of cpg over the appropriate temperature range is used in $Hbatch$.

Rules of Thumb Based on Laboratory Tests and Engineering Melters.

Crucible melting experiments for glass formulation and melter component engineering tests in small to pilot scale melters have resulted in some "rules of thumb" applicable to waste glass melters. These rules are only semi-quantitative, and exceptions can be found. None the less, the rules are reasonable summations of the results of diverse testing, and have generally held up in similar tests done in various laboratories, and at various times. In any event, they point out important melter design or operating criteria, or phenomena to be alert to so that operations reliable and small test conclusions can be used in general circumstances.

Well-Mixed Batch. All melter feed should be uniform on the scale of the largest feed particles, which are typically the silicon or glass frit particles. Local areas that are low in flux content can result in slowly melting areas, and formation of high melting temperature particles, such as spinels. This was graphically illustrated when calciners were coupled directly above a pilot melter. The calciner produced clinkers with high alumina and spinel content that did not mix well with the frit being added to the melter. This resulted in deep deposits of spinels accumulating at the bottom of the melter, where they could not be dissolved during normal melter operation.

1150 °C Nominal Temperature. The nominal melter temperature of waste glass melters is based on the empirical evidence of successful operations. This temperature is too close to the alloy melting point for reliable creep calculations of critical components submerged in the glass. Thus, Inconel components should only support their own weight. Temperature measurements in one area of the melter are only a general indication of the temperatures at particular locations. Cold batch and water are added at the same time that power is being applied to electrodes, and the highly-colored DOE waste glasses do not effectively transfer heat by thermal radiation. Where high stresses are applied, such as lid heaters or mechanical agitators, the nominal local temperature should be reduced to increase creep resistance.

Aqueous Slurry Feeds. Typical DOE waste glass feeds are waste, batch additives, and water or nitric acid solutions. The water provides a convenient way to slurry up and transfer the melter feed, but this is at a cost of melting efficiency. For typical melter feeds consisting of abut 50 wt% aqueous materials and less than 2 molar nitrates (or greater that 2 molar nitrates but without reducing agents), the melt rate is about one half of what it is for similar dry materials. The water absorbs heat on boiling, and further reduces the melter vapor space temperature by absorbing heat as the steam exits the melter. The water also quenches the surface of the melter making it difficult for partially melted material to spread over the surface of the melter. The water does non uniformly distribute feed by collecting in crater-shaped piles, which periodically over flow with fingers of the aqueous feed running on the surface of the melt towards the melter walls. Low melter feed concentration directly reduces the melt rate by requiring the same amount of delivered heat to produce less glass. The effect is not directly equivalent because the glass melting requires enthalpy with a higher thermodynamic quality (i.e. delivered at a higher operating temperature) than that required to evaporate the aqueous materials.

Auxiliary Vapor Space Heaters Resistance heating elements are used in the vapor space of melters for melter startup. If the heaters are operated while melting, they act as a supplemental source of heat to the electrodes directly heating the glass. The normal rule of thumb is that they can provide enough heat to boil off the aqueous slurry, compensating for the use of water. Recent analyses indicate that when the batch is very slow to melt, then the vapor space heaters supply a majority of the effective heat [5]. It is also expected that when the vapor space temperature is below 300oC the only major effect of the vapor space heaters is to dry the feed. At higher temperteres the heaters can initiate the redox reactions beween nitrate salts and reducing agents, and can suppl part of the energy needed to conduct the conversion of the feed to glass.

Air Inleakage Air inleakage reduces the quantity of usable heat from vapor space heaters, and reduces the quality of heat by dropping the vapor space temperature, slowing heat transfer to the top of the cold cap (batch pile).

Power Skewing The power distribution between upper and lower electrode pairs, called power skewing, has been a topic of debate. One school of thought is that power skewed to the bottom electrodes tends to increase overall circulation, increasing melt rates. The second school feels that delivering the majority of power to the top electrodes puts it where it is most needed, under the reacting. Test results are inconclusive, but tests indicate that adequate power is required at both elevations. Insufficient bottom power can result in stagnant conditions that can lead to accumulations of material at the bottom of the melter, and may lead to hot streaks in the upper melt pool, where only part of the surface of the melter is effective in melting. Too little power to the top electrodes can result in low temperatures throughout the top of the melter, slowing glass production [4,6].

Glass Viscosity Glass viscosity is indirectly related to the melt rate by electrode heating. High viscosity at the nominal melting temperature reduces the overall glass circulation, thickening the thermal boundary layer underneath the reacting glass feed, and slowing the delivery of heat from the joule effect to the batch.

Melt Pool Agitation The amount of mixing, or rate of circulation in the melt pool, has been shown to have the most dramatic effect on melt rate. Agitation reduces thermal boundary layers, delivers hot glass to the cold cap region, and can disperse feed reducing the distance that the heat must be transferred. Mechanical agitation by mixer blade has increased melt rate to 8 times that without the agitation. Simple sparger bubblers can increase the melt rate to double or triple that without sparging [7,8].

Cold Cap/ Melt Interfacial Area Cold cap coverage has a direct effect on melt rate. Melt rate is considered to be directly proportional to the area of the glass surface. For small melters the melt rate decreases because of the cooling and circulation damping effects of the walls and melter bottom. Very low viscosity or exothermic melts in small melters can be closer to that of large melters. In addition to general slowing of convection by friction, refractory walls tend to collect foam or viscous partially reacted glass because of the combined effects of cooling and frictional forces. For typical SRS feeds melters smaller than 8 inches in diameter are not reliable predictors of melt rate because of potential bridging of the melter by low temperature glass. Melters 8 to 16 inches in diameter show the same relative rates as larger ones. For a given SRS feed, the melt rate of 8 inch diameter melters has been typically one half of that of larger melters. Generally the coverage during SRTC pilot scale tests was 70-100% of the melt surface. Complete cold cap coverage (100%) is not desirable, since it can develop into overfeeding, and formation of bridges of batch over the melt. If bridging develops then vapors may accumulate under the cold cap separating the batch from the glass, and resulting in instability of gas pressure, melt rate and temperature/power demand.

Relative Melt Rates

Suite of Melt Rate Testing Methods

Waste Acceptance Product Specifications (WAPS) require the glass produced to pass a reference deionized water durability test, which effectively limits the amount and type of fluxes that can be used in the glass. The high temperature creep and melting point of the Inconel ™ alloys used restrict the nominal operating temperature of the melters to ~1150 °C. Further, the need to avoid filling the melter cavity or pouring area with undissolved material requires that the liquidus of the glass product be kept below the nominal operating temperature to assure that solids accumulations don't result from localized temperature variations, feed composition variations, or segregation during the melting operation. Still further, the need to pour the melted product restricts the viscosity to a manageable level in the pouring area, typically 300-1200 PaS (30-120 poise). The net result of these restrictions is to make the physical properties of the DOE waste glasses very similar, despite different waste origins [3]. The similarities in waste glass properties result in similar testing approaches for melt rate. Mini-bubbler testing was conducted in the Slurry Fed Melt Rate Furnaces (SMRF) installed at Aiken County Technology Laboratory (ACTL) and at Clemson Environmental Technology Laboratories (CETL). These furnaces were utilized to compare the melting behavior of different DWPF slurry feed formulations both with and without a mini-bubbler. The SMRFs were designed to mimic the heat transfer characteristics of a large-scale joule-heated melter. This was done by providing heating in one dimension through the bottom of an 8 inch diameter Inconel 690 crucible and insulating around the sides of the crucible in the melt pool area to minimize radial heat transfer to or from the melt pool and heat exchange with the plenum. Sketches of the ACTL and CETL furnaces are shown in Figures 2 and 3. A schematic depicting the SMRF system at both ACTL and CETL is shown in Figure 4.

Figure 2. ACTL Slurry Fed Melt Rate Furnace

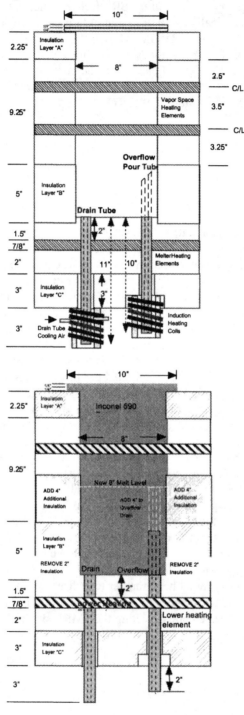

Figure 4. SMRF System Schematic

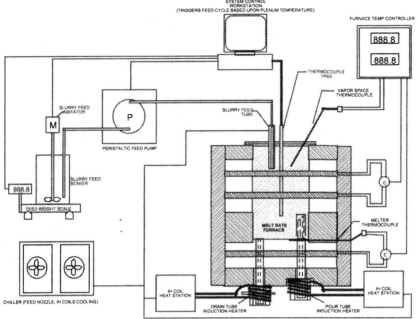

The two systems are identical with the exception of the melt pool depth and the method of heating the pour/drain tubes. The CETL SMRF has an 8" melt pool depth and a propane torch heats the pour/drain tubes. The ACTL SMRF has a 4" pool depth and induction coils heat the pour/drain tubes. The glass temperature is controlled by a thermocouple mounted on the bottom of the crucible and for these tests the setpoint was maintained between 1125 and 1175°C. Additional heating was applied to the plenum above the melt pool by Globar heaters that surround the top of the crucible. The plenum temperature was controlled by a thermocouple inserted into the vapor space of the crucible, and by changing the setpoint, different plenum conditions could be simulated. Feed additions to the melter were based on maintaining a plenum temperature set point. After each feed cycle, the controller waited for the melter to return to the vapor space set point temperature (typically between 600 and 800°C). Once the vapor space temperature setpoint was reached and the temperature was increasing, the feed cycle began again. As slurry was fed onto the melt surface, glass was continuously poured from the SMRF through the overflow pour tube. The break-over level for glass pouring from the SMRF required 3-1/2 inches of glass depth in the ACTL crucible and 8 inches in the CETL crucible. The poured glass was collected in a catch pan located beneath the pour tube discharge and was weighed. Melt rate was assessed by weighing the amount of glass poured over a test period and by measuring the mass decrease of the feed vessel over a test period.

The bubbler tests were conducted by feeding the slurry in controlled increments to the SMRF for a sufficient time to establish the cold cap and reach steady state operation, which typically was about two hours. Melt rate data and observations relating to cold cap and feed behavior were then obtained.

The CETL SMRF runs are summarized in Table 2. Each run is discussed in more detail below. For most of these tests the bubbler was positioned vertically at the glass/cold cap interface and a bubbler air flow rate of 5 cc/min was used. For the last two SMRF runs the bubbler was positioned to discharge approximately ¼" above the glass surface and a bubbler air flow of 25 cc/min was used. The bubbler was positioned at the location shown in the SMRF plan view of Figure 6.

TABLE 2. CETL SMRF Melt Rate Tests

DATE	FEED Solids (Wt%)	Bubbler Air Flow (cc/min)	Vapor Space Temp °C	Average Batch Feed Rate* (grams/min)	Average Glass Pour Rate* (grams/min)
8/30	52	0	750	24.87	9.86
9/3	52	5	750	51.67	19.63
9/4	41	5	650	30.47	11.22
9/4	41	0	650	27.07	9.75
9/4	41	0	750	46.02	18.32
9/5	41	5	750	47.92	19.15
9/27	41	0	750	44.20	16.34
9/30	41	25	750-893	55.81	19.75

*Feed Rate and Pour Rate averages represent values after initial 100 minutes of operation to assure that steady state conditions have been established

Baseline tests were conducted in the CETL SMRF using SB2 with Frit 320 both with and without use of a bubbler to compare melt rate and melting behavior. The initial series of tests were performed by feeding a slurry consisting of 52 wt% insoluble solids, a departure from the standard 41 wt% solids slurry typically used in melt rate testing in the SMRF. These runs revealed interesting insight into the importance of the behavior of the slurry when deposited onto the glass melt surface. It became evident that any melt rate advantage one would anticipate by reducing the amount of batch is overcome by the inability of the slurry batch to adequately flow across the glass melt surface and establish good heat transfer from the hot glass to the slurry batch. The incorporation of the bubbler, however, improved the melting behavior of the high batch solids feed such that feed and pour rates were nearly equal to that of the standard batch solids feed behavior.

The plenum temperature was controlled at 750°C, and the melt pool temperature was controlled to 1180°C during these two initial melt rate tests with the CETL SMRF. As the slurry feed was delivered onto the melt pool surface, it tended to mound directly beneath the feed tube due to its inability to flow across the melt pool surface. Subsequent feed additions tended to create a thicker mass of feed solids as its liquid was driven off by convective heat energy. The limited heat conducting into the feed mass from the melt pool, due to the reduced area of feed to pool contact, resulted in reduced rate of batch incorporation into the melt pool. The average feed and glass pour rates for the baseline run of high batch solids after the initial 100 minutes for steady state conditions to be established were 24.87 and 9.86 g/minute, respectively.

Figure 14. CETL Frit 320 / SB2 Baseline (High Wt% Solids) Run - 8/30/02

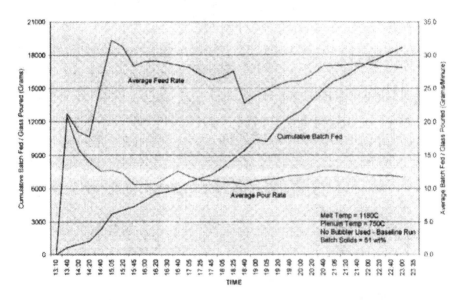

**Figure 15. CETL Frit 320 / SB2 Bubbler (High Wt% Solids)
Run - 9/3/02**

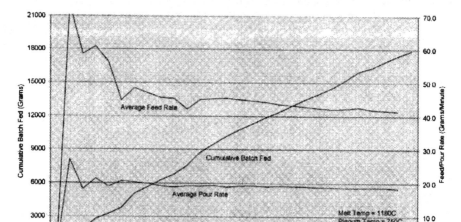

SEPT. 3 CETL SMRF AVERAGE FEED/POUR

The average feed and glass pour rates for the run containing high batch solids during which a bubbler was used after the initial 100 minutes for steady state conditions to be established were 51.67 and 19.63 g/minute, respectively. Use of the bubbler at 5 sccm air flow improved the average feed rate by 107.8%, while the average steady state glass pour rate increased by 99.1%.

CETL Lower Plenum Temperature Runs

Tests were conducted in the CETL SMRF using SB2 with Frit 320 at a reduced plenum temperature of 650°C both with and without use of a bubbler. While the initial series of tests were performed by feeding a slurry consisting of 52 wt% insoluble solids, the remaining tests were performed feeding the standard 41 wt% solids slurry typically used in melt rate testing in the SMRF. These runs revealed insight into the affects of reduced plenum temperature on the melting behavior of the slurry when deposited onto the glass melt surface. It became evident that melt rate is improved when the bubbler is used while feeding the melter under lower plenum temperature conditions. However, returning the plenum to the standard 750°C temperature improved the melting rate of the batch feed such that feed and pour rates were greatly improved without the use of a bubbler beyond that of operation of the melter at reduced plenum temperature both with and without bubbler use.

The plenum temperature was controlled at 650°C, and the melt pool temperature was controlled to 1180°C during the two melt rate tests with the CETL SMRF at reduced plenum temperature conditions. The bubbler was positioned into the melt pool with its upper discharge located just at the surface.

REFERENCES

[1]   D.F. Bickford, P. Hrma and B.W. Bowen, "Control of Radioactive Waste Glass Melters: II, Residence Time and Melt Rate Limitations", J. Am. Ceramic Soc., 73 [10] 2903-15 (1990).
[2]   L.D. Anderson, T. Dennis, M.L. Elliott, and P. Hrma, "Waste Glass Melting Stages", Env. & Waste Man. Issues in Ceramic Ind., Am. Ceramic Society, 213-220 ().
[3]   D.F. Bickford, A. Applewhite-Ramsey, C.M. Jantzen, and K.G. Brown, "Control of Radioactive Waste Glass Melters: I, Preliminary General Limits at Savannah River", J. Am. Ceramic Soc., 73 [10] 2896-2902 (1990).
[4]   S.R. Young and D.F. Bickford, "A Review of DWPF Pilot Melter Experience – Melt Rate and Associated Parameters", WSRC-TR-96-0405, NTIS (1996).
[5]   H.N. Guerrero, and D.F. Bickford, "Numerical Models of Waste Glass Melters Part I – Lumped Parameter Modeling of DWPF", these proceedings.
[6]   D.F. Bickford, R.C. Propst, and M.J. Plodinec, "Control of Radioactive Waste Glass Melters: III, Glass Electrical Stability", Advances in the Fusion of Glass, Am. Ceramic Soc. 19.1-19.17 (1988).
[7]   T.L. Allen, "Melt Flux Studies with the SCM-2 During Campaign I" DPST-83-746, NTIS (1983).
[8]   A.E. Hailey, "Melt Flux Studies with the SCM-2 During Campaign II", DPST-83-981 NTIS (1983)
[9]   J.A. Bunting and B.H. Bieler," Batch Free Time Versus Crucible Volume in Glass Melting", Ceramic Bulletin 48 [8] 781-85 (1969)
[10]   J.C. Hayes," Laboratory Methods to Simulate Glass Melting Processes", Advances in the Fusion of Glass, D.F. Bickford, et al eds., American Ceramic Society 14.1-14.8 (1988).
[11]   D.P. Lambert, T.H. Lorier, D.K. Peeler and M.E. Stone," Melt Rate Improvement for DWPF MB3: Summary and Recommendations", WSRC-TR-2001-00148, NTIS (2001).
[12]   A.D. Cozzi, D.F. Bickford, and M.E. Stone, "Slurry Fed Melt Rate Furnace Runs to Support Glass Formulation Development for INEEL Sodium-Bearing Waste", WSRC-TR-2002-00192, NTIS (2002).

[13]   H.N. Guerrero, D.F. Bickford and H. Naseri-Neshat, "Numerical Models of Waste Glass Melters Part II – Computational Modeling of DWPF", these proceedings

ANALYSIS OF FEED MELTING PROCESESS

J. Matyáš
Laboratory of Inorganic Materials
V Holešovičkách 41, Prague 8, Czech Rep.

P. Hrma and D-S. Kim
Pacific Northwest Nat. Laboratory
Richland, WA 99352

ABSTRACT

The vitrification process of two feeds that exhibited different rates of conversion was studied using thermal analyses, including evolved gas analysis with volume-expansion monitoring. Quantitative X-ray diffraction and scanning electron microscopy were performed on quenched samples. The difference in the conversion rates was attributed to different melt viscosities and crystal concentrations at the temperature at which the melt interfaces the cold cap. It is suggested that higher viscosity and the presence of crystals stabilizes foam under the cold cap, thus hindering heat transfer for melting.

INTRODUCTION

Melting with a cold-cap is the preferred method for vitrifying high-level waste (HLW). A cold cap reduces melt volatilization, but has three major disadvantages: 1) slow melting, 2) unsteady process, and 3) propensity for crystal generation and settling. This paper deals with the first issue, i.e., a slow rate of melting caused primarily by foaming. In small crucibles, where heat is rapidly delivered to the feed, the feed-to-glass conversion rate is fast. In large-scale melters, where heat is delivered to the cold cap from the pool of molten glass, the rate of melting is predominantly controlled by heat transfer. Heat is delivered to the cold cap by thermally-driven convection. If bubbles are present, they form a stagnant foam layer under the cold cap, and thus hinder heat transfer.[1]

Two sources of bubbles exist (Figure 1):[2-5] gas trapped within the cold cap at the final stage of melting (primary foam), and bubbles that ascend through the melt and accumulate under the cold cap (secondary foam). The source of primary foam is batch gases[2,5] and intermediate solid phases, such as sodalite[6] or nosean.[7] The source of secondary foam is transition metal oxides that gradually release oxygen as temperature increases.[8-10] A HLW glass with 11 mass% Fe_2O_3 liberates 2.5 m^3 O_2 per m^3 of glass.[11] These reactions are enhanced by H_2O.[12]

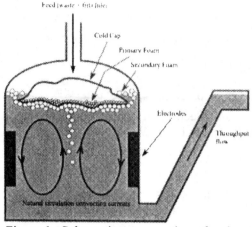

Figure 1. Schematic cross-section of a glass melter

Foam creates a low-density, thermally insulating layer under the cold cap that is virtually motionless and thus insulates the cold cap from the hot glass underneath. Commercial batches are formulated to prevent this.[13] Attempts to decrease bubble generation through reducing Fe_2O_3 in HLW feed faces major problems. Reductants must be used in large excess to overcome nitrates.[14] An excessive quantity of reductants could lead to an undesirable formation of spinels and sulfides.[8,15] Residual reductants can generate CO_2 bubbles, possibly at a larger quantity than that which they should prevent.[16]

An alternative option is a faster removal of foam. Mechanical bubblers drive a powerful forced convention that displaces the foam. Enhanced convection can also be achieved with a lower melt viscosity.[17,18] At the Defense-Waste Processing Facility (DWPF), a glass with a lower viscosity was formulated using Frit 320 (Frit II) and Macrobatch 3 (MB3) waste.[19,20] Though the main criterion used in developing Frit II was the viscosity-temperature function, the change in glass composition impacted the melting process in all its stages. The aim of this study is to characterize the effect of Frit II, as compared with the nominal Frit 200 (Frit I), on the conversion reactions and the overall cold-cap behavior.

EXPERIMENTAL PROCEDURE

This study was performed with the DWPF MB3 sludge simulant,[21] Frit I, and Frit II (feeds and glasses based on these frits are denoted as Feed I and II, and Glass I and II). The particle size of each frit was 74 to 177 μm. Frit compositions and major components of glasses are listed in Table I. Some minor components, i.e., (with mass% in glass) MnO_2 (0.79), NiO (0.31), ZrO_2 (0.20), Cr_2O_3 (0.09), ZnO (0.09), and BaO (0.07), appeared in solids detectable by XRD and SEM.

Table I. Nominal composition of frits and glasses in mass%

	SiO_2	Na_2O	B_2O_3	Li_2O	MgO	Fe_2O_3	Al_2O_3	CaO
Frit I	70.00	11.00	12.00	5.00	2.00	0.00	0.00	0.00
Frit II	72.00	12.00	8.00	8.00	0.00	0.00	0.00	0.00
Glass I	54.46	11.74	9.22	3.84	1.59	12.03	4.36	1.05
Glass II	55.99	12.51	6.14	6.14	0.06	12.03	4.36	1.05

The calcine factor for MB3-simulant sludge (15.06 mass%) was determined by heating dry slurry at 4°C/min to 900°C with 2-h dwell. The waste loading was 23.2 mass%. To make feeds, the sludge was mixed with a frit, homogenized in a closed glass beaker for 5 h using a stir bar, and dried at 110°C for 24 h.

The feed-to-glass conversion process was characterized by standard thermal-analysis techniques. The thermo-gravimetric analysis (TGA) and differential thermal analysis (DTA) (SDT 2960 Simultaneous TGA-DTA) were performed with the ambient flow rate of 65 cm^3/min He. Approximately 50-mg samples were heated at 4°C/min from room temperature to 1100°C. For evolved gas analysis (EGA) approximately 10-g samples of feed were heated at 4°C/min in a tall quartz-glass crucible in a furnace equipped with a quartz-viewing window. The change in feed volume was recorded on video. The off-gas carried in a 65-cm^3/min stream of pure He was analyzed with gas chromatography-mass spectrometry (GC-MS) (Hewlett Packard 5890A GC with 5971A MS).

Crystalline phases in feeds were determined with an X-ray diffractometer (XRD) (Scintag PAD-V) with Cu target and Peltier Detector. Samples of dried feed were heated up at 4°C/min to different temperatures, quenched in air, ground, and mixed with a 5 mass% internal standard (CaF_2) in the tungsten-carbide mill. The XRD-scan parameters were the step-size range of 0.04° 2-θ and the scan range of 5° to 70° 2-θ. The scans were analyzed with the Jade software and mass fractions of the phases were determined with the RIQAS 3.1 program. The microstructure of partly reacted feeds was investigated with scanning electron microscopy (SEM). The composition of crystals was analyzed with energy-dispersive spectroscopy (EDS). The oxidation-reduction equilibrium of Fe was obtained from wet chemical analysis and checked with Mössbauer spectroscopy.

The modified reboil test (MRT) was performed to determine how much gas is released from the glass after the completions of the feed-melting reactions. Approximately 10-g samples of feed were heated at 4°C/min in a tall quartz crucible and held at 870°C for 1.5 h to allow the melting reactions to complete and the melt to homogenize. Then the temperature was increased to 1200°C and held for 0.5 h. The gases were analyzed using MS. The feed volume was recorded on video. Glass viscosity was measured as a function of temperature with a rotating spindle viscometer (Brookfield Digital Viscometer Model LVTD).

RESULTS AND DISCUSSION
The results of simultaneous TGA-DTA for Feeds I and II are shown in Figure 2. Two broad indistinct exotherm peaks at temperatures below 200°C, associated with a mild mass loss, indicate possible solid-state reactions, probably too slow to reach a significant progress with the 4°C/min temperature-increase rate. Two pronounced endothermic peaks associated with substantial mass loss are seen at 200°C and 250°C. The first peak probably indicates the formation of $NaNO_2$/$NaNO_3$ eutectic melt and its reaction with NaCOOH, the second the melting of NaCOOH and its reaction with the nitrate-nitrite melt. Mass loss continued to ~750°C because of the continuing gas-releasing reactions of nitrates,

nitrites, formates, and carbonates as evidenced by EGA and XRD. Carbonate decomposition is suspected as the cause of the broad peak at ~670°C in Feed I and the doublet at 713°C and 728°C in Feed II. The average total mass losses were 9.36 mass% for Feed I and 8.46 mass% for Feed II.

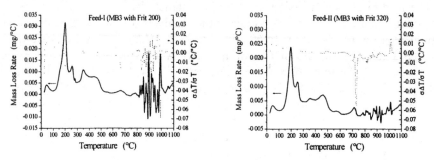

Figure 2. Comparison of TGA and DTA data for Feed I (left) and II (right)

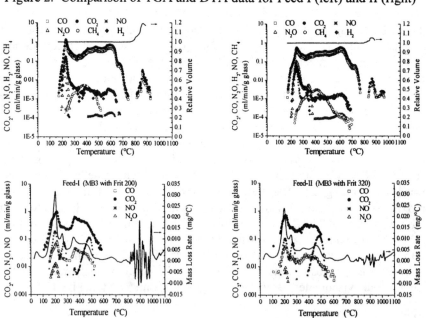

Figure 3. The rates of gas evolution and relative volume expansion by EGA (top) and of gas evolution and mass loss by TGA-EGA (bottom) from Feeds I and II

One remarkable feature of the gravimetric and calorimetry curves is fluctuations between ~800°C and 1000°C. By volume expansion data (Figure 3, top), the glass-forming phase becomes interconnected starting at 800°C. It is possible that bursting gas bubbles from the highly viscous, glass-forming phase caused the mechanical disturbances recorded by TGA. The reaction enthalpy fluctuation seen in the DTA curves show that a chemical reaction was involved in the process. These disturbances were attributed to the decomposition of sodalite crystals[6]. As Figure 4 (bottom) shows, sodalite begins to decompose at 800°C. Note that Feed I that produced a higher mass of sodalite exhibits a larger volume expansion (primary foam) and higher amplitude of the TGA disturbances.

The maximum off-gas rates of CO_2, CO, and H_2 occurred at ~240°C (Figure 3). This temperature roughly coincides with the temperature at which NaCOOH melts (253°C) and decomposes (~300°C). Continuous generation of CO_2, H_2, and CO below 780°C is probably due to the continued decomposition of residual NaCOOH (XRD data indicate its presence up to 500°C) and the initial decomposition of carbonates. A reaction of H_2 with CO or CO_2 produced CH_4 between 220 and 700°C.

In Feed-II maxima on the EGA curves for N_2O and NO (Figure 3, top right) occur at 240°C (both N_2O and NO), and at ~320°C for NO and 340°C for N_2O. The second maximum (absent in the Feed I) is most likely the result of molten $NaNO_2$ or $NaNO_2/NaNO_3$ eutectic melt reacting with NaCOOH. The decomposition of $NaNO_2$ may also generate NO at temperatures between 400°C and 700°C. These temperatures are somewhat lower on TGA-EGA curves (Figure 3, bottom). Whereas in the larger EGA samples, temperature gradients delay reactions, in small TGA samples, the atmospheric dilution of gaseous products by He accelerates reactions that generate gases.

Feed-volume expansion (Figure 3, top) began at ~800°C, indicating that glass-forming melt became interconnected at this temperature. When the open porosity closed, the release of gas temporarily stopped. The evolution of CO and CO_2 resumed above 800°C and continued to 950°C. This CO_x could not be produced from the primary melt. Its source could be a release from carbonate dissolved in the glass-forming melt or, more likely, a reaction of residual carbon from the pyrolysis of organics with Fe_2O_3. Apart from a higher concentration of sodalite, the higher expansion in Feed I was probably assisted by a higher melt viscosity.

Concentrations of all crystalline phases in feeds were low, rarely exceeding several mass%. Most of the feed was amorphous gels and glass frit. The XRD analysis (Figure 4) identified the initial feed minerals (baddeleyite, quartz, rock salt, etc.) and intermediate phases (hematite, spinel, sodalite, and nepheline) that eventually dissolve into the glass as the temperature increases. Sodium nitrate and nitrite decomposed partly by 300°C, continued to decompose or react, and disappeared from the feeds by 700°C. Quartz, baddeleyite, and sodium chloride disappeared from the feeds by 900°C. Quartz began to dissolve at ~700°C in Feed I and 600°C in the more reactive Feed II. Baddeleyite dissolved in a narrow temperature range, starting at 750°C in Feed I and at 700°C in Feed II. Sodium chloride began to disappear at ~600°C in Feed I and 800°C in Feed II. Calcium carbonate persisted to an unusually high temperature of 850°C. This was probably possible because dilute carbonates were preserved in the primary melt.

Not shown in Figure 4 are aluminum and iron hydroxides and sodium formate, which were also detected by XRD. Aluminum hydroxide disappeared by 400°C from Feed I and by 500°C from Feed II. Iron (oxy)hydroxides disappeared by 600°C from both feeds. The highest temperature at which sodium formate was detected in Feed I was 400°C and 500°C in Feed II. As in the case of calcium carbonate, the formate persisted to temperatures above the spontaneous decomposition of sodium formate, probably as a component of the primary melt.

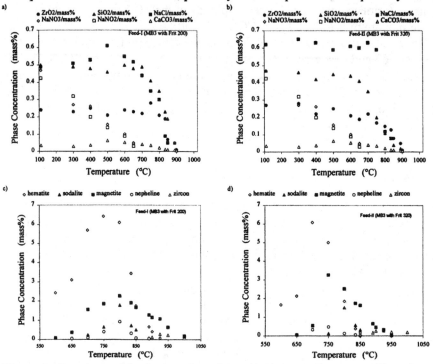

Figure 4. Fractions of feed minerals (a,b) and of intermediate crystalline phases (c,d) versus temperature in Feeds I (left) and II (right)

XRD detected several intermediate crystalline phases in both feeds: hematite, magnetite, nepheline, zircon, and sodalite (Figure 4, bottom). Hematite formed the largest fraction, more than 6 mass% at 750°C (Feed I) and 700°C (Feed II). Hematite was probably a product of the decomposition of amorphous iron hydroxide. Above 750°C (Feed I) and 700°C (Feed II), hematite was partly reduced to magnetite, a phase that peaked at 800°C (approximately 2 mass%), and disappeared by 900°C in Feed I and 850°C in Feed II. A small fraction of nepheline was detected above 600°C, peaking at 800°C with ~1 mass% in Feed I and at 750°C with ~0.3 mass% in Feed II. This was probably a result of the reaction between sodium aluminate and silicate melt. Nepheline disappeared from both feeds (dissolving in glass-forming melt) by 850°C. A small fraction of zircon (approximately 0.2 mass%) appeared above 900°C, the temperature at which baddeleyite disappeared. Traces of zircon were detected even at 1000°C. Sodalite ($6NaAlSiO_4 \cdot 2NaCl$) was produced above 700°C (Feed I) and 750°C (Feed II), peaking at 800°C in both feeds, and was gone between 850°C and 900°C. The sodalite peak coincides with the temperature at which the feed volume began to increase and also with the temperature at which solid sodium chloride began to disappear. Other phases detected (not shown in Figure 4) were trevorite ($NiFe_2O_4$) and zinc-chromite ($ZnCr_2O_4$). These spinels coexisted with magnetite, with which they probably formed a solid solution. Spinel persisted to 950°C in Feed II and 1000°C in Feed I. As mentioned before, a higher spinel concentration in Feed I at temperatures around 900°C, together with a higher viscosity, might stabilize foam and thus cause slower melting of Feed I.

The temperature of 800°C was chosen for SEM/EDS analyses because this is the temperature at which the key intermediate phases peaked (see Figure 4) and the feed began to expand by primary foaming (see Figure 3). The SEM micrographs showed heterogeneous areas embedded in a connected matrix of glass-forming melt. The heterogeneities consist of dissolving solids, insoluble inclusions, and gas bubbles. Baddeleyite was clearly detected as a conglomerate of dissolving crystals (3 to 20 µm) in both feeds. Silica (possibly quartz) was found as ~2 µm round gray objects. An elongated gray shape (~30 µm long) and conglomerates of submicron gray objects were suspected to be hematite. Submicron spinel (magnetite, trevorite, and jacobsite) were suspected in similar agglomerates, probably coexisting with sodalite. The extremely small sizes and diffuse shapes of some other phases detected with XRD precluded their clear identification with SEM. Barium sulfate, a minor component of the feed (not seen with XRD), was clearly detected as 10 to 30-µm agglomerates. Another minor ingredient, Pd, was seen as a ~12-µm nodulus of metallic Pd.

A cold-cap sample retrieved from the mini-melter run with Feed II exhibited similar features to laboratory samples quenched at 800°C. A continuous glass matrix contained both solid and gaseous inclusions. Iron-rich regions contained tiny solid inclusions, probably hematite with spinel and sodalite. A large (~40 µm) dissolving crystal (probably hematite), dissolving crystals of baddeleyite, and a dissolving grain of quartz were also found. Pd was not detected, but a loose agglomerate of RuO_2 was clearly seen. Unlike in laboratory samples, where crystals of sodalite were barely detectable, a large (~10 µm) dendrite of sodalite was observed in the melter sample. An iron-rich sodium aluminosilicate, possibly nepheline, was also identified.

A feature not seen in the laboratory sample was a darker amorphous silica-rich area, probably a frit residue that had not yet been fused with the rest of the glass-forming matrix. Another phenomenon not observed in laboratory samples was the presence of tridymite needles, twins, and star-shaped crystals that precipitated in the glass matrix. They probably formed from silica-rich glass (possibly frit) during an extended period at temperatures below the liquidus or during slow cooling.

Wet colorimetry was performed on two samples of each feed, both ramp-heated at 4°C/min to 870°C with 1.5-h dwell. Sample A was quenched, whereas Sample B was ramped from 870°C (after 1.5-h dwell) at 25°C/min to 1200°C with 8-min dwell. Approximately one third of the total Fe was reduced to Fe(II) in both feeds. The Fe(II)/Fe fraction was nearly constant during the temperature increase from 870°C to 1200°C in Glass I, but increased significantly in Glass II, suggesting that the redox reaction continued in Glass II after melting reactions were completed. This increase of Fe(II)/Fe fraction in Glass II is associated with the evolution of O_2 detected in Feed II in the MRT. Mössbauer spectroscopy was performed on a Glass I 870°C sample to check wet analysis. The spectral-area fraction derived from the RT spectrum was Fe(II)/Fe = 0.33, a value similar to that by wet colorimetry (see Table II).

Table II. Wet chemistry analysis of Fe(II) and Fe(III) in glasses I and II

Glass/Sample	T (°C)	Total Fe	Fe(II)/Fe
I/A	870	0.546	0.325
I/B	1200	0.505	0.328
II/A	870	0.491	0.308
II/B	1200	0.510	0.338

The MRT began at 870°C and was finished at 1200°C. The temperature of 870°C was selected based on XRD data showing that feed minerals were fully dissolved between 850°C and 900°C (Figure 4, top). Neither feed expanded in volume noticeably during the tests even though the temperature was increased at high rate of 25°C/min for the GS-MS study and 100°C/min for the MS study. This can be attributed to the low viscosity of the melt at 1200°C that results in low stability of foam. The amount of gas evolution was significant, though smaller than that observed in EGA for the temperature region from 800°C to 950°C. An approximately equal volume of CO and CO_2 gases evolved from both feeds (1.3 mL CO_2 and 0.8 ml CO per kg of Glass I and 1.4 mL CO_2 and 0.8 ml CO per kg of Glass II). The large volume of oxygen recorded only in Feed II (8.0 mL O_2 per kg of Glass II as compared to 0.6 mL O_2 per kg of Glass I) was corroborated with the redox measurement showing continuous reduction of Fe(III) to Fe(II) not observed in Feed I. To explain this unexpected high concentration of oxygen that could not be at equilibrium with a nearly identical concentration of CO_2 and CO at 1200°C, we can speculate that while residual carbon reacted with Fe(III) to produce CO_x (CO_2 could also be produced from residual carbonates), in other areas, with less carbon, Fe(III) simultaneously produced O_2.

The 2.1 mL of CO + CO_2 evolved per kg glass at $T > 870$°C corresponds to 0.2 L/h in the DWPF melter producing 100 kg glass/h. This quantity of gas can produce a foamy layer under the cold cap before entering the off-gas. However, the fast-melting Feed II produced 3.8 times larger total volume of gas at $T > 870$°C than Feed I. If the rate of melting is proportional to the thickness of the secondary foam, we must conclude that the foam produced in Glass II was less stable than the foam in Glass I. Hence, the dominant factor was foam stability rather than the gas flux from the melt.

Three factors for foam stability were operating:

1) A lower viscosity of Glass II destabilized the foam, allowing gas bubbles to coalesce easily. At 900°C, Glass I viscosity (101 Pa·s) was nearly twice as

high as that of Glass II (57 Pa·s). This difference decreased with increasing temperature. At 1150°C, Glass I viscosity was 5.9 Pa·s and Glass II viscosity was 4.3 Pa·s. The activation energy of was 18.9×10^3 K for Glass I and only 17.2×10^3 K for Glass II, which was deliberately formulated "longer" than Glass I to achieve as low viscosity at 900° as possible without undesirably affecting other glass properties.

2) A possible destabilizing effect was chemical imbalance. If CO_x bubbles coexist in the foam with O_2 bubbles, as was likely the case of Glass II, the foam was destabilized.

3) Finally, the foam layer of Glass I was likely stabilized by the presence of spinel crystals at higher concentration.

Unstable foam collapses into large bubbles that then merge into cavities. Cavities conduct radiative heat, insulating the cold cap less than spherical foam. Large cavities easily move and open to the atmosphere, either at cold-cap edges or into the vent holes. The result is a faster heat transfer from the melt to the cold cap and, consequently, a higher rate of melting. However, this scenario needs to be confirmed by further research.

CONCLUSIONS

The two main results from this research are that gases (both CO_x and O_2) can be produced in molten glass from a well-reduced feed, thus contributing to the foam formation under the cold cap, that this foam is destabilized by lowering glass viscosity at the melt interface with the cold cap (850°C to 900°C), and stabilized by the presence of crystals. However, direct evidence of this is lacking, and further research is needed to investigate foam behavior.

ACKNOWLEDGMENT

This study was funded by the U.S. Department of Energy's Office of Science and Technology (through the Tanks Focus Area). The authors wish to thank David Peeler from the Savannah River Technology Center for providing the waste sludge simulant and the frits and for his useful advice during the work. Mike Schweiger (he also reviewed this paper) and Jarrod Crum provided unwavering expertly laboratory support and Ravi Kukkadapu helped with Mössbauer spectroscopy. The authors thank Dennis Bickford and Troy Lorier for directing their attention to the stabilizing effect of crystals in the foam layer. Pacific Northwest National Laboratory is operated for the U.S. Department of Energy by Battelle under Contract DE-AC06-76RL01830.

REFERENCES

[1]Ungan A, WH Turner, and R Viskanta.. "Effect of Glass Bubbles on Three-Dimensional Circulation in a Glass Melting Tank," *Glastech. Ber.*, **56K** 125 (1983).

[2]Kim D-S, and P Hrma, "Volume Changes During Batch to Glass Conversion," *Ceram. Bull.*, **69** [6] 1039-1043 (1990).

[3]Darab JG, EM Meiers, and PA Smith, "Behavior of Simulated Hanford Slurries During Conversion to Glass," *Mat. Res. Soc. Proc.*, **556** 215-222 (1999).

[4]Gerrard AH, and IH Smith.. "Laboratory Techniques for Studying Foam Formation and Stability in Glass Melting," *Glastech. Ber.*, **56K** [1] 13-18 (1983).

[5]Kim D-S, and P Hrma, "Foaming in Glass Melts Produced by Sodium Sulfate Decomposition under Isothermal Conditions," *J. Amer. Ceram. Soc.*, **74** [3] 551-555 (1991).

[6]Izak P, P Hrma, BW Arey, and TJ Plaisted, "Effect of Batch Melting, Temperature History, and Minor Component Addition on Spinel Crystallization in High-Level Waste Glass," *J. Non-Cryst. Solids*, **289** 17-29 (2001).

[7]Emer, P, "On Foam Formation on Melt Pool Surface in Glass-Melting Furnaces," *Glastech. Ber.*, **42** [6] 221-228 (1969).

[8]Hrma P, P Izak, JD Vienna, GM Irwin and M-L Thomas, "Partial Molar Liquidus Temperatures of Multivalent Elements in Multicomponent Borosilicate Glass," *Phys. Chem. Glasses*, **43** (2) 128-136 (2002).

[9]Schreiber HD, CW Schreiber, MW Riethmiller, and JS Downey, "The Effect of Temperature on the Redox Constraints for the Processing of High-Level Nuclear Waste into a Glass Waste Form," *Material Research Society Symposium Proceeding*, **176** 419–426 (1990).

[10]Lucktong C and P Hrma, "Oxygen Evolution during $MnO-Mn_3O_4$ Dissolution in a Borosilicate Melts," *J. Amer. Ceram. Soc.*, **71** [5] 323-328 (1988).

[11]P. Hrma, J. Matyáš, and D.-S. Kim, "The Chemistry and Physics of Melter Cold Cap," *9th Biennal Int. Conf. on Nucl. and Hazardous Waste Management, Spectrum '02*, American Nuclear Society, Reno, CD-ROM (2002).

[12]Goldman DS, DW Brite, and WC Richey, "Investigation of Foaming in Liquid-Fed Melting of Simulated Nuclear Waste Glass," *J. Am. Ceram. Soc.*, **69** [5] 413-417 (1986).

[13]Morelissen HW, AHM Rikken, and AJM Van Tienen, "Pelletized Batch: Its Manufacture & Melting Behavior," *The Glass Industry*, **61** 16-20 (1980).

[14]Ramsey WG, CM Jantzen, and DF Bickford, "Redox Analyses of SRS Melter Feed Slurry; Interactions between Nitrate, Formate, and Phenol-Based Dopants," *Ceram. Trans.*, **23** 259-265 (1991).

[15]Bickford DF, and RB Diemer, Jr, "Redox Control of Electric Melters with Complex Feed Compositions, Part I: Analytical Methods and Models," *J. Non-Cryst. Solids*, **84** 276-284 (1986).

[16]Hrma P, "Melting of Foaming Batches: Nuclear Waste Glass," *Glastech. Ber.*, **63K** 360-369 (1990).

[17]Kim D-S, and P Hrma, "Laboratory Studies for Estimation of Melting Rate in Nuclear Waste Glass Melters," *Ceram. Trans.*, **45** 409-419 (1994).

[18]Vienna JD, PA Smith, DA Dorn, and P Hrma, "The Role of Frit in Nuclear Waste Vitrification," *Ceram. Trans.*, **45** 311-325 (1994).

[19]Peeler DK, TH Lorier, DF Bickford, DC Witt, TB Edwards, KG Brown, IA Reamer, RJ Workman, and JD Vienna. "Melt Rate Improvement For DWPF Macrobatch 3: Frit Development and Model Assessment (U)," WSRC-TR-2001-00131, Westinghouse Savannah River Company, Aiken, SC (2001).

[20]Lambert DP, TH Lorier, DK Peeler, and ME Stone, "Melt Rate Improvement for DWPF Macrobatch 3: Summary and Recommendations (U)," WSRC-TR-2001-00148, Westinghouse Savannah River Company, Aiken, SC (2001).

[21]Stone ME, and DP Lambert, "Melt Rate Improvement for DWPF Macrobatch 3: Feed Preparation," WSRC-TR-2001-00126, Westinghouse Savannah River Company, Aiken, SC (2001).

ELECTRON EQUIVALENTS REDOX MODEL FOR HIGH LEVEL WASTE VITRIFICATION

C.M. Jantzen, D.C. Koopman, C.C. Herman, J.B. Pickett, and J.R. Zamecnik
Savannah River Technology Center
Westinghouse Savannah River Co.
Aiken, SC 29808

ABSTRACT

Control of the REDuction/OXidation (REDOX) state in High Level waste (HLW) glass melters is critical in order to eliminate the formation of metallic species from overly reduced melts while minimizing foaming from overly oxidized melts. To date, formates, nitrates, and manganic (Mn^{+4} and Mn^{+3}) species in the melter feeds going to the Savannah River Site (SRS) Defense Waste Processing Facility (DWPF) have been the major parameters influencing melt REDOX. The sludge being processed for inclusion in the next DWPF Sludge Batch (SB-3) contains several organic components that are considered non-typical of DWPF sludge to date, e.g., oxalates and coal. A mechanistic REDOX model was developed to balance any reductants (e.g., oxalate, coal, sugar, formates) and any oxidants (e.g., nitrates, nitrites, and manganic species) for any HLW melter feed. The model is represented by the number of electrons gained during reduction of an oxidant or lost during oxidation of a reductant. The overall relationship between the REDOX ratio of the final glass and the melter feed is given in terms of the transfer of molar Electron Equivalents, ξ.

INTRODUCTION

High-level nuclear waste (HLW) is being immobilized at the Savannah River Site (SRS) by vitrification into borosilicate glass at the Defense Waste Processing Facility (DWPF). A similar HLW vitrification program has just been completed at West Valley Nuclear Services (WVNS) and another facility is being built to process HLW at the Hanford Waste Treatment and Immobilization Plant (WTP). The REDOX equilibrium in a HLW melter must be controlled to prevent the following:

- liberation of oxygen which can cause foaming from decomposition of Mn^{+4} or Mn^{+3} because the $MnO_2 \rightarrow MnO + \frac{1}{2} O_2$ reaction liberates oxygen at the melt temperature
- liberation of NO_x and oxygen caused by decomposition of nitrate species via reactions such as $NO_3 \rightarrow NO + O_2$
- retardation of melt rate due to foaming from nitrates and manganic species
- reduction of metallic species such as $NiO \rightarrow Ni^\circ + \frac{1}{2} O_2$ and $RuO_2 \rightarrow Ru^\circ + O_2$ which can fall to the melter floor and cause shorting of electrical pathways in the melt and accumulations which may hinder glass pouring
- reduction of sulfate ($SO_4^=$) to sulfide ($S^=$) which can complex with Ni° and/or Fe° to form metal sulfides which can fall to the melter floor and cause shorting of electrical pathways and/or hinder glass pouring
- overly reduced glasses which can be less durable than their oxidized equivalents [1].

Controlling the HLW melters at a REDuction/OXidation (REDOX) equilibrium of $Fe^{+2}/\Sigma Fe \leq 0.33$ [2, 3] prevents the potential for conversion of $NiO \rightarrow Ni^\circ + \frac{1}{2} O_2$, $RuO_2 \rightarrow Ru^\circ + O_2$, and $2SO_4^= \rightarrow S_2 + 4O_2$ during vitrification. Control of foaming due to deoxygenation of manganic species is achieved by having 66-100% of the MnO_2 or Mn_2O_3 species converted to MnO [4] during pretreatment in the Sludge Receipt Adjustment Tank (SRAT). At the lower redox limit of $Fe^{+2}/\Sigma Fe \sim 0.09$ about 99% of the manganic species are converted to Mn^{+2} [2, 3]. Therefore, the lower REDOX limit eliminates melter foaming from deoxygenation of manganic oxides.

BACKGROUND

During melting of HLW glass, the REDOX of the melt pool cannot be measured. Therefore, the $Fe^{+2}/\Sigma Fe$ ratio in the glass poured from the melter must be related to melter feed organic and oxidant concentrations to ensure production of a high quality glass without impacting production rate (e.g., foaming) or melter life (e.g., metal formation and accumulation).

Most REDOX models developed to date only include one oxidant and one reductant. For example, at SRS the first REDOX model developed balanced formic acid [F] and nitric acid [N] with a 1:1 stoichiometry, e.g., $Fe^{2+}/\Sigma Fe = -0.8 + 0.87\{[F]-[N]\}$, $R^2 = 0.80$ [5, 6, 7, 8]. The data used to develop the $\{[F]-[N]\}$ relationship was revisited in 1997, and glass quality and REDOX measurement criteria were developed to screen the experimental data used for modeling [9]. This redefined the population of glasses used for modeling by excluding those below the $Fe^{+2}/\Sigma Fe$ measurement detection limit of 0.03 and those that precipitated metallic and/or sulfide species.

Regression of the redefined data demonstrated that the $\{[F]-[N]\}$ parameter was a less accurate predictor ($R^2=0.68$) of waste glass REDOX than had previously been calculated ($R^2=0.80$). The regression of the redefined data showed that there was an {[F]-3[N]} relationship between the feed reductants, oxidants, and the glass REDOX ratio, e.g., $Fe^{2+}/\Sigma Fe = 0.217 + 0.253[F]-0.739[N]$}, $R^2=0.89$ where the F and N concentrations are normalized to a feed that is 45 wt% solids. Both the {[F]-[N]} and the {[F]-3[N]} REDOX models assumed that the melter feeds were properly formated and refluxed to ensure that 66-100% of the Mn^{3+} and Mn^{4+} were converted to Mn^{+2} as $Mn(COOH)_2$ during preprocessing, e.g., before the melter feed entered the melter.

Investigations were also performed at West Valley Nuclear Services (WVNS) to determine the effect of total solids and the concentrations of oxidants and reductants on the REDOX state of iron in glass. The major difference between the WVNS and SRS feeds is that the reductant in WVNS feed is sugar instead of formate. Preliminary investigations by WVNS indicated that the logarithm of the REDOX state, e.g., $\log(Fe^{2+}/Fe^{3+}$, in a glass can be predicted from the feed using the Index of Feed Oxidation (IFO) which is defined as [10]:

$$IFO \equiv \frac{(1-\phi)[NO_3]_{ppm}}{[TC]_{ppm}}$$

where ϕ is the fraction of solids and TC is Total Carbon. While the IFO parameter predicted the REDOX of WVNS glasses it did not predict the REDOX of SRS glasses very well [9]. Conversely, it was shown that the WVNS data fit the {[F]-3[N]} model when the formate coefficient was multiplied by a factor of two in order to account for the differences in the oxidation state of carbon in formic acid and sugar* [9].

PNNL had also developed an iron REDOX model which was similar to that developed by WVNS. The iron REDOX index (ri) suggested by PNNL models two feed oxidants and two feed reductants. The $\log(Fe^{2+}/Fe^{3+})$ is related to ri computed from the concentrations in M/L and normalized to 130g waste oxides/L using either of the following stoichiometrically based relationships depending on whether or not the feed was formated [11].

$$ri = \begin{cases} \dfrac{[COOH^-]-[NO_2^-]+[TC]}{[NO_3^-]} & \textit{formated} \text{ samples} \\[2em] \dfrac{[TC]}{[NO_2^-]+[NO_3^-]} & \textit{unformated} \text{ samples} \end{cases}$$

The effects of manganic species as oxidizers was noted during this study but not incorporated into the model. A separate REDOX model was developed by PNNL for oxalated feeds at WVNS. In this study the $Fe^{+3}/\Sigma Fe \propto g\ NO_3^- / g\ TOC$ [12]

---

* the ratio of the number of electrons transferred during oxidation of C in sugar divided by number of electrons transferred during oxidation of C in formic acid

where TOC is Total Organic Carbon. While the PNNL models were each fit to the data generated in the respective studies, these models did not predict the REDOX of SRS glasses very well [9].

Thus there was a need for a mechanistic REDOX model that could account for all oxidizers (nitrates, nitrites, soluble and insoluble manganic species) and reductants (formates, sugar, coal, oxalate). In addition the model needed to be able to account for the relative oxidizing and reducing power of each species.

EXPERIMENTAL

Twenty-nine simulated SB3 melter feeds were tested in sealed crucibles at four different waste loadings with two different frits, e.g., a high sodium frit (F320) and a low sodium frit (F202). Detailed preparation and analyses of the refluxed feeds used and the sealed crucible studies are given elsewhere [13]. The feeds varied in formic, nitrate, nitrite, oxalate, coal, and manganese. The noble metals were varied from 10 wt% to 100 wt% of the amount calculated to be present in the actual sludge, e.g., 0.0511 wt% Rh, 0.183 wt% Ru, 0.0275 wt% Pd and 0.0005 wt% Ag. Manganese varied from 2.92 wt% Mn to 3.87 wt% Mn. The feeds contained Sm (as a Pu surrogate) and Gd. The Sm was present at 0.024 wt% while Gd varied from 0.037-0.061 wt%. Mercury was added at a constant 0.076 wt%.

Over 200 sealed crucible melts were performed. Sealed crucible vitrification was achieved by sealing Al_2O_3 crucibles with a nepheline ($NaAlSiO_4$) gel that melts at a temperature lower than that at which the slurry vitrifies. This causes the crucible to seal before the slurry vitrifies so that air inleakage does not occur during vitrification. This is extremely important as air inleakage will alter the glass REDOX ratio, $Fe^{2+}/\Sigma Fe$, and allow oxidizers and reductants to escape, rather than reacting with the transition elements in the glass. The $Fe^{+2}/\Sigma Fe$ analyses were performed by the Baumann colorimetric method [14].

Fifty three glasses for modeling were selected out of the 200 glasses melted by applying the following criteria used in previous modeling studies [9]:

- Vitrified material must be visibly black and homogeneous; that is, it must contain no brown discoloration due to metallic copper and/or no crystalline or other metallic material as these species make both reliable REDOX ratio and cation measurements difficult
- The iron REDOX ratio must be greater than or equal to the measurement detection limit of $Fe^{2+}/\Sigma Fe = 0.03$
- Both REDOX and feed chemistry measurements must be available for the same sample
- Measured or as-made total solids information must be available: measured total solids are preferred to minimize modeling error.

REDOX REACTIONS IN THE MELTER COLD CAP

During melter feed-to-glass conversion, multiple types of reactions occur in the cold cap and in the melt pool. The REDOX reactions occur in the cold cap along with feed decomposition and calcination. In the melt pool, further degassing and homogenization occur primarily by additional REDOX reactions. The gaseous products from the cold cap and the volatile feed components further react with air in the vapor space. In order to represent the gradual nature of the feed-to-glass conversion, a 4-stage cold cap model was developed by Choi [15] which approximates the melting of feed solids as a continuous, 4-stage counter-current process [16]. In Stage 1 formated salts such as NaCOOH, are decomposed to CO, CO_2 and H_2. The CO subsequently gets oxidized by the air diffusing into the cold cap from the top and by the oxygen being liberated during the Stage 2 denitration reactions (at further depth in the cold cap). Thus the overall decomposition and calcination reactions occurring in Stages 1 and 2 can be represented by the combined equation:

$$2NaCOOH + 2NaNO_3 \rightarrow CO\uparrow + CO_2\uparrow + H_2\uparrow + N_2\uparrow + 2.5O_2\uparrow + 2Na_2O \rightarrow$$

MelterFeed1   *melterFeed*   *Stage1*   *Stage2*   *Stage1+2*

$$2CO_2\uparrow + H_2O\uparrow + N_2O\uparrow + 4Na_2O$$

plenum   *plenum*   *plenum*   *glass*

(1a)

Multiple oxides begin to form during Stage 3. These oxides are assumed to form solid solutions such as spinels which coexist with the REDOX species in the same phase. Stage 4 represents the final fusion where the oxides formed in Stage 3 dissolve in a silica-rich matrix to form silicate groups in the melt, e.g., Fe_2SiO_4 and Na_2SiO_3. In order to represent all four stages of cold cap reaction simultaneously and include terms for reduced and oxidized iron and silica one can assume a generalized form of the reactions as follows:

$$Fe_2O_3 + 5SiO_2 + 6NaCOOH + 2NaNO_3 \rightarrow Fe_2SiO_4 + 6CO_2 + N_2 + 4Na_2SiO_3 + 3H_2O \quad (1b)$$

Equation 1b assumes that Fe^{3+} enters the melter as Fe_2O_3 and that the reductant $COOH^-$ and the oxidizer NO_3^- enter as sodium formate and sodium nitrate, respectively. The formated and nitrated salts react with glass formers such as SiO_2 to form Fe^{+2} and Na_2SiO_3 components in the glass and liberate CO_2, N_2O and H_2O vapors to the melter plenum (Equation 1a).

ELECTRON EQUIVALENTS MODEL

For simplicity, the generalized REDOX cold cap reaction (Equation 1b) can be rewritten in terms of Fe^{2+} and Fe^{3+} instead of the iron oxides, and the SiO_2 term can be omitted. In addition, the vapor species generated in Stage 1 and Stage 2 of the cold cap, where these reactions occur, are used as the product phases rather than the vapor species measured in the plenum (Equation 1a). This generates

Equation 2 below as one of the controlling REDOX reactions, the one between reducing formate salts and oxidizing nitrated salts, in the cold cap:

$$2Fe^{+3} + 6Na\overset{+2}{C}OOH + 2Na\overset{+5}{N}O_3 \rightarrow 2Fe^{2+} + 6\overset{+4}{C}O_2 + \overset{0}{N_2} + 3Na_2O + 3H_2O + 2Na^+$$

with annotations: $-2e^-/C$ or $-12e^-/6$ formate; $+2e^-$; $+5e^-/NO_3$ or $1x(+10e^-/2NaNO_3)$

(2)

The oxidation/reduction equilibrium shown in the Equation 2 between nitrate and formate indicates that one mole of nitrate gains 5 electrons when it is reduced to N_2 while one mole of carbon in formate loses 2 electrons during oxidation to CO_2. This is an oxidant:reductant ratio of 5:2 which indicates that nitrate is approximately 2½ times as effective an oxidizing agent as formate is a reducing agent (when nitrogen gas is the reaction product).

The oxidation/reduction equilibrium shown in Equation 3 between coal and the oxidized nitrated salts indicates that one mole of nitrate gains 5 electrons when it is reduced to N_2 while one mole of carbon in coal loses 4 electrons during oxidation to CO_2. This is an oxidant:reductant ratio of 5:4 which indicates that nitrate is only 1¼ times as effective an oxidizing agent as coal is a reducing agent (when nitrogen gas is the reaction product).

$$2Fe^{+3} + 3\overset{0}{C} + 2NaN^{+5}O_3 \rightarrow 2Fe^{+2} + \overset{0}{N_2} + 3\overset{+4}{C}O_2 + 2Na^+$$

with annotations: $-4e^-/C$ or $-12e^-/3$ coal; $2x(+1e^-/Fe)$; $+5e^-/NO_3$ or $+10e^-/2NaNO_3$

(3)

The oxidation/reduction equilibrium between the oxalate and nitrate salts is given in Equation 4. This reaction, written in the format of the preceding cold cap reactions (Equations 2 and 3), indicates that one mole of nitrate should gain 5 electrons when it is reduced to N_2 while one mole of carbon in oxalate should lose 1 electron during oxidation to CO_2. This is an oxidant:reductant ratio of 5:1 which indicates that nitrate is 5 times as effective an oxidizing agent as the carbon in oxalate is a reducing agent (when nitrogen gas is the reaction product).

$$2Fe^{+3} + 6Na_2\overset{+3}{C_2}O_4 + 2Na\overset{+5}{N}O_3 \rightarrow 2Fe^{+2} + \overset{0}{N_2} + 12\overset{+4}{C}O_2 + 2Na^+ + 6Na_2O$$

with annotations: $-1e^-/C$ or $-2e^-/$oxalate or $-12e^-/6$ oxalates; $2x(+1e^-/Fe)$; $+5e-/NO_3$ or $1x(+10e-/2NaNO3)$

(4)

During REDOX modeling the data indicated that oxalate appeared to be twice as strong a reductant as indicated by Equation 4. During further investigation of the apparent increase in the reducing power of oxalate, data became available that demonstrated that oxalate salts convert to oxalic acid and then disproportionate to formic acid and CO_2 during SRAT processing [17] via the following equation:

$$HC_2^{+3}O_4^- \rightarrow HC^{+2}O_2^- + CO_2 \uparrow \qquad (5)$$

Experimentally, it was found that between 8-37% of the oxalate present in the SRAT was determined to disproportionate during processing into HCOOH and CO_2 gas [17].

Therefore, it was assumed that additional disproportionation occurs in the cold cap when the liquid slurry impacts the melt pool surface. The pertinent oxidation/reduction equilibrium for oxalate, including the disproportionation, would then be as expressed in Equation 6. Note that this equation includes the decomposition of the oxalate into formic acid and CO_2. Only half the oxalate is acting as a reductant (the half that disproportionates does not affect the REDOX equilibrium). Hence, the reduction potential of oxalate is doubled.

-1e⁻/C or -2e⁻/oxalate = -12e⁻/**12x0.5** oxalates**

$$2Fe^{+3} + 12Na_2C^{+3}{}_2O_4 + 2NaN^{+5}O_3 \rightarrow 2\ Fe^{+2} + N^0{}_2 + 12C^{+4}O_2 + 2Na^+ + 12Na_2O + 12C^{+4}O_2$$

2x(+1e⁻/Fe)

+5e⁻/NO or 1x(+10e⁻/2NaNO3)

$$\qquad (6)$$

A similar equation can be written for the reduction of manganese by any carbon containing species, for example:

-4e⁻/C

$$2Mn^{+4}O_2 + C^0 \rightarrow 2Mn^{+2}O + C^{+4}O_2$$

2x+2e⁻/Mn

$$\qquad (7)$$

Equations 2, 3, 4 and 7 demonstrate that the relative factors for the electrons exchanged upon oxidation and reduction are the following:

- 4 for the number of moles of coal
- 2 for the number of moles of formate
- 4 for the number of moles of oxalate
- 5 for the number of moles of nitrate
- 2 for the number of moles of manganese.

The signs for the oxidation of the reductants are positive while the signs for reduction of the oxidants is negative indicating gain and loss of electrons.

The effectiveness of the oxidants and reductants depends on their concentrations relative to the other slurry components. Therefore, the molar Electron Equivalents term must be multiplied by the factor 45/T, where T is the total solids (wt%) content of the slurry. This factor puts all concentrations on a consistent basis of 45 wt% total solids. The normalized molar Electron Equivalents, ξ, are then:

$$\xi \left(\frac{mol/kg\ feed}{@\ 45\ wt\%\ solids} \right) = (2[F]+4[C]+4[O_T]-5[N]-2[Mn])\frac{45}{T} \qquad (8)$$

Therefore, the basis for the relation of REDOX to electron equivalent transfers, ξ, is

$$\frac{Fe^{2+}}{\Sigma Fe} = f\left[(2[F]+4[C]+4[O_T]-5[N]-2[Mn])\frac{45}{T}\right] = f[\xi] \qquad (9)$$

where f = indicates a function
         $[F]$ = formate (mol/kg feed)
         $[C]$ = coal (carbon) (mol/kg feed)
         $[O_T]$ = oxalate$_{Total}$ (soluble and insoluble) (mol/kg feed)
         $[N]$ = nitrate + nitrite (mol/kg feed)
       $[Mn]$ = manganese (mol/kg feed)
         T = total solids (wt%)

$$\xi = (2[F]+4[C]+4[O_T]-5[N]-2[Mn])\frac{45}{T}$$

When the REDOX data generated in this study and the data from the 1997 study [9] are then fit as a linear function of ξ:

$$\frac{Fe^{2+}}{\Sigma Fe} = 0.1942 + 0.1910\xi \qquad (10)$$

which is the DWPF Electron Equivalents REDOX model (see Figure 1) with an adjusted R^2 of 0.8037 and a Root Mean Square Error of 0.0690 for 120 data observations (53 from the current study and 67 from the 1997 study.

If sugar were used in HLW processing then the form of the model would be changed to include Equation 11:

$$8\overset{+3}{Fe} + \overset{0}{C}_{12}H_{22}O_{11} + 8Na\overset{+5}{N}O_3 \rightarrow 8\overset{+2}{Fe} + 4\overset{0}{N}_2 + 12\overset{+4}{C}O_2 + 11H_2O + 8Na^+$$

$$-4e^-/C \text{ or } -48\ e^-/\text{sucrose}$$
$$8x(+1e^-/Fe)$$
$$4x(+10e^-/2NaNO_3)$$

(11)

The Electron Equivalents term becomes:

$$\xi = \left(2[F] + 4[C] + 4[S] + 4[O_T] - 5[N] - 2[Mn]\right)\frac{45}{T}$$

where $[S]$ = sugar (carbon) (mol/kg feed).

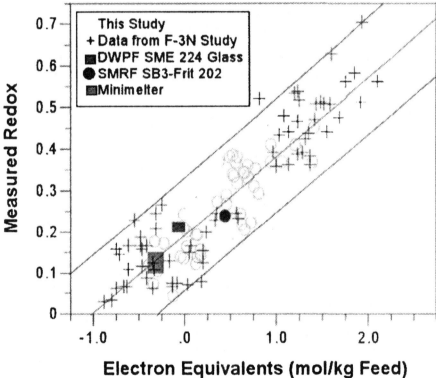

Figure 1. REDOX model with formate, oxalate, coal, nitrate, and manganese normalized for 45 wt% solids, where $\dfrac{Fe^{2+}}{\Sigma Fe} = 0.1942 + 0.1910\xi$.

VALIDATION OF ELECTRON EQUIVALENTS MODEL

Data for three glasses produced in various SRS melters, including the DWPF, were analyzed for REDOX. The validation data are shown in Figure 1 (see solid symbols).

- The DWPF sample REDOX was from a sample pulled from the pour spout of the DWPF during radioactive operation and analyzed at SRTC in the Shielded Cell Facility (SCF). The feed to the melter was comprised mostly of melter feed from SME Batch 224. The target REDOX based on the Electron Equivalents model was ~0.2.
- Samples of minimelter feeds were vitrified in closed crucibles and preliminary analysis of the redox indicated that the target REDOX using the Electron Equivalents model was achieved in the closed crucible tests. A REDOX of 0.12 was achieved after continuous feeding in the minimelter due to air inleakage.
- Lastly, Slurry-fed Melt Rate Furnace (SMRF) tests were performed at a target REDOX of 0.22 based on the Electron Equivalents model. The resulting measured REDOX values shown in Figure 1 indicate an average REDOX of 0.239 (ten REDOX values measured after the SMRF had achieved steady state conditions).

The data from these validation melter tests are shown in Figure 1 (see solid symbols) along with the "Model Data" and the fitted model. The Electron Equivalents term is fitted using the SME analyses. All three data points from actual melters fall well within the 95% confidence interval of the Electron Equivalents model.

The Electron Equivalents REDOX model was also validated against REDOX and feed data from the SRS Integrated DWPF Melter System (IDMS), which was a one-ninth scale (of DWPF) facility used to test various aspects of DWPF operation [13]. Agreement between the Electron Equivalents model and data generated in quartz crucible and pilot-scale melter runs at Pacific Northwest National Laboratory (PNNL) [18,19,20] was also excellent (see reference 13). The PNNL quartz crucible tests used both formate and sugar as reductants, while nitrate and nitrite provided the oxidant.

Crucible studies of the effect of formate and nitrate on glass REDOX adjustment were performed by PNNL in 1996 [21]. These studies were performed on feeds that were very similar to the DWPF, since at that time the Hanford Waste Vitrification Project (HWVP) process was similar to the DWPF process. Nineteen crucible melts were made with varying amounts of nitrate and formate. This data was also used to validate the Electron Equivalents model. Detailed discussions are given elsewhere [13].

Lastly, REDOX and feed data from WVNS was assessed against the Electron Equivalents model. The data included crucible melts [22] and data from the operation of a 1/10th scale test melter [10]. This melter was a 1/10th scale prototype of the joule-heated, ceramic-lined melter used to vitrify wastes stored at the West Valley Demonstration Project (WVDP). Tests were run by doping simulated waste slurries with varying amounts of nitric acid to simulate WVDP flowsheet levels of nitrate, and sucrose was used as a reductant. All data were then normalized to 45 wt% solids. The WVNS data show that the Electron Equivalents model handles sugar as a reductant quite well [13].

CONCLUSIONS

Glasses used in REDOX modeling must be produced from refluxed melter feed material to ensure conversion to nitrate and formate species. Vitrified material must be visibly (10X magnification) black and homogeneous for reliable REDOX ratio and cation measurements. The iron REDOX ratio (i.e., $Fe^{2+}/\Sigma Fe$) should be measured using a highly reproducible and accurate method such as the Baumann colorimetric technique, which was recommended, for use in DWPF in 1989 [23]. Use of other REDOX measurement techniques has been shown to give less reliable measurements [23]. REDOX values must be greater than or equal to the method detection limit of $Fe^{2+}/\Sigma Fe \leq 0.03$. Both REDOX and feed chemistry measurements must be available for the same sample to decrease modeling error. Measured or as-made total solids information must available.

Reduction makes an atom or molecule less positive by electron transfer. Oxidation makes an atom or molecule more positive by electron transfer. The number of moles of electrons transferred for each REDuction/OXidation reaction are weighted by the number of electrons transferred providing an Electron Equivalents term for each reductant and oxidant species defined. The weighted Electron Equivalents are then summed (oxidation reactions have a positive sign and reduction reactions have a negative sign:

$$\frac{Fe^{2+}}{\Sigma Fe} = f\left[\left(2[F] + 4[C] + 4[O_T] - 5[N] - 2[Mn]\right)\frac{45}{T}\right] = f[\xi]$$

where f = indicates a function
  [F] = formate (mol/kg feed)
  [C] = coal (carbon) (mol/kg feed)
  [O_T] = oxalate$_{total}$ (soluble and insoluble) (mol/kg feed)
  [N] = nitrate + nitrite (mol/kg feed)
  [Mn] = manganese (mol/kg feed)
  T = total solids (wt%)
and ξ(mol/kg feed) = Electron Equivalents

In the presence of sugar the Electron Equivalents term becomes

$$\xi \text{ (mol/kg feed)} = \left(2[F] + 4[C] + 4[S] + 4[O_T] - 5[N] - 2[Mn]\right)\frac{45}{T}$$

The REDOX data generated in this study were fit along with previous model data as a linear function of ξ:

$$\frac{Fe^{2+}}{\Sigma Fe} = 0.1942 + 0.1910\xi$$

with an $R^2 = 0.80$ and a RMSE $= 0.0690$.

The $\dfrac{Fe^{2+}}{\Sigma Fe}$ predictions from the Electron Equivalents model were validated against (1) REDOX data generated from the DWPF melter from SME Batch 224, (2) data generated by the SRTC mini-melter and (3) data from the SRTC Slurry-fed Melt Rate Furnace (SMRF). All the data from these melters fell within the 95% confidence bands of the Electron Equivalents REDOX model developed in this study. Validation data from SRS pilot scale melters, Pacific Northwest Laboratory testing, and West Valley Nuclear Fuel Services testing agreed with the Electron Equivalents model better than all previous REDOX models.

ACKNOWLEDGEMENTS

This paper was prepared in connection with work done under Contract No. DE-AC09-96SR18500 with the U.S. Department of Energy.

REFERENCES

1   C.M. Jantzen, J.B. Pickett, K.G. Brown, T.B. Edwards, D.C. Beam, "Process/Product Models for the Defense Waste Processing Facility (DWPF): Part I. Predicting Glass Durability from Composition Using a Thermodynamic Hydration Energy Reaction MOdel (THERMO)," U.S. DOE Report WSRC-TR-93-0672, Westinghouse Savannah River Co., Aiken, SC, 464p. (Sept. 1995).

2   H.D. Schreiber, and A.L. Hockman, "Redox Chemistry in Candidate Glasses for Nuclear Waste Immobilization," J. Am. Ceram. Soc., Vol. 70, No. 8, pp. 591-594 (1987).

3   C.M. Jantzen and M.J. Plodinec, "Composition and Redox Control of Waste Glasses: Recommendation for Process Control Limit," U.S. DOE Report DPST-86-773, E.I. duPont deNemours & Co., Savannah River Laboratory, Aiken, SC (November, 1986).

4   M.J. Plodinec, "Foaming During Vitrification of SRP Waste," U.S. DOE Report DPST-86-213, E.I. duPont deNemours & Co., Savannah River Laboratory, Aiken, SC (January, 1986).

5   W.G. Ramsey, C.M. Jantzen, and D.F. Bickford, "Redox Analyses of SRS Melter Feed Slurry; Interactions Between Formate, Nitrate, and Phenol Based Dopants," Proceed. of the 5th Intl. Symp. on Ceramics in Nuclear Waste Management, G.G. Wicks, D.F. Bickford, and R. Bunnell (Eds.), American Ceramic Society, Westerville, OH, 259-266 (1991).

6   W.G. Ramsey, T.D. Taylor, K.M. Wiemers, C.M. Jantzen, N.D. Hutson, and D.F. Bickford, "Effects of Formate and Nitrate Content on Savannah River and Hanford Waste Glass Redox" Proceed.of the Advances in the Fusion and Processing of Glass, III, New Orleans, LA, D.F. Bickford, et.al. (Eds.) Am. Ceramic Society, Westerville, OH, 535-543 (1993).

7   W.G. Ramsey, N.M. Askew, and R.F. Schumacher, "Prediction of Copper Precipitation in the DWPF Melter from the Melter Feed Formate and Nitrate Content," U.S. DOE Report WSRC-TR-92-385, Westinghouse Savannah River Co., Aiken, SC (Nov.30, 1994).

8   W.G. Ramsey, and R.F. Schumacher, "Effects of Formate and Nitrate Concentration on Waste Glass Redox at High Copper Concentration." U.S. DOE Report, WSRC-TR-92-484, Westinghouse Savannah River Co., Aiken, SC (October 23, 1992).

9   K.G. Brown, C.M. Jantzen, and J.B. Pickett, "The Effects of Formate and Nitrate on REDuction/OXidation (REDOX) Process Control for the Defense Waste Processing Facility," U.S. DOE Report WSRC-RP-97-34, Westinghouse Savannah River Co, Aiken, SC (Feb. 1997).

10 B.W. Bowan, "**A Redox Forcasting Correlation Developed Using a New One-Tenth Area Scale Melter for Vitrifying Simulated High-Level Radioactive Wastes**," M.S. Thesis, Alfred University, Alfred, New York (1990).

11 P.A. Smith, J.D. Vienna, and M.D. Merz, "**NCAW Feed Chemistry: Effect of Starting Chemistry on Melter Offgas and Iron Redox**," U.S. DOE Report PNL-10517, PNL, Richland, WA (March 1995).

12 G.K. Patello, R.L. Russell, G.R. Golcar, H.D. Smith, G.L. Smith, and M.L. Elliott, "**Processing Simulated Oxalated High Level Waste Through a Vitrification Feed Preparation Flowsheet**," Ceram. Trans. v. 93, Am.Ceram. Soc., Westerville, OH, 163-170 (1999).

13 C.M. Jantzen, J.R. Zamecnik, D.C. Koopman, C.C. Herman, J.B. Pickett, "**Electron Equivalents Model for Controlling REDuction-OXidation (REDOX) Equilibrium During High Level Waste (HLW) Vitrification**," U.S. DOE Report WSRC-TR-2003-00126, Rev. 0, Westinghouse Savannah River Company, Aiken, SC (May, 2003).

14 E.W. Baumann, "**Colorimetric Determination of Iron(II) and Iron(III) in Glass**," Analyst, 117, 913-916 (1992).

15 A.S. Choi, "**Validation of DWPF Melter Off-Gas Combustion Model**," U.S. DOE Report WSRC-TR-2000-00100, Westinghouse Savannah River Co., Aiken, SC (June 23, 2000).

16 A.S. Choi, "**Prediction of Melter Off-Gas Explosiveness**," U.S. DOE Report WSRC-TR-90-00346, Westinghouse Savannah River Co., Aiken, SC (January 22, 1992).

17 D.C. Koopman, C.C. Herman, N.E. Bibler, "**Sludge Batch 3 Preliminary Acid Requirements Studies with Tank 8 Simulant**," U.S. DOE Report WSRC-TR-2003-00041, Westinghouse Savannah River Co., Aiken, SC (January 31, 2003).

18 P.A. Smith, J.D. Vienna, and M.D. Merz, "**NCAW Feed Chemistry: Effect of Starting Chemistry on Melter Offgas and Iron Redox**," U.S. DOE Report PNL-10517, Pacific Northwest Laboratory, Richland, Washington (March 1995).

19 D.R. Jones, W.C. Janshiki, and D.S. Goldman, "**Spectroscopic Determination of Reduced and Total Iron with 1,10-Phenalthroline**," Anal. Chem., 53, 923-924 (1981).

20 R.W. Goles, R.K. Nakaoka, "**Hanford Waste Vitrification Program Pilot-Scale Ceramic Test Melter 23**," U.S. DOE Report PNL-7142, UC-721, PNL, Richland, Washington (1990).

21 K. D. Weimers, "**The Effect of HWVP Feed Nitrate and Carbonate Content on Glass REDOX Adjustment**," U.S. DOE Report, PNNL-11044, PNL Laboratory, Richland, WA (March 1996).

22 D.S. Goldman and D.W. Brite, "**Redox Characterization of Simulated Nuclear Waste Glass**," J. Am. Ceram. Soc., 69 [5], pp. 411-413, (1986).

23 C.M. Jantzen, "**Verification and Standardization of Glass Redox Measurement for DWPF**," U.S. DOE Report DPST- 89-222, E.I. du Pont de Nemours & Co., Savannah River Laboratory, Aiken, SC (1989).

SULFATE RETENTION DURING WASTE GLASS MELTING

P. Hrma, J. D. Vienna,
and W. C. Buchmiller
Pacific Northwest National Laboratory
Richland, WA 99352

J. S. Ricklefs
Department of Chemical Engineering
University of California
Davis, CA 95616

ABSTRACT

Sulfate segregation significantly increases the cost for vitrifying low-activity waste (LAW) and, therefore, is a major concern in vitrifying sulfate-rich LAWs. The fraction of sulfate that is not dissolved during early stages of melting is transported in gas bubbles to the glass surface, where it remains segregated. Tests with crucible melts and runs of experimental melters of various scales indicate that the physicochemical makeup of an LAW feed, LAW glass formulation, and melter design and operation can be optimized to increase SO_3 retention in glass.

INTRODUCTION

Sulfate is a common component of low-activity waste (LAW) at Hanford. In 2001, Li et al.[1] collected SO_3 solubility data for both commercial and waste glasses and showed that the SO_3 solubility in terms of SO_3 mass fraction, $w_{SO_3}^{eq}$, is a power-law function of a single parameter, $\varsigma = c_{NBO}^2/c_{NB}$, where c_{NBO} and c_{NB} are molar concentrations of the nonbridging and bridging oxygen, respectively:

$$\ln w_{SO_3}^{eq} = a + b \ln \varsigma \qquad (1)$$

with $a = -5.02$ and $b = 0.22$. This equation describes data for borosilicate and soda-lime glasses except those with more than 2 mass% P_2O_5. These glasses dissolved more SO_3 than Equation 1 would predict. Rong et al. studied and explained the interaction between sulfate and phosphate in borosilicate glass.[2]

Not all SO_3 up to its solubility limit can be dissolved in glass during continuous melting. As reported by Pegg et al.,[3] sulfate segregates at an SO_3 concentration fairly below the solubility limit when glass is processed in all-electric continuous melters equipped with bubblers. Segregated sulfate accumulates on the top of the melt as a corrosive molten salt. According to Pegg et al.,[3] the maximum allowable SO_3 in all-electric continuous melters equipped with bubblers is

$$w_{SO_3}^{max} = 5 \times 10^{-4} / w_{Na_2O} \qquad (2)$$

where w_{Na2O} is the Na_2O mass fraction in glass. Equations (1) and (2) are compared in Figure 1, which shows that for $\zeta > 1$, only a small fraction of soluble SO_3 is retained in the glass processed.

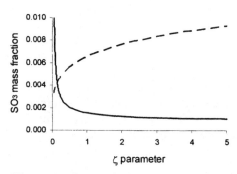

Figure 1. Comparison between the SO_3 solubility limit (broken line) and the maximum allowable SO_3 fraction (solid line) for LAW glass melters

How a well-dispersed easily soluble component can segregate at such low levels of concentration was not understood until recently. As was recently reported,[4] molten sulfate is carried to the melt surface in gas bubbles. Though sulfate was observed in bubbles in sulfate-fined commercial glasses for decades, bubbles were not recognized as vehicles that transport sulfate to the melt surface.

Figure 2 shows fracture surfaces of partly melted LAW feeds. The fracture passes either through a bubble or leaves a shell of sulfate intact, thus evidencing the mechanism of segregation.

How does sulfate get into the bubbles? At early stages of melting, sulfate is a component in the primary—predominantly nitrate—melt that appears in the feed at temperatures already below 300°C. As the dominant nitrate component decomposes by ~700°C, the remaining sulfate melt spreads on the available internal surfaces of the melting feed. These internal surfaces eventually, at temperatures just below 800°C, collapse into bubbles, trapping sulfate inside. This explains the observation made by Li et al.[1] that sulfate segregation begins at the melting process as soon as the connected glass phase is formed.

Some sulfate dissolves in the glass melt before bubbles are formed and continues to dissolve from the surfaces of bubbles as they ascend to the top surface of the melt.   The following equation was derived for the final concentration of the dissolved sulfate:[4]

$$c_{dis} = \frac{4\pi}{k_s k_L} nh\Delta c \left(\frac{D\eta}{\Delta\rho g} \right)^{\frac{2}{3}}$$ (3)

where k_s is the Stokes constant, k_L the Levich constant, n the bubble-number density, h the bubble-path length, D the diffusion coefficient, η the melt viscosity, $\Delta\rho$ the difference between the melt density and the average bubble density, g the acceleration due to gravity, and $\Delta c = c_{sat} - c_{dis}$; here c_{sat} is the SO_3 solubility.

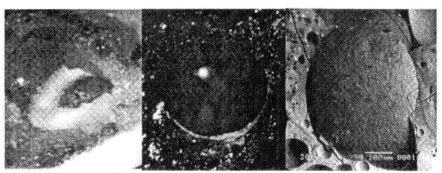

Figure 2. Sulfate in a bubble at 900°C (left); scanning electron microscopy (SEM) micrographs of a crescent of sulfate in a bubble at 950°C (middle) and of a spherical layer of sulfate in a bubble at 875°C (right)

The term in parentheses in the right-hand side of Equation 3 changes little with melt composition and temperature.   Although viscosity and diffusion coefficient change exponentially with composition and temperature, their product, by the Stokes-Einstein equation, is nearly constant.   Therefore, neither viscosity nor diffusion coefficient significantly affect sulfate retention.

The claim by Pegg et al.[3] that sulfate retention is controlled by diffusion is based on the assumption that sulfate is incorporated in melt through diffusion from the top surface of the melt, i.e., through redissolving sulfate that has been segregated.   However, segregated sulfate, though it may redissolve during melter idling, is unlikely to get appreciably redissolved during a normal melter operation.

Based on observations of pilot-scale melters, Pegg et al.[3] reasoned that sulfate dissolution is a slow process, and thus very slow melting would prevent segregation.   As stated above, the dissolution of segregated sulfate is indeed slow. However, as crucible experiments demonstrate, rapid melting actually increases sulfate retention.   Hence, sulfate dispersed in the feed dissolves more rapidly and segregates less if the feed temperature is rapidly increased.   The opposite trend observed in melters may be associated with bubbling.   An increased rate of bubbling leads to an increased rate of melting and, simultaneously, leads to sulfate removal at a faster rate.

EXPERIMENTS

Experimental studies were performed with Hanford LAWs and sodium-bearing Idaho wastes of similar composition. Crucible melts were always carried out with simulated feeds heat treated at 4°C/min, a rate deemed similar to that experienced in a continuous melter. Melter studies were conducted using a quartz-glass-tube melter (QTM) described by Darab et al.[5] This tiny melter with ~5 cm^2 melt area allows continuous feeding and visual observation of the melt during processing. A research-scale melter (RSM)[5,6] was also used. This Joule-heated melter has a melt area of ~150 cm^2.

RESULTS AND DISCUSSION

Using the numerical values of constants and approximate values of glass properties, i.e., $k_s \approx 0.19$, $k_L \approx 2.62$, $\eta \approx 10$ Pa·s, $D \approx 10^{-11}$ m^2/s, $\Delta\rho \approx 2\times10^3$ kg/m^3, and $g \approx 10$ m/s^2, Equation 3 yields

$$c_{dis} = \frac{c_{sat}}{1+13.4/(nh)} \tag{4}$$

where n is in mm^{-3} and h in mm. Equation 4 is represented in Figure 3.

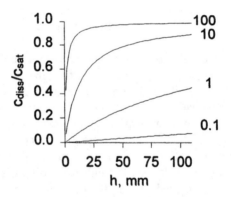

Figure 3. The effect of h and n
(values at the lines are in mm^{-3}) on
SO$_3$ retention in a LAW glass

The n and h values are mostly affected by the melter design and operation, though n can also be affected by the feed makeup. Figure 4 compares the sulfate retention achieved with the RSM as compared with the limit defined by Equation 2 for large melters with bubblers. The length of the bubble residence time in the melt, or the length of the bubble trajectory, h, can be increased by mechanical stirring that would thus help dissolve more sulfate. On the other hand, mechanically-introduced large bubbles are likely to coalesce with smaller sulfate-containing bubbles and thus remove sulfate from the melt at a rapid rate.

Sulfate segregates at temperatures from just below 800°C to just above 950°C. At these temperatures, the composition of molten glass changes as some solid components of the feed continue to dissolve. Therefore, the c_{sat} value in Equation 4 must be considered at the glass composition and temperature when the bubbles move through the melt.

As, by Equation 1, c_{sat} increases with increasing ζ-parameter, which could be considered as an expression for glass basicity, sulfate retention can be increased if major acid glass components, such as B_2O_3, Al_2O_3, or SiO_2, are either at lower concentrations or their incorporation into the melt is delayed. As Figure 5 shows, larger grains of silica, which take a longer time to dissolve, enhanced sulfate retention in the QTM (results of only those melter runs are shown that were free from segregated sulfate). Although other parameters were changed from run to run, the positive effect of silica grain size on the SO_3 retention is clearly plausible. An even more dramatic impact was observed when the QTM was charged with a feed in which Al, Ca, Mg, and Zr silicates were used instead of oxides; SO_3 retention increased to 0.8 to 1.1 mass% for the same feed.

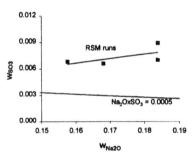
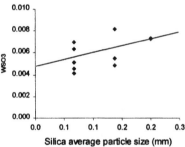

Figure 4. SO_3 retention achieved in RSM compared to the allowable SO_3 fraction for large melters

Figure 5. SO_3 retention achieved in QTM runs versus silica particle size

A dramatic increase of both waste loading and SO_3 retention was accomplished by an improved formulation of glass as shown in Table 1. The waste loading of LAWC22 glass was 16.2 mass%. By Equation 2, the allowable SO_3 fraction in this glass would be 0.34 mass%. However, it was possible to melt LAWC22 glass in a QTM without forming a sulfate layer with additional sulfate at 0.82 mass% SO_3. Glass LAWN represents an improved formulation. The waste loading in this glass was 22.2 mass%, 6 mass% higher than in glass LAWC22, corresponding to 0.25 mass% of allowable SO_3. Moreover, the SO_3 level actually achieved during sulfate-free continuous operation in a QTM was

1.09 mass% SO_3 in glass. Figure 6 depicts the difference in added glass components other than silica. Less B_2O_3, Al_2O_3, TiO_2, and ZrO_2, and additional K_2O increased glass basicity. Adding P_2O_5 and V_2O_5 also increased SO_3 retention.

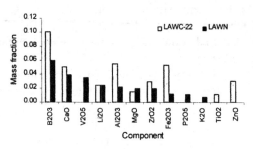

Figure 6. Added oxides in two glasses formulated for an identical LAW

Table 1. Two LAW glasses formulated for an identical LAW

	LAWC22		LAWN	
	Waste	Additives	Waste	Additives
Al_2O_3	0.0057	0.0552	0.0078	0.0222
B_2O_3		0.1005		0.0600
CaO	0.0004	0.0506	0.0005	0.0395
Cl	0.0008		0.0011	
Cr_2O_3	0.0002		0.0003	
F	0.0015		0.0020	
Fe_2O_3	0.0017	0.0539	0.0024	0.0126
K_2O				0.0075
Li_2O		0.0250		0.0250
MgO		0.0151		0.0200
Na_2O	0.1458		0.2000	
NiO	0.0003		0.0004	
P_2O_5	0.0012		0.0016	0.0114
PbO	0.0002		0.0002	
SO_3	0.0038		0.0052	
SiO_2		0.4660		0.5253
SrO		0.0000		
TiO_2		0.0114		
V_2O_5				0.0350
ZnO		0.0306		
ZrO_2		0.0301		0.0200
Sum	0.1615	0.8385	0.2215	0.7785

CONCLUSIONS

Adjusting feed makeup, glass composition, and possibly melter design and operation can accelerate sulfate dissolution and thus increase SO_3 retention in sodium-alumino-borosilicate waste glasses. Using additive feed components, such as coarse silica or less-reactive silicate minerals, to decrease the basicity of glass-forming melt at temperatures below 900°C, helps sulfate dissolve faster. Formulating glass of higher basicity, balancing alkali and earth alkali oxides, and increasing the level of P_2O_5 as a minor addition can substantially increase SO_3 retention. Elements of melter design and operation, such as bubblers versus mechanical stirrers, also influence SO_3 retention rate.

ACKNOWLEDGMENT

This study was funded by the U.S. Department of Energy's Office of Science and Technology (through the Tanks Focus Area). The authors wish to thank David Peeler from the Savannah River Technology Center for his interest and useful advice during the work. Mike Schweiger and Jarrod Crum provided indispensable and friendly laboratory support. Pacific Northwest National Laboratory is operated for the U.S. Department of Energy by Battelle under Contract DE-AC06-76RL01830.

REFERENCES

[1]H. Li, P. Hrma, and J. D. Vienna, "Sulfate Retention in Simulated Radioactive Waste Bororsilicate Glasses," *Ceram. Trans.* **119**, 237-245 (2001).

[2]C. Rong, K.C. Wong-Moon, H. Li, P. Hrma, and H. Cho, "Solid-State NMR Investigation of Phosphorus in Aluminoborosilicate Glasses," *J. Non-Cryst. Solids* **223**, 32-42 (1998).

[3]Pegg I. L., H. Gan, I. S. Muller, D. A. McKeown, and K. S. Matlack, *Summary of Preliminary Results on Enhanced Sulfate Incorporation During Vitrification of LAW Feeds*, VSL-00R3630-1, Vitreous State Laboratory, The Catholic University of America, Washington, DC (2001).

[4]P. Hrma, J. D. Vienna, and J. S. Ricklefs, "Mechanism of Sulfate Segregation during Glass Melting," *Mat. Res. Soc. Proc.* In print (2003).

[5]Darab, J. G., D. D. Graham, B. D. MacIsac, R. L. Russell, H. D. Smith, J. D. Vienna, and D. K. Peeler, *Sulfur Partitioning During Vitrification of INEEL Sodium Bearing Waste: Status Report*. Pacific Northwest National laboratory, PNNL-13588 (2001).

[6]Goles R. W., J. M. Perez, B. D. MacIsaac, D. D. Siemer, and J. A. McCray, *Testing Summary Report INEEL Sodium Bearing Waste Vitrification Demonstration RSM-01-1*, PNNL-13522, Pacific Northwest National Laboratory, Richland, Washington (2001).

THE CHARACTERIZATION AND DISSOLUTION OF HIGH LEVEL WASTE CALCINE IN ALKALI BOROSILICATE GLASS

S. Morgan, P.B. Rose, R.J. Hand, N.C. Hyatt, W.E. Lee
Immobilisation Science Laboratory,
Department of Engineering Materials, University of Sheffield,
Mappin Street, Sheffield, S1 3JD, UK

C. R. Scales
British Nuclear Fuels plc
Research and Technology
Sellafield
Seascale, CA20 1PG, UK

ABSTRACT

The decomposition and dissolution of high level waste calcine in alkali borosilicate glass was examined using simulated calcine from a full scale inactive trial and mixed alkali borosilicate glass. The initial, poorly crystalline calcine material contained $LiNO_3$ and $Sr(NO_3)_2$, and a fluorite-type phase. Decomposition took place in 4 distinct stages. Dehydration occurs between room temperature and 125°C; loss of bound water between 125°C and 400°C ; denitration and formation of several secondary RE fluorite- type phases between 400°C and 700°C and from 800-1000°C a single RE fluorite phase evolves along with $CsLiMoO_4$ and additional minor phases. Dissolution of the calcine into the glass occurs in 2 stages, the first involving rapid migration of Cs and Mo into the melt, forming a low density $CsLiMoO_4$ composition liquid on the melt surface. The second stage involves more extensive migration of RE (Ce, Nd, Gd) elements and Zr , accompanied by disappearance of the surface layer of liquid $CsLiMoO_4$, due to increased convection currents within the melt. Unreacted calcine is still present after 16 minutes of melting.

INTRODUCTION

In the UK, High Level Waste (HLW) arising from nuclear fuel reprocessing is vitrified in a lithium sodium borosilicate glass matrix.[1,2] The waste vitrification process begins when spent nuclear fuel (SNF) rods are sheared and dissolved in

---

hot nitric acid. The resultant Highly Active Liquor (HAL) is then treated to recover U and Pu and the remaining raffinate is calcined to produce a granular solid. The calcination process is complicated by the volatilisation of Ru and the formation of refractory oxides which are difficult to incorporate in the subsequent glass melt. Therefore, prior to calcination, the raffinate is combined with a solution of lithium nitrate to complex Fe and Al species thus preventing the formation of refractory compounds of these metals. Volatilisation of Ru is suppressed by adding sucrose to react with excess nitrate, thereby preventing the nitrate mediated oxidation of Ru to volatile RuO_4. The effective operating temperature of the calciner is <500°C. The granular solid arising from calcination is discharged into an induction heated Inconel melter, together with a quantity of alkali borosilicate glass frit. The calcine / glass batch (~250kg) is maintained at 1050°C for several hours and sparged with air to homogenise the melt prior to pouring into stainless steel canisters. These canisters are allowed to cool and are subsequently sealed and decontaminated before being transferred for interim storage.

This study is concerned with characterisation of a simulant waste calcine and the dissolution of this material during the early stages of the melting process. The composition of the calcine, given in Table I, was formulated to simulate the HAL derived from reprocessing of UK Oxide and Magnox fuels, blended in a 75:25 ratio (on an oxides basis, by weight). This material was obtained from full scale inactive trials at the Sellafield WVP (Waste Vitrification Plant) and, as such, represents a close approximation to the active waste material, with the exception that Ru was omitted from the composition of this material on economic grounds. To better understand the calcination process the phase evolution of the calcine to 1000°C has been investigated in a series of laboratory experiments, using powder X-ray diffraction (XRD), Thermogravimetric Analysis (TGA), Differential Thermal Analysis (DTA) and Fourier Transform Infra-Red (FT-IR) spectroscopy. The interaction of the simulant calcine with molten glass was studied by analysing specimens quenched during the early stages of calcine dissolution, which were cross-sectioned and examined by Scanning Electron Microscopy (SEM) coupled Energy Dispersive X-ray spectroscopy (EDX).

Table I: Simulant (non- radioactive) HLW composition

Oxide	Weight % in simulant calcine	Oxide	Weight % in simulant calcine	Oxide	Weight % in simulant calcine
SiO_2	0.06	CeO_2	4.97	P_2O_5	0.27
TiO_2	0.01	Cs_2O	4.86	Cr_2O_3	0.61
Al_2O_3	3.74	Gd_2O_3	8.15	ZrO_2	7.48
Fe_2O_3	2.81	La_2O_3	2.35	HfO_2	0.10
CaO	<0.01	MoO_3	6.00	BaO	0.71
MgO	4.02	Nd_2O_3	7.60	SrO	1.60
K_2O	<0.01	NiO	0.40	B_2O_3	<0.05
Na_2O	<0.10	Pr_6O_{11}	2.35		
TeO_2	0.86	Sm_2O_3	1.26		
Li_2O	5.53	RuO_2	<0.01		

EXPERIMENTAL

The simulant HLW calcine (see Table I) was prepared from a solution of the appropriate nitrate reagents mixed in stoichiometric ratio. This solution was passed through a full scale rotary calciner during commissioning of vitrification lines at the Sellafield WVP. The calciner is divided into four heating zones, dehydration of the HAL occurs in the first two zones with partial denitration occurring in the latter zones. The maximum estimated temperature in the calciner during operation is <500°C. The composition of this material was determined by inductively coupled plasma spectroscopy (Table I). The Loss on Ignition (determined from thermogravimetric analysis to 1000°C) was 45wt%.

Thermal analysis of the simulant calcine was undertaken using a Perkin Elmer Pyris 1 TGA and a Perkin Elmer DTA 7. Specimens were prepared by placing ~15mg of well ground powder, sieved to < 100µm, in alumina crucibles and heating at a rate of 2°C/min to 1000°C in air. To investigate the thermal events apparent in the TGA and DTA data, larger specimens, ~2g, were heated in a muffle furnace at various temperatures between 80°C and 1000°C, at a rate of 2°C/min. Specimens were removed from the furnace at designated temperatures and analysed as described further in the Results section.

The phase composition of the calcine material, as a function of temperature, was examined using X-ray powder diffraction. Specimens (ground and sieved to < 100µm) were front loaded into aluminium sample holders and analysed with a Philips PW1373 X-ray Powder Diffractometer employing Cu/Kα radiation and operating in reflection mode.

FTIR spectroscopy was performed on pellets pressed from a well ground mixture calcine material (2mg) and KBr (200mg), using a Perkin Elmer Spectrum 2000 FT-IR in transmission mode (range 4000– 600cm^{-1}).

The interaction between the calcine and glass melt during the early stages of vitrification was examined by heating a well mixed batch of glass frit and simulant calcine in an Inconel 601 crucible (75ml capacity) at 1050°C. The glass frit employed was a sample of that supplied to the Sellafield WVP, with the composition given in Table II. The batch composition was calculated to yield a calcine waste loading of ~25wt%, equivalent to that employed on the full scale vitrification lines at the Sellafield WVP. Samples were heated for periods of 5, 10 and 16 minutes at 1050°C, in air, and subsequently quenched in water. Specimens suitable for analysis by SEM/EDX were prepared from crucibles filled with cold-setting epoxy resin, cross sectioned using a diamond saw. The cross sections were polished to a finish of 1μm, using an oil based lubricant, to prevent hydration of the residual calcine. Specimens were carbon coated and analysed using CamScan and JEOL JSM 6400 scanning electron microscopes, equipped with Oxford Analytical Link EDX systems.

Table II: Borosilicate glass frit

Component	Wt%
SiO_2	62.9 ± 1.0
B_2O_3	23.0 ± 0.8
Na_2O	11.4 ± 0.5
Li_2O	2.7 ± 0.4

RESULTS

Heat treatment of simulant calcine

Analysis of powder XRD data acquired from the simulant calcine, Figure 1, revealed the presence of $LiNO_3$, $Sr(NO_3)_2$ and a rare earth fluorite-type phase (LnO_{2-x}) characterised by broad reflections. The broad nature of these reflections indicate that this phase is poorly crystalline and also reflects the presence of several rare earth cations (La, Nd, Ce, Sm, Gd) with a range of ionic radii (1.18 – 1.08A, for 8-fold co-ordination[3]) incorporated within the fluorite structure. Strong absorption bands at ~3430cm[-1] and 1387cm[-1], were apparent in the infra-red spectrum of this material. These bands attributed, respectively, to the O-H and N-O stretches of H_2O molecules and the NO_3^{2-} oxyanion, confirmed a high water and nitrate content in this material.

Thermal analysis of the simulant calcine, Figures 2a and 2b, revealed several thermal events arising between room temperature and 1000°C. DTA revealed an intense endotherm arising between room temperature and 125°C associated with a weight loss of 12.0%. FT-IR analysis of a sample heated to 140°C revealed a decrease in the intensity of the absorption band associated with O-H stretch of water. This thermal event is therefore attributed to dehydration of the calcine material and the loss of sorbed water. Between 125 – 400°C several endothermic events are observed in the DTA data of the calcine. The sharp intense endotherm at 246°C is attributed to the melting of $LiNO_3$, consistent with published data for this compound.[5] The weak endotherms observed at 325°C and 375°C and the small weight loss of ~4% may be associated with the loss of chemically bound water molecules ("water of hydration"), consistent with a further reduction in the intensity of the O-H absorption band in the infra-red spectra. Thermogravimetric analysis reveals a significant weight loss, ~30wt%, between 400-700°C, concomitant with a decrease in the intensity of the absorption band associated with the N-O stretch of the NO_3^{2-} species. In addition, this absorption band is resolved into two components, a broad band (at 1550-1250cm^{-1}) and a sharp band at 1385cm^{-1}. The broad band is observed to decrease in intensity above 400°C, whereas the sharp band remains essentially unchanged. Powder diffraction data acquired from calcine samples heated to 400-700°C reveal the decomposition of $Sr(NO_3)_2$, between 400-500°C, and $LiNO_3$, between 500–700°C. Sharp reflections, characteristic of at least two additional rare-earth fluorite phases, appear in the diffraction pattern of a sample of calcine heated to 700°C (Fig. 1). These reflections are superimposed on the broad reflections also attributed to the rare-earth fluorite type phase, in the diffraction pattern acquired from the calcine prior to heat treatment. Thus, between 400-700°C denitration of the calcine occurs leading to the formation of secondary rare-earth fluorite phases together with decomposition of $LiNO_3$ and $Sr(NO_3)_2$. This is accompanied by change in colour of the calcine material from pale to deep brown. Between 700 – 1000°C no significant weight loss is observed, however, the powder diffraction patterns of calcine samples heated to 800°C and 1000°C reveal that the constituents of the calcine undergo a series of reactions to yield a single, highly crystalline, rare-earth fluorite phase together with $CsLiMoO_4$, MgO, $Sr_3Al_2O_6$ and other, as yet, unidentified phases (Fig. 1).

Figure 1: X-ray powder diffraction data from heat treated calcine

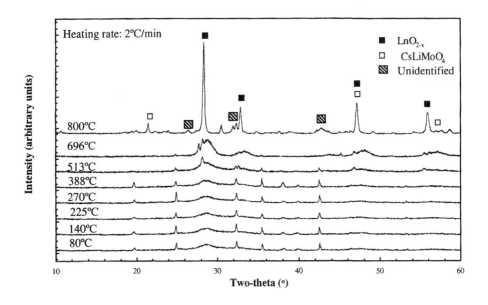

Figure 2a: DTA data from as- received simulant calcine

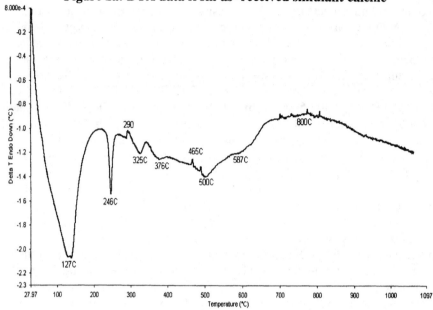

Figure 2b: TGA data from as- received calcine

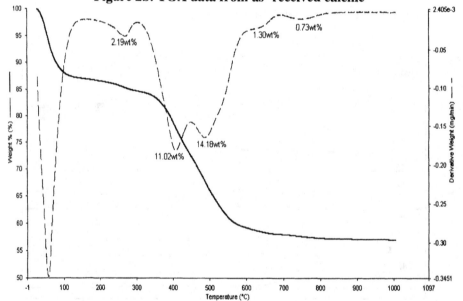

HLW/ Borosilicate glass interactions

Examination of calcine / glass interaction revealed a continuous yellow/green layer was observed on the surface of the melts quenched after 5 and 10 minutes.

This material was identified as $CsLiMoO_4$ by X-ray powder diffraction, Figure 3. Alkali molybdates of this type are known to be highly soluble in water[4] and this was confirmed by the rapid dissolution of a small amount of this material in distilled water. This material was also observed on the surface of the sample quenched after 16 minutes but in localised patches, several square millimetres in size.

In cross section, the melts quenched after 5 and 10 minutes appeared inhomogeneous, regions of transparent glass were observed together with particles of calcine material surrounded by a yellow / green zone of reaction. In contrast, the glass matrix appeared more homogeneous in the cross section of the melt quenched after 16 minutes, with no regions of transparent colourless glass apparent, although particles of undissolved calcine were observed within a uniform green glass matrix.

As shown in Figures 4a-c, the calcine particles (C), glass matrix (M) and reaction zones (R) were observed as regions of differing contrast in low magnification backscattered electron images. The effect of convection currents in the melt quenched after 5mins at 1050°C is apparent in Figure 4a. The surface layer of this sample, ~4µm thick, was rich in Cs and Mo as revealed by X-ray dot maps (Figures 5a and 5b), consistent with the powder diffraction data discussed above. Calcine particles, 100 – 1000µm in size, appear distributed thoughout the melt as regions of bright contrast, see Figure 4a. These particles were rich in rare-earth cations (Ce, Nd, Gd) and Zr (Figures 5c – d). Diffuse reaction zones, of intermediate brightness (Figure 4a), were observed around the perimeters of calcine particles and extending ~ 200µm into the glass matrix. X-ray dot-mapping revealed these zones to contain rare earth elements (Ce, Nd, Gd) together with Zr, Cs and Si. The Si X-ray map, Figure 5e, shows clear regions of silicon enrichment and depletion (that were fluid at 1050°C) suggesting that during the early stage of melting, phase separation of the vitreous matrix occurs under these conditions.

The backscattered electron images of the samples quenched after 10mins, Figure 4b, and 15mins, Figure 4c, show more extensive dissolution of the calcine, as expected. This is apparent from the reduced size of the calcine particles, <300µm, and the increased extent of the reaction zones, of intermediate brightness. In addition, the effects of thermal convection are clearly apparent in the backscattered electron image of the sample quenched after 15mins, Figure 4c. X-ray dot mapping indicated that the reaction zones were rich in rare-earth elements (Ce, Nd, Gd) together with Zr, Cs and Si as observed in the sample quenched after 5mins.

Figure 3: XRD of melt surface - CsLiMoO$_4$

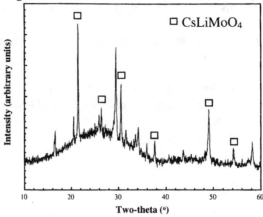

Figure 4a: Quenched after 5 mins

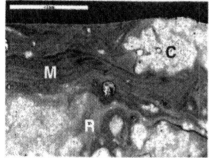

Figure 4b: Quenched after 10 mins

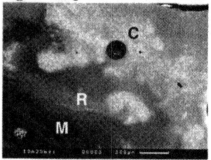

Figure 4c: Quenched after 16 mins

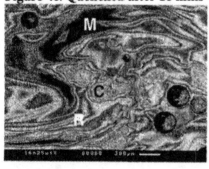

C= Calcine particle
M= Glassy matrix
R= Reaction zone around calcine
particle within glassy matrix

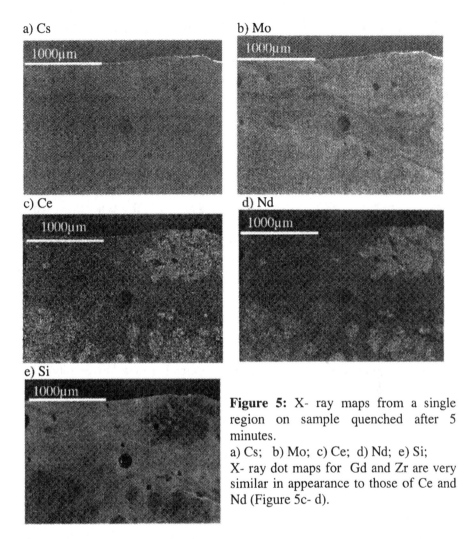

a) Cs b) Mo c) Ce d) Nd e) Si

Figure 5: X-ray maps from a single region on sample quenched after 5 minutes.
a) Cs; b) Mo; c) Ce; d) Nd; e) Si;
X-ray dot maps for Gd and Zr are very similar in appearance to those of Ce and Nd (Figure 5c-d).

DISCUSSION

Characterisation of the initial (inactive) calcine material, obtained from the full scale rotary calciner, has demonstrated that this material is composed of a poorly crystalline rare-earth fluorite phase together with $LiNO_3$ and $Sr(NO_3)_2$. Importantly, this material is heavily hydrated (12-16wt%) and contains a high level of residual nitrate (~30wt%). Further decomposition of this material occurs in four distinct stages. Between room temperature and 125°C dehydration of the calcine proceeds; at temperatures between 125 and 400°C, loss of chemically-bound water molecules is thought to occur; denitration, together with the formation of secondary rare-earth fluorite phases, takes place between 400-700°C; with formation of various crystalline phases arising between 700 – 1000°C. These results indicate that temperatures in excess of 500°C are required to effectively denitrate the calcine material. More complete denitration of the calcine would be beneficial since this would avoid the evolution of nitrous oxides in the melter and the extraction of heat from the glass melt to effect denitration *in situ* which is accompanied by a drop in the melt temperature.

Dissolution of simulant calcine in alkali borosilicate glass occurs in two stages. The first involves rapid migration of Cs and Mo from the calcine into the glass melt, these elements combine with Li to form $CsLiMoO_4$ which separates, as a fluid of lower density, on the surface of the melt. During this stage, which covers the first 5 minutes of dissolution, some limited migration of rare-earth elements (Ce, Nd, Gd), together with Zr and Fe into the glass melt also occurs. In the second stage of dissolution (10 – 16 minutes of reaction), more extensive migration of rare-earth elements, Zr and Fe takes place and this is accompanied by the disappearance of the continuous surface layer of $CsLiMoO_4$ and the formation of a more homogeneous glass. However, unreacted calcine material is still present after 16 minutes of melting, indicating that a longer reaction time, or stirring of the glass, is needed to incorporate all the elements of the waste into the matrix in the melting process.

CONCLUSIONS
Calcination takes place in four distinct stages and for effective denitration of the raffinate temperatures >700°C are needed. Dissolution of calcine into the glass occurs in two stages, the first involving the formation of a low-density alkali molybdate which floats to the surface of the glass, and secondly, extensive migration of RE (Nd, Ce, Gd) elements into the glass accompanied by removal of the molybdate surface layer via convective mixing.

ACKNOWLEDGMENTS
The authors thank Ms Jagdish Roopra for heat treatment of the calcine, and the EPSRC and BNFL for funding this work.

REFERENCES
[1] G. V. Hutson, *Waste Treatment – Chapter 9, The Nuclear Fuel Cycle*, ed. P.D. Wilson, (Oxford University Press, Oxford, UK, (1996)
[2] W. Lutze, *Silicate Glasses* pp1- 192, *Radioactive Wasteforms for the Future*, ed. W. Lutze, R.C. Ewing, North Holland, Amsterdam, (1988).
[3] R. D. Shannon, *Acta Cryst.* A32 751(1976).
[4] F.A. Cotton, G. Wilkinson *Advanced Inorganic Chemistry* 5th ed, (Wiley, New York, 1988)
[5] *The Elements and Inorganic Compounds- CRC Handbook of Chemistry and Physics* p114, 63rd edition, ed R.C. Weast, CRC Press, London, 1982

SUMMARY OF RESULTS FROM 786-A MINIMELTER RUN WITH MACROBATCH 3 (SLUDGE BATCH 2) BASELINE FEED USING FRIT 320

*Michael E. Smith and Donald H. Miller
Westinghouse Savannah River Company
Savannah River Technology Center
*Building 773-42A
Aiken, SC 29808

ABSTRACT

During the vitrification of high level waste at the Savannah River Site's Defense Waste Processing Facility (DWPF) Melter, melt rates have never consistently achieved the design basis of 103 kg/hr. Frit 200 has been the frit added to the high level waste sludge to make the waste glass since radioactive operations were begun. Frit 200 was used as it was expected that coupled feed (sludge and salt solution) operations were going to begin soon after DWPF was started. Since unforeseen problems in processing the salt solution are now delaying for several years the addition of the salt solution that contains additional alkalies (which helps increase melt rate), a program was initiated to develop a new sludge-only frit. Lab scale tests led to the development of Frit 320 (higher alkali content than Frit 200) and the conclusion that this frit would substantially increase the DWPF melt rate and produce an acceptable final glass waste form. The final verification that Frit 320 increased melt rate were larger scale melter tests using Frit 200 and Frit 320 in the 786-A Mini-Melter. This paper discusses the results of these Mini-Melter tests.

INTRODUCTION

During the vitrification of high level waste at the Savannah River Site's Defense Waste Processing Facility (DWPF) Melter, melt rates have never consistently achieved the design basis of 103 kg/hr. Frit 200 has been the frit added to the high level waste sludge to make the waste glass since radioactive operations were begun. Frit 200 was used as it was expected that coupled feed (sludge and salt solution) operations were going to begin soon after DWPF was started. Lab scale tests led to the development of Frit 320 (higher alkali content than Frit 200) and the conclusion that this frit would substantially increase the

DWPF melt rate and produce an acceptable final glass waste form[1]. This work has become increasingly important with the DOE thrust for the accelerated cleanup of the sites. Each five percent increase in DWPF melt rate would reduce processing time by one year and cut life cycle costs by approximately $430 million. The final verification that Frit 320 increased melt rate were two larger scale melter tests (the first using Frit 200 and the second Frit 320) in the 786-A Mini-Melter. The general objectives of the run relative to melt rate were to 1) determine a relative melt rate for the mini-melter using Frit 320 and 2) establish any significant operational differences between the use of Frit 200 and Frit 320.

TEST SETUP

The 786-A Mini-Melter (see Figures 1 and 2) is joule heated and has a one-foot diameter Carborundum Monofrax K-3 refractory pot. The vertical electrodes and plenum are made of Inconel 690. There are two vertical Kanthal™ lid heaters that are capable of supplying 5000 watts each. The overflow spout is heated by a split clam shell 1500W resistance heater. The melter is kept under partial vacuum with an air eductor and pressure is controlled with the addition of air. The off-gas passes through a quencher/scrubber and then through a mist eliminator prior to exiting a stack. Sample ports allow the off-gas to be sampled at the melter exit and after the condensate tank. Two gas chromatographs (GC) are used for off-gas analysis.

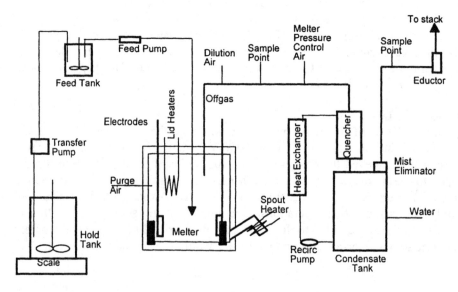

Figure 1. 786-A Mini-Melter schematic

Figure 2. 786-A Mini-Melter

Details of the 786-A Mini-Melter Frit 200 run performed in midyear 2001 are documented elsewhere[2]. At the start of Frit 320 run, the melter was still full of the Frit 200 glass made in this previous run. Therefore the melter had to be flushed with Frit 320 glass. The specifications for Frit 200 and Frit 320 are shown in Table I. The main difference between the two frits is the higher alkali (Li_2O and Na_2O) in Frit 320. The melter was initially fed for 5 hours at a rate of about 50 cc/min. This was the standard rate used during previous testing and would represent normal operation during a transition between different types of feed. The rate was gradually increased until approximately one melter volume had been poured. Feed samples were taken at least once a day, usually during the transfer from the hold tank to the feed tank. Glass samples were taken at the end of the last pour each day. Feed rates up to 67 cc/min were used during this transition period. The feed rate was subsequently reduced during the continuous feed portion of the testing. Typical analyses of the two feed batches with waste loading targeted at 25% are shown in Table II.

Table I. Frit 200 and Frit 320 specifications

Oxide	Frit 200 (wt %)	Frit 320 (wt %)
B_2O_3	12	8
Li_2O	5	8
Na_2O	11	12
SiO_2	70	72
MgO	2	0
Total	100	100

Table II. Compositions for Frit 200 and Frit 320 simulated feeds

Element	Frit 200 Feed Weight %	Frit 320 Feed Weight %
Al	2.77	2.67
B	2.66	1.87
Ba	0.092	0.084
Ca	0.924	0.688
Cr	0.924	0.083
Cu	0.046	0.030
Fe	7.85	8.61
K	0.143	<0.010
Li	1.56	2.73
Mg	0.915	0.026
Mn	0.866	0.828
Na	8.33	9.22
Ni	0.433	0.469
P	0.035	0.029
Pb	0.062	0.058
Pd	0.011	<0.010
Rh	0.006	<0.010
Ru	0.012	0.017
Si	23.1	25.9
Zn	0.126	-
Zr	0.202	0.165

SHORT TERM MELT RATE

The short term melt rate test was conducted by feeding 150 cc/min of Frit 320 feed for 5 minutes and visually determining the time required for the cold cap to burn off. A VCR recorded the surface for comparison to other tests. The duration of volatile generation was also recorded. The melt pool and plenum temperatures were controlled automatically with setpoints of 1150°C and 850°C respectively. The purge air flow was zero and the dilution flow was 160 standard liters per minute. The burn off was determined visually by the absence of any remaining feed on the surface. There is a residual texture to the surface for a long period after feeding, but this was not considered during this test. The feed used during this testing had 46.9 wt % solids. This is 0.5% lower than that used in the previous run with Frit 200 but was not deemed significant enough to skew the test results. When all the feed was introduced, there was nearly complete cold cap coverage.

The volatile concentrations were measured using the gas chromatographs (GC). The sample point selected was directly after the addition of the dilution air. Due to low concentrations and short sample times the absolute value of the reading is probably not accurate. Durations were counted for the period that the concentration was above the background value. The durations for the visual cold cap burn off times are shown in Table III below. The average feed rate is based on the data recording system.

Table III. Results from short term melt rate test

Test #	Feed Rate cc/min	Visual Cold Cap Burn Off (min)
1	147	18
2	144	24
3	150	27
Average	147	23
Frit 200		
Run Average	142	19.3

The results of the short term testing were difficult to interpret due to problems detecting the volatile components during short sample times. The volatile concentration tests results were too erratic to be of any use and will probably not be used in future mini-melter melt rate tests. In addition, the times for visual cold cap burn off were so varied that they were not deemed as a reliable measure of melt rate. From past experience, longer duration (several hours) melt rate tests are a much better technique to use.

MELT RATE VERSUS PLENUM TEMPERATURE

In addition to the short term melt rate testing described above, two additional runs were made to determine relative melt rate at different plenum (vapor space) temperatures. The plenum temperature was initially lowered to 600°C by adding purge air. Once the target temperature was reached, the lid heaters were energized with a set point of 600°C. The feed rate was initially high to build up the cold cap. The goal during the testing was to maintain a consistent cold cap coverage of about 90%. This is a subjective measurement but generally the presence of several vent holes or a small area of glass constitutes normal coverage. Once steady state was obtained, the plenum temperature was raised to 700°C and the test was repeated.

The testing indicated that a steady rate of about 44 cc/min could be maintained with a plenum temperature of 600°C and the glass pool in automatic control with a set point of 1150°C. The increase to a 700°C plenum temperature yielded an estimated steady state feed rate of about 58 cc/min.

The feed rate vs. plenum temperature for this test is shown in Figure 3. There were feed problems experienced during the testing, especially at the lower temperature. Replacing a pump hose seemed to solve the problem. Enough feed time was achieved to get a good approximation of the maximum feed rate at plenum temperatures of 600 and 700°C.

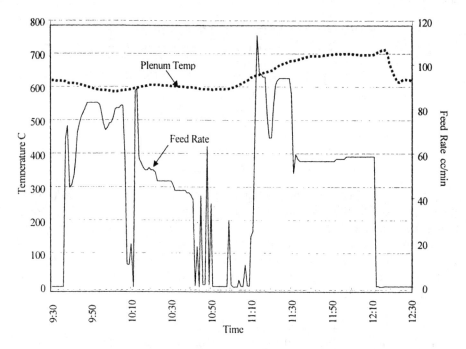

Figure 3. Melt rate versus plenum temperature for Frit 320 feed

EXTENDED MELT RATE

Continuous operation was started by initially feeding with a melter pressure of -4 inwc. Feeding continued until glass began to drip from the pour spout. The pressure was changed to + 5 inwc to initiate a steady stream. The set point was then placed at -0.5 inwc and feeding continued for 6 more hours. No problems were observed using the new feed. The cold cap was maintained with similar coverage in both runs. The melter was capable of sustaining a feed rate of ~64 cc/min using the feed made with Frit 320. This is an improvement over the last campaign using Frit 200, where a maximum sustainable feed rate was ~54 cc/min. Similar feed rates for the Frit 320 were also observed during the melter turnover portion of the test. Operating conditions for both runs are shown in Table IV. The glass pool and plenum temperatures were maintained in automatic mode for both tests. Due to an error in the previous report[2], the plenum temperature for the Frit 320 test was held at 850°C rather than 800°C. This is not believed to have a major effect on the melt rate comparison. This is based on the decreasing slope in melt rate when plotted against increasing plenum temperature. A rate of 63 cc/min at 800°C is estimated (see Figure 4) when graphing the rates at different temperatures. These values correlate to a higher melt rate of about 17 percent for Frit 320 versus Frit 200. This extended melt rate test should be the best test used to date when comparing the relative melt rates of Frit 320 and Frit 200 feed.

Table IV. Melt rate conditions

Variable	Unit	Frit 320	Frit 200
Feed Rate	cc/min	64	54
Feed Density	g/cc	1.21	1.21
Wt % Solids	%	47	46.1
Calcine Ratio		0.92	0.915
Glass Melt Rate	lb/hr	4.4	3.6
Glass Temperature	°C	1150	1150
Plenum Temperature	°C	850	800
Feed Duration	hr	6	5

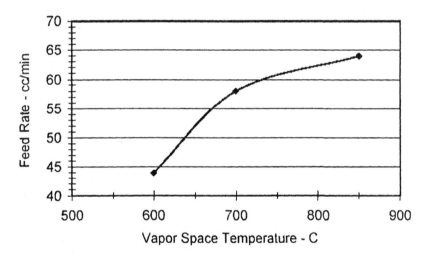

Figure 4. Frit 320 melt rate versus vapor space temperature with extended melt rate data at 850°C

PROCESS PARAMETERS

Data collection occurred automatically during testing. Additional information was recorded in the laboratory notebook, including setpoint and output changes. Table V represents average values during normal and idle operation for both frits. The idle parameters were taken from a 24 hour period several days after the completion of testing for each frit. The feeding parameters were taken during the extended melt rate portion of each test. The electrode power settings required during idle and feeding periods indicate a difference in electrical resistivity between the two glasses. Glass made with Frit 320 exhibited a lower resistance as indicated by the lower voltage and higher current that was required. Glass samples from both the Frit 200 and Frit 320 Mini-Melter runs were subsequently measured for glass resistivity at various temperatures. At 1141°C, the resistivities for Frit 320 and Frit 200 glasses were 2.9 and 4.4 ohm-cm respectively. The lower Frit 320 resistivity is still well within the range specified for DWPF and the DWPF Melter has sufficient additional available electrode amperage for Frit 320 glass.

Table V - Average melter parameters

Variable	Unit	Frit 200 Idle	Frit 320 Idle	Frit 200 Operating	Frit 320 Operating
Electrode	Kw	4.2	4.4	3.9	4.7
Electrode	Amp	141.5	174.7	143.6	186.3
Electrode	Volt	29.4	25.2	27.2	25
Glass Pool	°C	1158	1141	1150	1150
Lid Heater	Kw	NA	NA	6.1	6.1
Lid Heater	Amp	NA	NA	76.1	76.6
Lid Heater	Volt	NA	NA	79.5	79.9
Plenum Temp	°C	780	804	800	850
Feed Rate	cc/min	NA	NA	55	64
Date of Data		7/30/01	4/02/02	7/26/01	3/14/02
Duration	Hours	24	24	5	6

PREDICTED GLASS PROPERTIES

The analyses from 4 pour samples of the run were used to predict the processing properties of the glass. The results indicate that the estimated values for liquidus, viscosity, and durability (NL – normalized leach rate) all fall within the accepted ranges using the DWPF Process Acceptability Region (PAR) criteria. The maximum allowable liquidus temperature is 1050°C. The acceptable range for viscosity at 1150°C is between 20 and 100 poise. The maximum allowable predicted NL for boron is 1.222 g/L. The results for all properties are shown in Table VI. The increasing sample number indicates transition between Frit 200 and Frit 320 glass. The values predicted by the liquidus model indicate that higher sludge loading could be achieved with Frit 320 and still stay within the limits.

Table VI. Predicted glass properties for Frit 320 glass samples

Sample ID	Viscosity (Poise)	*Predicted NL [B(g/L)]	Predicted Liquidus(°C)
MMG 020	92.90	0.375	1004
MMG 022	71.47	0.577	920
MMG 025	62.09	0.686	900
MMG 028	56.08	0.832	908

* Predicted normalized leach rate of boron in the Product Consistency Test (PCT)

GENERAL OPERATIONAL OBSERVATIONS

The use of the new larger inner diameter feed tube (0.457cm versus 0.333cm) during the testing greatly improved the reliability of the system. Long runs could be made without the feed tube plugging. Continuous pouring also allowed the operation of the melter with very few high glass pool level alarms. Operation for long periods at low plenum temperature still continues to be a problem. Two water leaks were discovered during the run. One was on the #1 electrode clamp, which required the water to be shut off during most of the testing. Occasionally during the high feed rate test, the water was turned on briefly to bring the clamp temperature below the alarm point. There was a second leak in the rotometer supplying the feed tube flush water. This valve had to be opened manually prior to and after each feed initiation or stoppage.

CONCLUSIONS

The results indicate that glass made with Frit 320 melts at a rate about 17% faster than that using Frit 200. This is in line with results from smaller scale testing. The run also showed that there were no significant operational problems associated with the processing of Frit 320 glass with regards to cold cap behavior, feed pumping, glass pouring, etc. The glass produced has predicted properties within the acceptable processing range. A lower electrical resistance was indicated with Frit 320 glass via lower voltage and higher current melter electrode requirements during both idle and feeding conditions. This was verified by resistivity measurements made on Frit 200 and Frit 320 glass produced by the 786-A Mini-Melter. This lower glass resistivity is within the operational range specified by DWPF.

Frit 320 produced higher melt rates in all lab scale tests and in the 786-A Mini-Melter. The reason for this faster melt rate is the higher alkali content of Frit 320 (Li and Na). A WSRC paper study reviewed DWPF pilot melters and discussed various parameters that can impact melt rate[3]. The parameters include air in-leakage, lid heater power, electrode power, melt pool temperature, melt pool agitation, melt foaming, glass viscosity, and feed composition. It is believed that the sludge-only nitric/formic acid flowsheet currently being used at DWPF may result in a high amount of gas being generated during glass melting. The viscous cold cap would then trap these bubbles and would act as a very effective, foamy insulating layer between the glass and the cold cap[4]. This foam layer would then exacerbate the melt rate problem even more by causing the cold cap/melt pool interface to become even more viscous. It is therefore conceivable that this foamy layer would continue to get thicker with extended periods of operation. Visual inspection of the DWPF Melter surface during idling has shown that some type of interface layer is indeed on top of the melt pool. By making this cold cap/melt pool interface less viscous with the use of Frit 320, more of the gas bubbles would be able to escape and therefore minimize this foamy insulating layer.

As the DOE complex is pushed for an accelerated cleanup of the various sites, this work could prove to be very significant. A ten percent increase in DWPF melt rate over the production life of the plant translates to a two year reduction in DWPF processing time and a cost savings of about $860 million for the entire HLW process. Tailoring frits to achieve the maximum melt rate for each DWPF macrobatch batch will be required. It is uncertain, however, at this time whether the melt rate increases gained with using a sludge-only frit for sludge-only processing can be continued when coupled feed processing is done. This process improvement technique could be used at other sites such as the vitrification plants currently under construction at the Hanford site. Other process changes being considered by SRS's Savannah River Technology Center (SRTC) such as airlift bubbling and melt pool agitation could result in even greater gains in DWPF Melter glass production rates. Efforts to increase waste loading can lead to additional cost savings. One SRTC study[5], however, indicated that increasing waste loading in DWPF feed too much may actually begin to decrease the waste treatment throughput by negatively impacting melt rate. These two competing process changes will have to be balanced to achieve the fastest possible treatment of the high level waste at the DWPF.

REFERENCES
1. D. P Lambert, T. H. Lorier, D. K. Peeler, M. E. Stone, "Melt Rate Improvement for DWPF MB3: Summary and Recommendations (U)", USDOE Report WSRC-TR-2001-00148, WSRC.
2. D. H. Miller, "Summary of Results for Simulated Tank 8/40 Blend Run in Minimelter (U)", USDOE Report WSRC-TR-2001-00530, WSRC.
3. S. R. Young and D. F. Bickford, "Review of DWPF Pilot Melters – Melt Rate and Associated Parameters (U)", USDOE Report WSRC-TR-96-0405, WSRC.
4. A. G. Fedorov and R. Vistanta, "Radiation Characteristics of Glass Foams", Journal American Ceramic Society, 83 [11] 2769-2776 (2000).
5. T. H. Lorier and P. L. McGrier, "Melt Rate Improvement for the DWPF: Higher Waste Loading (U)", USDOE Report WSRC-TR-2002-00344, WSRC.

NUMERICAL MODELS OF WASTE GLASS MELTERS
PART I – LUMPED PARAMETER ANALYSES OF DWPF

Hector N. Guerrero and Dennis F. Bickford
Savannah River Technology Center
Westinghouse Savannah River Co.
Aiken, SC 29808

ABSTRACT
Defense Waste Processing Facility melter production data from three waste batches were analyzed using a lumped parameter approach which separates effects of melter feed, heater temperature, and power on melt rate under various modes of operation. A detailed distribution of power inputs and heat consumption pathways, as provided by the lumped parameter model, evaluated possible causes of melt rate reduction and other operational data. Theoretical aspects of the steady state analysis, as well as transient analysis, are presented. The lumped model complements the more detailed multi-dimensional computational models by providing boundary conditions for such models, and is the only practical way of predicting transients.

INTRODUCTION
The Department of Energy's Defense Waste Processing Facility (DWPF) melter has operated for six years at varying rates. An examination of the power, materials flow and thermal data from radioactive operations was conducted to provide trend analyses and isolate sources of variation. Actual DWPF production data was evaluated using a lumped parameter approach which separates effects of melter feed, heater temperature, and power on melt rate under various modes of operation. This model can be used to:

- Facilitate the analyses of existing melter problems from a thermal perspective to determine possible improvements.
- Provide a rapid, disciplined way of evaluating relative effects of proposed methods of increasing melt rate and other changes to melter operation.
- Isolate and diagnose changes in the behavior of the melting process.

- Provide the boundary conditions such as shell heat losses, radiant heat fluxes to the cold cap and upper plenum for more detailed Computational Fluid Dynamics (CFD) analyses of the glass melt and cold cap.
- Obtain a transient analysis of melter operation and to acquire an insight into the physical mechanisms occurring during transients.

The lumped parameter method provides a steady state balance between melter power inputs and various power consumption pathways under continuous feeding/pouring or idling conditions. These various power losses include heat lost through the melter shell, evaporation of feed water, heating of steam and in-leakage air, chemical reactions, and heating of glass. These are not constant and depend on melter operating conditions such as feed rate, solids ratio, feed batch and net pool circulation. Through this model, the separate effects of engineering and physical chemistry batch effects can be obtained for quick evaluation of their effects on melt rate. The thermal model also provides a framework where laboratory determination of batch effects, e.g., specific heat, viscosity, thermal conductivity, etc. can be varied and the relative melt rate effect predicted.

The lumped parameter model is inherently limited in that space variations in glass and cold cap temperatures and heat transfer coefficients for glass to cold cap convection are not accounted for. A 3-dimensional model using computational fluid dynamics is currently being developed which may in the future provide more accurate averaged parameters for this lumped parameter model.

OVERVIEW OF HISTORIC DWPF MELTER OPERATION
In this section, the performance of the DWPF Melter to date with three different feed compositions, termed macrobatches, is discussed. Daily DWPF Melter power data as a function of feed rate for the period, 11/97 to 2/98, representative of Macrobatch 1 feed, are summarized in Figure 1. The scatter of the total power data points around the linear trend-line is much tighter than for the corresponding data points for the electrode and dome heater powers. This suggests that it is the total power that is important and the trend-lines for the electrode power and dome heater power should also be linear. Using the equations for the trend-lines, it is apparent that the electrode power was initially high (173 kW) at zero feed and decreased slightly to 163 kW at 0.75 gpm. The dome heater power increased linearly from 103 kW at 0 gpm feed rate to 278 kW at 0.75 gpm. In DWPF operation, glass melt pool and dome heater temperature limits dictate the above power settings. These conditions resulted in a melt mass flux of 8 lbs/hr-ft^2 at the maximum feed of 0.75 gpm and 49% solids ratio. This was only 87 % of the value achieved in the SGM and IDMS runs. This may be attributable to scale-up effects; or, the current nitric acid flow sheet feed is slower melting than the formic acid flow sheet used for most of the pilot scale studies.
A similar set of daily power data for the period, 1/99 to 8/99, representing Macrobatch 2 runs, indicate that for Macrobatch 2, the melt rate decreased

approximately 20% from that of Macrobatch 1. The maximum feed attained for Macrobatch 2 was 0.6 gpm, while the maximum for Macrobatch 1 was 0.75 gpm. The electrode power decreased for Macrobatch 2 (143 kW) relative to Macrobatch 1 (163 kW). For Macrobatch 3, the electrode power decreased to 101 kW, and the Macrobatch the melt rate decreased 27% from Macrobatch 1. The maximum feed for Macrobatch 3 was 0.5 gpm.

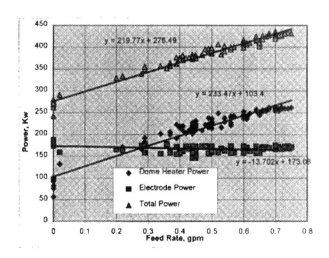

Figure 1 DWPF Melter Daily Averaged Power Data from 11/1/97 to 2/18/98 - Macrobatch 1

Under Macrobatch 1 and 2 conditions, the available power from the electrodes transferred to the cold cap directly from the melt pool or via the upper plenum was 90 kW and 68.5 kW, respectively. It was much less for Macrobatch 3, at 25.7 kW. The total power, or the sum of electrode and dome heater powers, for the three batches did not vary significantly since the energy to melt the glass was only a small fraction of the total energy input. The decrease in electrode power under feeding conditions may be attributed to the possible presence of a thermally resistant layer in the cold cap. The waste glass batches are known to have foaming characteristics. Under feeding conditions, this thermally resistant layer reduced convection heat transfer from the melt pool to the cold cap, thus decreasing electrode power requirement in order not to exceed the maximum glass temperature limit.

Under idling conditions, Macrobatch 3 electrode power decreased compared to Macrobatches 1 and 2, which were similar. The vapor space temperature also decreased by as much as 27C. Calculations show that the radiant heat from the glass surface decreased, as well as the effective cold cap surface temperature.

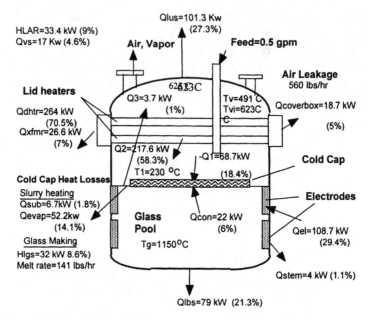

Figure 2   Distribution of Power Inputs and Power Consumption
for Macrobatch 3

This implies the presence of a thermally resistant layer, which may be due to persistent foam layer or accumulation of melt resistant layer.

DETAILED POWER INPUTS AND HEAT LOSS DISTRIBUTIONS

A detailed distribution of power consumption for Macrobatch 1 conditions may be determined by using the lumped parameter model to estimate component losses, which were not directly measured. Under feeding conditions of 2.65 lpm, the electrode power required was 163 kW or 39.7% of the total power. The dome heater power was 266.8 kW. The energy required to eliminate subcooling and evaporate the water was19.7% (87 kW) of the total energy input and the actual amount of energy required to melt the glass was only 10.7% (47 kW). The shell heat losses amounted to about 41% of the total energy input and the energy lost by gas mass flux due to air in-leakage, steam and calcined gas was about 14.6%. The remainder was miscellaneous losses. The resulting melt mass flux was 39.7 Kgs/hr-m^2, which was 11 % less than assumed for the design conditions.

Power inputs and losses for Macrobatch 2 full feed conditions of 2.08 lpm were similar to Macrobatch 1. The melt rate was 30.7 kg/hr for a solids ratio of 49%, which was 30% less than the design value. The electrode power had dropped down to 138.4 kW (from 163 kW for Macrobatch 1) under full feed conditions.

This suggests that something happened with the cold cap, perhaps significantly more foaming in Macrobatch 2, as compared to Macrobatch 1.

The power consumption for Macrobath 3 full feed conditions of 1.89 lpm, required an electrode power of 108.7 kW or 29.4% of the total power input. The dome heater power was 264 kW. The energy required to eliminate subcooling and evaporate the water was14.1% (52.2 kW) of the total energy input and the actual amount of energy required to melt the glass was only 8.6% (32 kW). The shell heat losses amounted to about 48.6% of the total energy input and the energy lost by gas mass flux due to air in-leakage, steam and calcined gas was about 13.6%. The resulting melt mass flux was 27.9 Kg/hr-m^2, which was 37 % less than assumed for the design conditions.

With the lumped parameter model, it was possible to estimate the amount of heat directly transferred to the cold cap from the melt pool, 22 kW, which was much lower than in Macrobatches 1 and 2. The radiant heat absorbed by the cold cap from the dome heaters and plenum walls was 68.7 kW. Through the radiant heat exchange method (to be discussed later), the surface temperature of the cold cap was estimated as 230°C. The corresponding values for Macrobatch 2, which has close to the same feed rate are: 400°C and 52.7 kW. The cold cap temperature is an effective temperature, which implies that a large proportion of the cold cap is covered with wet slurry. It is clear that a highly insulating cold cap layer reduced heat addition to the cold cap from the melt pool. To make up for this, the dome heater power had to be increased, 264 kW (up from 243 kW for Macrobatch 2 for close to the same feed rate). This required a low cold cap surface temperature of 230°C.

Under idling conditions, the indicated vapor space temperature (Tvi) decreased to 800°C (from 892°C in Macrobatch 1 and 883°C in Macrobatch 2). The vapor temperatures, corrected for radiation heating effect, were 730 °C, 730 °C, and 649 °C. Using the lumped model for idling conditions, the glass surface temperature and the radiant flux from the glass surface can be predicted, to be 747°C and 25.7 kW, respectively. The corresponding values for Macrobatches 1 and 2 are 900°C, 877°C and 90 kW, 87 kW, respectively. This appears to confirm the premise that a thermally resistant upper glass layer has formed, e.g., spinels.

ANALYTICAL METHOD

This analysis uses a lumped parameter approach for simplicity and to provide an overall perspective of a very complex process. The averaged or lumped parameters of the model can come from more detailed 3-dimensional computational fluid dynamics models currently in progress, from experiments, or from DWPF data. First a steady state heat balance of the melter is calculated. This calculation borrows heavily from the 1988 analysis of Yoshioka, (Ref. 1) including property parameter relations. The BASIC program archived for this work did not seem to correspond to the logic presented in Reference 1, and running that program did not provide the same results as in the report. In that report, the vapor temperature was assumed known, 680°C at design conditions, and the demanded (required) electrode power and dome heater powers were calculated. Due to high actual upper shell heat losses in the melter, the measured vapor temperatures are much lower, typically 493°C (with radiation correction) at 0.7 gpm (Macrobatch 1). Consequently, the heat balance calculation was redone, using most of the same equations in the Basic program. The present calculation however differs from 1 in that the radiation heat transfer in the melter upper plenum accounts for radiation exchange among all the surfaces in the upper plenum and includes the steam as participating media. The results of the steady state heat balance closely resemble the results of Ref. 1 if the same inputs are used. The present analysis however assumes the electrode power, dome heater power, and feed rate are known functions of time. Further, estimates of the shell heat losses based on actual DWPF Melter power data, which are almost twice as much as those in Ref. 1, are used. Then the glass, vapor space, and dome heater temperatures, and internal heat distributions are then calculated.

Steady state equations

By performing a heat balance on the cold cap, Equation 1below follows. Here, the sum of the convective heat from the glass pool, Qcon, and the net radiant heat absorbed by the cold cap, -Q1, is used to evaporate the feed water, melt the glass, and raise its temperature to the operating temperature.

$$Qcon-Q1= SFR*cpw*(100-25) + SFR*\Delta Hevap + MR*Hbatch \quad\quad [1]$$

Where,

$$Hbatch = (\int_{20°C}^{1150°C} c_p dT + \int_{20°C}^{1150°C} \Delta H_r dT) = cpg*(Tsoft-25) + \Delta Hmelting + cpg*(1150-$$
554), $\Delta Hmelting$ is estimated to be 120 cal/gm for endothermic reaction heat and –80 cal/gm for exothermic reaction, which includes 20 cal/gm for silica melting. SFR is the water (or steam) feed rate. $\Delta Hevap$ is the heat of evaporation of water, cpw and cpg are the specific heats of water and glass, respectively. The average value of cpg over the appropriate temperature range is used in $Hbatch$.

The vapor space, lid heater, and plenum wall temperatures and the wall heat loss, radiant heat absorbed by the cold cap, and lid heater power are calculated by radiant heat exchange equations of the form,

$$Q_k = A_k \sigma [T_k^4 - \sum_{i=1}^{N} F_{ik} \tau_{ik} T_i^4 - \varepsilon_v T_v^4] \quad\quad [2]$$

assuming all surfaces are black. Here, σ is the Stefan-Boltzman constant, A_k is the area of the surface k, F_{ik} is the view factor from surface i to surface k, τ_{ik} is the transmittance, and ε_v is the emissivity of the vapor. The sum of all surface radiant heats into the plenum goes into heating the steam. The radiant heat absorbed by the walls equals the sum of the heat lost through the walls and transferred to the in-leakage air.

The steady state equations were programmed in an Excel spreadsheet. While the number of equations equals the number of variables, direct solution of the equations is very difficult because of the nonlinear nature of the equations, where the radiation terms involve temperatures to the 4th power. Yoshioka used an iterative method to obtain convergence on the cold cap coverage and glass surface temperature. He also assumed a cold cap surface temperature of $100°C$. For this work, a set of 5 nonlinear equations of the form [5] for the glass, cold cap, dome heaters', side wall, and top lid surface radiant heats were written. These equations consisted of temperature terms to the 4th power and linear terms. These were solved by iteration. By comparison to Yosioka's assumption of $100\,°C$, the calculated cold cap upper surface temperature ranges from $450°$-$477°C$, which is in the film boiling regime for water.

The transient analysis focuses on the melter glass pool, the cold cap, and the steam/air temperature responses. Transient heat balance equations are written for the rate of increase in temperature of the glass pool and the air/vapor mass in the plenum. The cold cap coverage is variable and highly dependent on the difference between the total feed rate and the actual melt rate. Here, a constant cold cap height and porosity is assumed (if changes are slow enough) so that the cold cap expands if the feed rate exceeds the melt rate, and vice-versa. The radiation view factors are functions of the cold cap area and thus vary with time. These result in three simultaneous first order differential equations for the melter glass temperature, the vapor temperature, and the uncovered glass surface area. No time lags are assumed due to radiant heat transfer, boiling, and melting. Additional relations are included for the convective heat flux from the glass pool to the cold cap, the radiant heat flux from the glass surface, the heat flux from the lid heaters, the heat absorbed by the cold cap from the upper plenum. The steam is fully participating in the radiant heat exchange. (In Yosioka's analysis, steam was not included in the radiation exchange.)

BENCHMARKING WITH POWER DATA
To benchmark the lumped parameter model, a calculation was done for two specific Macrobatch 1 conditions provided by the correlations of dome and electrode powers, one at zero feed and the other at 0.7 gpm. The run at zero feed represents a case of the glass surface completely uncovered. The run at 0.7 gpm represents a case of a cold cap area covering 88.8% of the available glass surface area. The calculations use the measured dome heater temperature ($950°C$) and the measured vapor temperature ($892°C$) corrected for radiant heating.

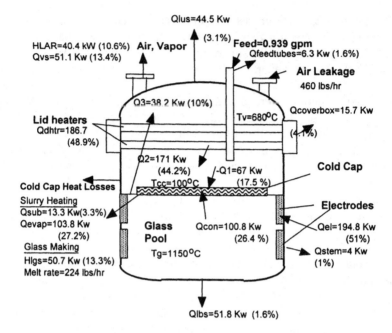

Figure 2　Theoretical Distribution of Power Inputs and
Power Consumption for Design Basis Case

However, in his comparison with DWPF data, the above correlation under-predicted the DWPF data by as much as 50°C. This correlation was used in the calculations but was adjusted upwards by 30°-50°C, which resulted in better heat balance.

An uncontrolled air in-leakage rate of 45.4 kg/hr (estimated by DWPF) is also used in addition to the known controlled air in-leakage of 209 kg/hr. Also, data for dome heater transformer bus bar cooling are used, as well as natural convection cooling, to add to the heat losses to the lid heater.

In the case of the zero feed run, the total power of 276.4 kW goes into heating the in-leakage air of 254.4 Kg/hr (47.5 kW), heat loss through the shell of 185.4 kW, and 33.5 kW for miscellaneous losses. The heat loss through the shell was calculated as the difference between the power input and the leakage air loss plus miscellaneous losses.

Design Basis Case – Melt Rate of 224 Lbs/hr
The steady state heat balance is solved using an Excel spreadsheet. Results for the nominal case, considered by Yoshioka (feed rate=0.939 gpm, melt rate of 224 lbs/hr, vapor temperature=680°C, electrode power=194.8 kW, lid heater power=186.7 kW, total air in-leakage flow=460 lbs/hr), are summarized in Figure

19. He assumed a cold cap area coverage of 89% and a glass surface area of 24.7 ft2, taking into account the area taken by the electrodes. The glass melting term includes sensible heating, a silica glass heat of melting (20 cal./gm) and an endothermic reaction heat of -100 cal/gm for the present glass formulation

The component heat losses responsible for slurry heating and evaporation, glass heating, and vapor superheating amount to 218.9 kW or 57.4% of the total power. The total heat losses through the shell (96.3 kW), heating of leakage air (40.4 kW) and miscellaneous losses (26 kW) make up the difference. However, this calculation did not include a number of heat loss sources such as transformer bus bar cooling, radiation into the off-gas outlet flange and other shell penetrations, which significantly increases the total heat loss as evidenced by actual power data.

CONCLUSIONS

A lumped parameter steady and transient thermal analysis model of the DWPF melter has been completed. The steady state analysis has been benchmarked against actual DWPF Melter data. The difference between the design basis predictions and the actual data can be attributed to:
(1) larger heat losses through the melter shell than can be accounted for in the analysis;
(2) scale up effects; and
(3) a larger thermal resistance between the cold cap and the glass melt pool, probably due to a foam layer present with the actual waste, or current nitric acid based feed.
This thermal resistance results in a larger cold cap area for the same feed rate (with less venting) than experienced in the SGM and IDMS runs. Therefore, the design feed rate of 0.939 gpm can not be achieved due to almost complete cold cap coverage at 0.7 gpm for Macrobatch 1 feed and 0.55 gpm for Macrobatch 2 feed.

From these results, it is also concluded that the radiant heat incident on the cold cap is insufficient to completely evaporate the slurry water. Additional heat is required from the glass pool. This heat which passes through the glass/foam layer significantly affects the melt rate.

Other conclusions from the steady state analysis are the following:

• The decrease in electrode power from Macrobatch 1 to Macrobatch 2 and then Macrobatch 3 is consistent with an interface layer buildup on top of the glass of foam or crystalline deposits which reduce heat transfer from the melt pool to the cold cap and hence melt rate.

- A decrease in electrode power for Macrobatch 3 feed under idling conditions is also consistent with an interface layer buildup which reduce heat transfer to the plenum. The lumped model predicts a decreased glass surface temperature and reduced radiant heat to the plenum, as a result of the reduced measured vapor space temperature.

Recommendations

This lumped parameter model is another step forward after Yosioka's analysis. Understandably, there are still many areas that can be improved upon since large gaps in understanding of many physical processes in the melter still exist. However, the model provides a framework where experimental values or good averages from 3-dimensional CFD analysis of the following parameters can be inserted in place of current assumptions. Improvements in the model should include:

- Convective heat transfer coefficient between cold cap and melt pool for new macrobatches, possibly from bench top slurry melt rate furnace tests,
- Re-evaluation of linearity of cold cap area vs. feed rate, especially at low feed rates, and during transients,
- Values of Hbatch for different macrobatches from bench top experiments,
- Relation between measured and true average dome heater temperatures,
- Average cold cap surface temperature from 3D analysis,
- More accurate determination of the dome heater view factors,
- More accurate determination of glass thermal time constant that includes mixing.

Thus, with input from bench top experiments and iteration with 3-dimensional CFD analysis, a good simple lumped parameter model can be developed for use in evaluating melter performance with new waste macrobatches and also for transient analysis of melter operation.

REFERENCES

[1]. M. Yoshioka to M.D. Boersma, "Prediction of Joule-Heated Glass Melter Operation", NTIS DPST-88-660, 1988.

[2]. Hector G. Guerrero and D.F. Bickford, "Steady and Transient Thermal Analysis of the DWPF Melter Operation", NTIS WSRC-TR-2002-00159, 2002.

NUMERICAL MODELS OF WASTE GLASS MELTERS
PART II - COMPUTATIONAL MODELING OF DWPF

Hector N. Guerrero and Dennis F. Bickford
Savannah River Technology Center
Westinghouse Savannah River Co.
Aiken, SC 29808

H. Naseri-Neshat
School of Engineering
Technology and Science
South Carolina State University
Orangeburg, SC 29917

ABSTRACT
Computational fluid-dynamics numerical models are developed for joule-heated
slurry fed waste glass melters, such as the Defense Waste Processing Facility
Melter. An important feature of the analyses is the simulation of the cold cap
region with its thermally resistant foamy layer. Using a simplified model which
describes the foam void fraction as a function of temperature, based on laboratory
sample testing, characteristic features of the cold cap are simulated. Two- and
three-dimensional models are presented.

INTRODUCTION
The performance of radioactive waste glass melters, such as the Defense Waste
Processing Facility, would benefit from a detailed thermal-hydraulic analysis
using computational methods in order to improve the prediction of glass melt
rates and thus glass production rate of the melter. Specific goals of such an
analysis would be:
- to predict the melt rate for known thermo-physical properties of various waste
  batches,
- to optimize melter operating conditions for maximum glass production, and
- to evaluate the effect of melt rate enhancing devices such as bubblers.
By providing predicted temperature distributions within the bulk glass, melter
temperature readings can be interpreted properly with respect to operational limits
to avoid local glass overheating, or reaching lower glass transition temperatures.
Further, knowing potential flow distributions may lead to strategies for increasing
convection to the melting cold cap.

The model presented in this paper is a simple one but still incorporates important features of the melt pool and cold cap by which some insights can be made on the melter temperature and flow distributions. This is first illustrated with a two-dimensional model, which is useful for parametric studies. Finally, a three-dimensional model was generated to demonstrate the effect of a full three-dimensional geometry. The computational model complements the Lumped Parameter approach (Ref. 1) by providing the average effect of nonuniform temperature and flow distributions on heat transfer coefficients to the cold cap. On the other hand, the Lumped Parameter model, based on melter operating data provides known boundary conditions for the computational model.

The DWPF Melter, a joule-heated glass melter, uses a slurry feed of glass frit, water and waste which forms a semi-solid cold cap on top of the glass. The thermal analysis of the melter is complicated by the interplay of three separate heat transfer problems. First, the glass pool is electrically heated by passing direct current through the molten glass. Temperature gradients set up by the introduction of a glass slurry on top of the glass surface produce natural convection currents within the bulk glass. Further, these temperature gradients set up non-uniform electrical heat generation due to the positive temperature coefficient of electrical conductivity of the glass. Second, the cold cap formed by evaporation of the slurry water, is heated from above by radiant heat from the dome heaters and plenum walls and from below by convection from the bulk glass. The cold cap consistency, its radial thickness variability, and foaming characteristics are largely unknown. Third, the radiant heat impinging on the cold cap depends on radiant heat exchange between the cod cap, the dome heaters' surfaces, the walls, and the steam, which participates in the exchange.

ANALYTICAL METHOD
The approach taken here is to analyze the glass and cold cap as a unit and utilize the radiant heat absorbed by the cold cap as a boundary condition. The radiant heat problem in the plenum would be solved separately, where the boundary conditions at the cold cap surface (temperature and heat flux) must match the assumed boundary conditions in the glass/cold cap model.

The computational fluid-dynamics software package, FIDAP, was used to predict the temperature and flow distributions in the electrically-heated glass pool. FIDAP has the advantage that joule heating is already incorporated in the code. The differential equations solved are:

Continuity and Momentum Equations
$$\nabla . u = 0 \tag{1}$$

$$\rho \, u \, . \nabla u = -\nabla P + \rho \, g \, \beta \left(T - T_r \right) + \mu \nabla^2 u \tag{2}$$

Energy

$$\rho c_p \, u \cdot \nabla T \;=\; k \, \nabla^2 T \;+\; \rho \, \alpha_E \, (\nabla \Phi)^2 \qquad\qquad [3]$$

Joule Heating

$$\rho \, \alpha_E \, \nabla^2 \Phi = 0 \qquad\qquad [4]$$

where u is velocity; T, temperature; ϕ, electric potential; P, pressure; ρ, density; g, gravitational constant, β, volumetric coefficient of expansion, μ, viscosity, k, thermal conductivity; c_p, specific heat; and α_e, electrical diffusivity.

Modeling Considerations

Previous modeling efforts have utilized simple boundary conditions (B.C.) in place of the cold cap, such as constant heat flux or constant temperature boundary conditions. These assumed boundary conditions however resulted in contradictory results where flow is upward in the center of the melter for constant temperature B.C. and downward for constant heat flux B.C. To eliminate this seeming contradiction, the present work prototypically includes a cold cap, which is characterized by variable thermal conductivity, electrical conductivity, specific heat, and viscosity, which are all functions of temperature. In addition, the latent heat of melting of glass and heats of chemical reactions are included in the energy equation.

One of the major problems in modeling the cold cap is how to represent the foamy layer. This thermally resistant layer is clearly important, as demonstrated by significant melt rate decreases observed with certain DWPF feed batches (Ref. 1). Theoretically, the best approach is to model the growth of bubbles by diffusion of gases from chemical reactions and the transport of these bubbles by convective and diffusive forces. However, the pertinent experimental data on diffusion coefficients for molten glass are unavailable. Consequently, our approach is to assume that a steady foam void fraction forms in the cold cap at various temperatures similar to the experimental results of batch sample tests performed in laboratory. Here, samples of waste glass frit are loaded into ceramic crucibles, melted and held at various temperatures for 2 hour sand then pulled out of the oven. Samples sectioned show that for Frit 200, bubbles do not form below 700°C, then go to a maximum value of about 50% at 750°C, and then disappear above 800°C.

The effective thermal conductivity of the cold cap, k_{eff} is then the liquid thermal conductivity, k, multiplied by the liquid fraction, or,

$$k_{eff} = k(1 - \alpha) \qquad\qquad [5]$$

where α is the void fraction and is estimated from steady state melt sample testing in the following temperature ranges as:

$$\alpha=0 \qquad\qquad\qquad T<700\ ^\circ C \qquad\qquad\qquad\qquad [6]$$
$$\alpha=0.5(T-1023) \qquad\qquad 700<T<750\ ^\circ C$$
$$\alpha=0.5[1-(T-1023)] \qquad 750<T<800\ ^\circ C$$
$$\alpha=0 \qquad\qquad\qquad T>800\ ^\circ C$$

The specific heat of glass is shown in Figure 5. On the same plot is a curve for the specific heat of the cold cap. This includes the effect of heat of reaction of chemical reactions in the temperature range, 973-1173°K. This specific heat is defines as,

$$Cp_{cc} = \frac{\int (Cp_g + L)dT}{\int dT} \qquad\qquad\qquad [7]$$

where L is the sum of heats of reactions.

TWO-DIMENSIONAL MODEL DESCRIPTION

The 2D model is basically a rectangular region with upper and lower electrodes located along the vertical sides. The width is 1.42 m, which simulates the actual distance between electrodes in the DWPF Melter. The height of the glass, 0.914 m, is also the same as in the melter. A cold cap region, which is included in the 0.914 m dimension, is 1.18 cm high. The outlet pipe is located at the bottom of the melter, since for this 2D model, placing it on the side-wall would interfere with the electrodes at that wall. The depth of the model can be considered to be 1.02 m which is the electrode depth in the melter. The model had 20,000 cells.

The cold cap area is assumed to cover the entire top surface of the melter model. The boundary conditions at the cold cap top surface is a parabolic temperature distribution, starting at 300°C at the center (where the feed-pipe for the slurry feed is located) and increasing to 850°C at the edges. This cold cap surface temperature variation is an initial guess to connect a cold cap temperature (300°C) close to film boiling conditions where the slurry feed is injected to a an exposed glass temperature at the edges calculated in the Lumped Parameter report (Ref. 1).

A flow boundary condition at the cold cap is based on a total feed rate of 68 Kg/hr. This flow is assumed to have a parabolic profile across the cold cap surface such that the flow is maximum at the edges and zero at the center. This flow distribution is based on the premise that the foam underneath the cold cap is thinnest at the edges and thickest at the center of the cold cap as the chemical reaction gases migrate to the periphery of the cold cap. During small-scale melter tests, gases were observed to vent at the cold cap periphery. Consequently, the melt rate would be highest at the edges and lowest at the center. The rest of the boundary conditions are convective heat transfer coefficients on the side wall and the bottom surface, based on a reference temperature of 940°C, which is a measured DWPF Melter electrode temperature. These coefficients were chosen so that the calculated heat fluxes (losses) after the solution was obtained would be equal to the melter heat losses obtained from the Lumped Parameter Model.

The method of solution is as follows: First, the electrode potentials are guessed and a solution is obtained. Then the heat losses through the cold cap, side wall and bottom surface are calculated. The convective coefficients are iterated until the calculated and prescribed heat losses match.

The prescribed heat flux at the cold cap represents the heat transferred from the melt pool through the cold cap, including the foam layer, after supplying heat to raise the glass temperature from the cold cap surface temperature to the bulk glass temperature, as well as any chemical reaction heat. This left over heat is available for evaporating slurry water, in addition to radiant heat from the dome heaters and plenum walls. The boundary condition for the cold cap, as stated above, is a temperature boundary condition and not a heat flux. After the solution, the heat loss through the cold cap is calculated and compared with the prescribed heat loss above. A solution is achieved if the calculated flux matches the required flux. Otherwise, the convection coefficient is changed and another iteration step is done. A converged solution is one with an error less than 10^{-4}.

A summary table of results is given for an electrode potential drop of 75 volts. The side wall surface flux is higher than in the actual melter because for this 2D model, there are no heat losses to the front and back sides. The total electrode density translate to a melter power of 140 kW, taking into account the actual melter volume.

Table 1 Summary Table for Two-Dimensional Model

Two Dimensional Model	Electrode Potential Drop=75 volts
Ave. Melt Pool Temp.	1381°K
Ave. Melt Pool Velocity	0.000836 m/s
Cold Cap Surface Flux	17.1 kW/m^2
Side Wall Surface Flux	16.1 kW/m^2
Bottom Surface Flux	3.2 kW/m^2
Glass Melting Power	8.5 kW/m^2
Total Electrode Power Density	66 kW/m^3

Temperature Distribution

The predicted temperature distribution with an electrode potential difference of 75 volts is shown in Figure 1. The average pool temperature (excluding the cold cap) is 1381°K (1108°C). The temperature distribution of the cold cap shows a region of low temperatures (less than 1358°K or 1085°C) that is thick at the center and tapers to a thin section near the walls. The temperature distribution in the center of the melt pool is quite uniform, which is about 1417°K (1144°C). The temperature then decreases in the lower quarter of the melt pool.

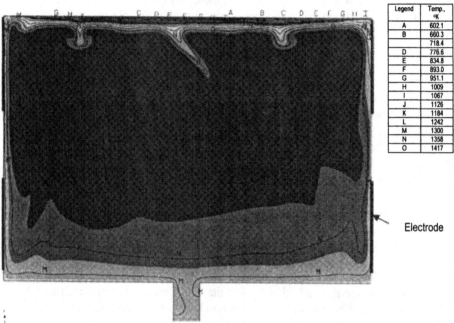

Legend	Temp., °K
A	602.1
B	660.3
	718.4
D	776.6
E	834.8
F	893.0
G	951.1
H	1009
I	1067
J	1126
K	1184
L	1242
M	1300
N	1358
O	1417

Electrode

Figure 1 Temperature Contour Map for Two-Dimensional Melter
Model at 75 Volts Potential Drop

Velocity (Streamline) Distribution

Figure 2 gives the streamline contour map, where circulation cells show up better than in a velocity vector map. Streamlines are lines tangent to the direction of the velocity vectors. The distance between two streamlines is inversely proportional to the flow, such that the closer the streamlines, the higher the velocities. Figure 2 shows basically two pairs of large counter-rotating cells. A dominant cell structure is not evident, which may be due to the relatively low average pool velocity.

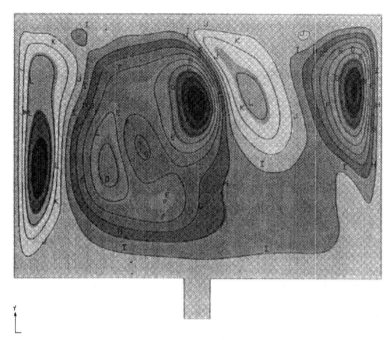

Figure 2 Streamline Contour Map for Two-Dimensional Melter
Model at 75 Volts Potential Drop

THREE-DIMENSIONAL MODEL

The features of the three-dimensional model are shown in Figure 3, which also
provides the temperature contour map. The upper and lower electrode pairs are
0.304m high and 1.015m wide. The distance between electrode pairs is 1.42m.
The melter diameter is 1.76m, and the glass depth is 0.86m. The cold cap is
assumed to be 76 mm high. The top elliptical section is an artifact of another
modeling effort and does not affect the results. The outlet riser leads to the pour
spout, where glass is poured into canisters. The model had 400,000 cells.

The glass properties are the same as in the 2D model. The void fraction in the
cold cap is also assumed to be a temperature function as given by Eq. 6. The
temperature BC is assumed to be radially symmetrical, 300°C at the center and
850°C at the edges. The inlet flow is a parabolic profile, zero at the center and
maximum at the edges, so that when integrated , the total flow is 68 Kg/hr. The
convective coefficients at the side wall, bottom, and cold cap surface are iterated
to give similar heat losses as in the Lumped Parameter model for Macrobatch 2
conditions. Solution of this 3D model was performed on a Dell WHL530 machine
with 2 GHz Pentium IV processor, 2 Gigabyte of RDRAM. Convergence was

severely hampered by the very sensitive variation of thermal and electrical conductivity with temperature.

Figure 3 shows the temperature contour map along a plane that connects the left and right electrodes for a potential drop of 75 volts. This plane would be analogous to the 2D model. Also, as in the 2D model, a region of severe temperature gradients appear in the cold cap region. It is generally thin at the edges and thickest at the center. There are also localized thick sections associated with junctures of circulation cells in the glass. In general, the glass temperature in the main pool is fairly uniform, with an average temperature of 1100°C. The velocity vector map is given in Figure 4. There are a number of small cells that are not as well organized as in the 2D model. This is because of the degree of freedom for flow in the perpendicular direction. The exiting flow to the outlet riser also affects the overall flow pattern. However, there appears to be one large cell directly below the center of the cold cap. Figures 5 and 6 give the temperature contour and velocity vector maps for a plane at the center, perpendicular to those in Figures 3 and 4, or parallel to the electrodes.

CONCLUSIONS

Two-dimensional and three-dimensional numerical models of the DWPF melter were obtained using computational fluid-dynamics methods. These featured a simplified cold cap model that assumed the foam layer void fraction is a function of temperature, as provided by laboratory batch sample testing at various temperatures. Although simple, the results showed cold cap region, thick at the center and thin at the edges, which is believed to be representative of actual melters.

An area of improvement would be the modeling of the cold cap to include the generation of bubbles and their venting or transport away from the cold cap. This would require two-phase flow modeling and basic experimental data required by the two-phase model. Experimental data on the cold cap consistency, thickness, surface temperature and radial movement are required in order to achieve an accurate melter model.

REFERENCE
1. "Numerical Models of Waste Melters, Part I Lumped Parameter Analysis of DWPF" by H. N. Guerrero and D.F. Bickford, this Proceedings.

Prepared for the U.S. Department of Energy under Contract DE-AC09-96SR18500. The authors are grateful for the helpful comments of Douglas Witt.

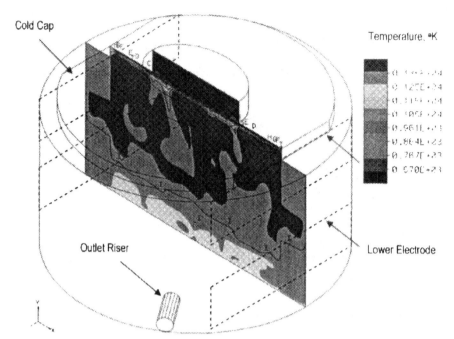

Figure 3 Temperature Contour Map for 3D Model at 75 Volts Potential Drop
- Plane Transverse to Electrodes

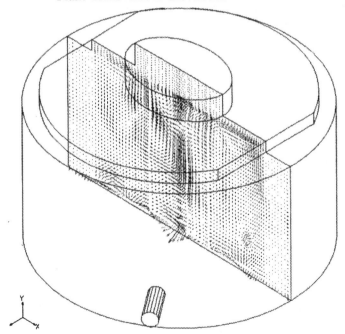

Figure 4 Velocity Vector Map for 3D Model at 75 Volts Potential Drop
- Plane Transverse to Electrodes

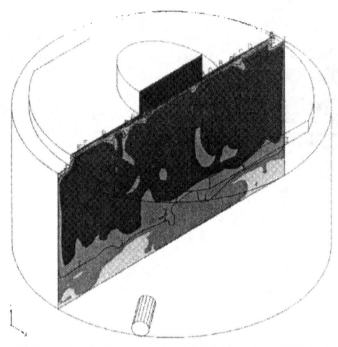

Figure 5 Temperature Contour Map for 3D Model at 75 Volts Potential Drop
-Plane Parallel to Electrodes

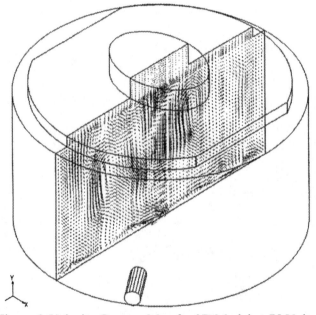

Figure 6 Velocity Contour Map for 3D Model at 75 Volts Potential Drop
-Plane Parallel to Electrodes

DISCLAIMER

This report was prepared as an account of work sponsored by an agency of the United States Government. Neither the United States Government nor any agency thereof, nor any of their employees, makes any warranty, express or implied, or assumes any legal liability or responsibility for the accuracy, completeness, or usefulness of any information, apparatus, product or process disclosed, or represents that its use would not infringe privately owned rights. Reference herein to any specific commercial product, process or service by trade name, trademark, manufacturer, or otherwise does not necessarily constitute or imply its endorsement, recommendation, or favoring by the United States Government or any agency thereof. The views and opinions of authors expressed herein do not necessarily state or reflect those of the United States Government or any agency thereof.

TAILORED ELECTRICAL DRIVING AS A MEANS OF CONTROLLING HEAT DISTRIBUTION AND CONVECTION PATTERNS IN JOULE-HEATED WASTE GLASS MELTERS

J.A. Fort and D.L. Lessor
Pacific Northwest National Laboratory
902 Battelle Boulevard
Richland, WA 99352

INTRODUCTION

In electrically heated glass melters, favorable convection patterns may mitigate gas layer buildup under the unmelted "cold cap", enhance heat transfer to the batch, and possibly accelerate batch reactions, thereby increasing melt rate and glass throughput. Favorable control of convection patterns may also improve homogenization, avoid cold spots, and provide other operational benefits. Convective patterns in an electrically heated melter are strongly influenced by Joule heat distribution and thermal boundary conditions for a given melter design and geometry. Electrical driving control, in particular, interactive control of electrical potentials connected to distinct electrode pairs, can be used to vary the Joule heat generation. One parameter affecting the Joule heat distribution is the "overlap" of the driving voltages waveforms. The "overlap" due to harmonic driving is determined by the relative phase. For electrical driving using waveforms chopped by Silicon Controlled Rectifiers (SCRs), the chopping influences the "overlap." Our study demonstrates with a model that tailored electrical driving is a means to control heat distribution and convection patterns in joule-heated waste glass melters.

THEORY

Consider a medium such as a glass melt into which a pair of electrodes is inserted, each pair being separately driven by an external time-dependent emf source. The current distribution within the medium can be represented by a superposition:

$$\mathbf{J}(\mathbf{r},t) = I_1(t)\mathbf{J}_1(\mathbf{r}) + I_2(t)\mathbf{J}_2(\mathbf{r}) \tag{1}$$

The space dependent current densities per unit of driving current, $\mathbf{J}_1(\mathbf{r})$ and $\mathbf{J}_2(\mathbf{r})$, can be obtained from solutions of Poisson's equation, each solution having one member of a driven electrode pair as a current source and the other as a current sink. The resulting time- or cycle-averaged Joule heating power per unit volume P_V will be:

$$P_V = \langle \mathbf{J} \bullet \mathbf{J} \rangle / \sigma = [\langle I_1^2 \rangle \mathbf{J}_1 \bullet \mathbf{J}_1 + \langle I_2^2 \rangle \mathbf{J}_2 \bullet \mathbf{J}_2 + 2\langle I_1 I_2 \rangle \mathbf{J}_1 \bullet \mathbf{J}_2] / \sigma \qquad (2)$$

Here the enclosing angle brackets $\langle \ \rangle$ indicate a time or cycle average. From Equation 2 we see that the time dependence of the two source currents from the two emf sources influences the Joule heat distribution though a parameter which we have labeled the "overlap" of the driving currents, namely:

$$Overlap = \frac{2\langle I_1 I_2 \rangle}{\langle I_1^2 \rangle + \langle I_2^2 \rangle} \qquad (3)$$

The generalization to more electrode pairs and independent emf sources is obvious.

TEST PROBLEM

The test problem geometry is a simple 2-D rectangular enclosure with two end-mounted electrode pairs. This geometry is shown in Figure 1. Figure 2 shows the electrical connections for parallel firing across the melt between opposing electrodes.

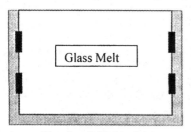

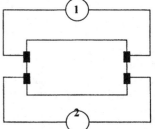

Figure 1.   Simple 2-Field Melter Geometry

Figure 2.   Electrode Connections for Two Fields

Figure 2 also shows electrical connections with two emf sources separately controlled for amplitude and waveform. Phase control in sinusoidal driving is a subset of possible control measures. We will set electrical driving to limiting cases to show magnitude of variability. The option specifying electrical driving in TEMPEST [1] is the set of cycle-averaged current products $\langle I_i(t) I_j(t) \rangle$ (or parameters to determine them) for each pair of source-sink currents from the external emfs. For a pair of sinusoidal currents with phase angles ϕ_1 and ϕ_2, $I_1(t) = I_{1,0} \cos(\omega t + \phi_1)$ and $I_2(t) = I_{2,0} \cos(\omega t + \phi_2)$, the current product input to TEMPEST is

$$\langle I_1(t) I_1(t) \rangle = I_{1,0}^2 / 2 = I_{1.rms}^2 \qquad (4)$$

$$\langle I_2(t) I_2(t) \rangle = I_{2,0}^2 / 2 = I_{2.rms}^2 \qquad (5)$$

$$\langle I_1(t) I_2(t) \rangle = 1/2\, I_{2,0} I_{2,0} \cos(\phi_1 - \phi_2) = I_{1.rms} I_{2.rms} \cos(\phi_1 - \phi_2) \qquad (6)$$

For fixed current amplitudes $I_{1,0}$ and $I_{2,0}$, the limiting cases occur when $\cos(\phi_1 - \phi_2) = \pm 1$, or more specifically, when the difference in phase angles is zero (in phase) and π (180 degrees out of phase). Thus

$$\max\langle I_1(t)I_2(t)\rangle = I_{1,rms}I_{2,rms} \qquad (7)$$

$$\min\langle I_1(t)I_2(t)\rangle = -I_{1,rms}I_{2,rms} \qquad (8)$$

which in the following text are referred to as the 'positive' and 'negative' limits, respectively. A third case was included

$$\langle I_1(t)I_2(t)\rangle = 0 \qquad (9)$$

This is referred to as the 'zero overlap case,' which requires $\cos(\phi_1 - \phi_2) = 0$ for full sinusoidal currents. Zero overlap could also be achieved by chopped waveforms from an SCR.

We note that switching between the $\cos(\phi_1 - \phi_2) = \pm 1$ cases can usually be done by exchanging electrical lead pairs, and that $\cos(\phi_1 - \phi_2) = 0$ can be achieved from a three-phase source using the Scott-T configuration. We will not here go further into electrical engineering questions, or attempt to describe systems in which the "overlap" is continuously variable, though SCR control probably permits this, subject to some design constraints.

TEST RESULTS

Baseline tests examined two opposing electrode pairs fired in parallel (Figure 2). Diagonal firing tests (bottom left electrode connected to top right electrode, etc.) are discussed later.

Initial runs used constant properties without flow, representing Joule heating of a solid of uniform electrical conductivity. Current levels and resulting Joule heat were arbitrarily set as only Joule relative heat distribution was of interest. Predicted Joule heating results for the three electrical driving cases are shown in Figure 3. Note that plots contain only the computational domain inside the electrodes (see Figure 1 'Glass Melt' portion).

These results demonstrate that changes in electrical driving can dramatically alter Joule heat distribution. Maximum heat distribution in Figure 3 always occurs adjacent to electrodes, with minima of heat generation moving from the melt corners (positive limit) to the melt centers (negative limit). This ability to control Joule heat distribution may offer advantages. The first impact on melter operation is that change in Joule heat distribution changes local temperature. This effect is moderated somewhat by thermal conduction and convection within the melt and to the boundaries, but the highest temperatures are generally in the highest Joule heat regions. Melt convection is proportionate to local density difference and hence to temperature difference, so convective flow patterns and velocities are also affected by changing Joule heat distribution.

To investigate this effect, the test problem is altered to simulate a melter. The simple 2-D geometry is retained (dimensions are 0.64m between electrodes, 0.46m glass depth), but with real glass properties, realistic boundary conditions, and viscous fluid flow. Properties are taken from the West Valley glass, and the top surface boundary conditions are the same as the feed condition used on the West Valley melter model [2]. Because real glass properties are used, electrical conductivity rises strongly with increasing glass temperature above the melting point; therefore Joule heat distribution is changed somewhat from the constant electrical conductivity cases presented in Figure 3. Figure 4 shows Joule heat and glass convective velocity distributions for this case.

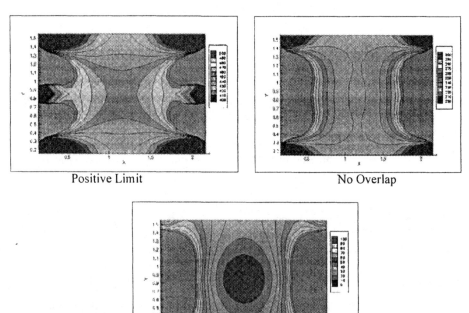

Positive Limit   No Overlap

Negative Limit

Figure 3. Joule Heat Distribution for Parallel Firing – Constant Electrical Conductivity

Velocity maps clearly show that electrode firing can change melt convection. For the positive limit case (Figure 4, top), Joule heat in the middle of the melt is the highest of the three cases and locally upward glass velocities are the predicted result. The negative limit (Figure 4, bottom) presents the contrasting case where mid-melt Joule heat is minimum and glass flows down in the middle of the melt. Peak glass velocities for all three cases are approximately the same, roughly 5 mm/s.

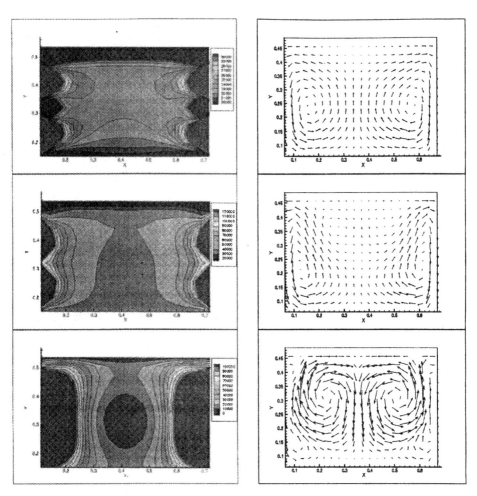

Figure 4. Joule Heat Distribution and Glass Velocities for Positive Limit (top), No Overlap (middle) and Negative Limit Cases

Finally, a sample result is included for diagonal firing. Joule heat distribution and melt convection are not significantly different from the parallel-fired cases at positive limit and zero overlap. The change in Joule heat distribution was most changed for the negative limit case. This result is shown along with predicted glass velocities in Figure 5 While the Joule heat distribution is changed from that of the parallel-fired case in Figure 4, the minimum Joule heat still occurs in the center of the melt and there is no visible change in the glass convection. It should be noted that, quite generally, the regions where the Joule heat generation can be changed most by changing overlap are those where the current densities per unit driving current $J_1(r)$ and $J_2(r)$ are nearly parallel or antiparallel.

The results presented here for this simplified case suggest that melt convection can be controlled by tailored use of electrical driving. This also suggests that melt convective patterns could be changed for operating purposes, for example and as mentioned previously, to help mitigate gas layer buildup under the cold cap or to discourage/encourage noble metal deposition in certain areas of the melter floor.

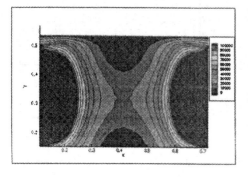 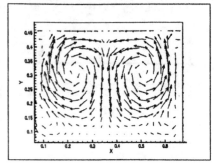

Figure 5. Joule Heat Distribution and Glass Velocities for Diagonal Firing - Negative Limit

Whether or not the glass velocities are adequate to result in these desired changes is not investigated here. Direction changes alone may not suffice as shear forces induced by the low velocity glass flow may be unable to overcome buoyant forces and surface interaction with gas layer trapped under the cold cap. Additional work is required but results presented in this study encourage further investigation.

CONCLUSIONS AND RECOMMENDATIONS
    Simulations of a simple 2-D rectangular melter have shown that changes in electrical driving can have a strong effect on Joule heat distribution and melt convection pattern. While peak glass velocities are not significantly changed, their direction can be reversed. This predicted ability to control melt convection should be further investigated with models of real melters and realistic electrode geometries as a means of improving operating performance, especially as related to mitigation of gas layer buildup beneath the cold cap. More detailed models including 3-D geometry and detailed treatment of boundaries and cold cap region will be required in this follow-on work.

REFERENCES
    [1]D.S. Trent and L.L. Eyler, *TEMPEST: A Computer Program for Three-Dimensional Time-Dependent Computational Fluid Dynamics*, PNL-8857, Vols. 1-4, Pacific Northwest National Laboratory, Richland, Washington, 1993.
    [2]J.A. Fort, D.L. Lessor, T.E. Michener, K.P. Recknagle, and M.L. Elliott, "Model Predictions of the West Valley Demonstration Project Melter with Nobel Metals and Alternative Electrode Configurations," WVSP 02-09, Pacific Northwest National Laboratory, 2002.

EFFECTS OF POLY(ACRYLIC ACID) ON THE RHEOLOGICAL PROPERTIES OF AQUEOUS MELTER FEED SLURRIES FOR NUCLEAR WASTE VITRIFICATION

Hong Zhao, Isabelle S. Muller, and Ian L. Pegg
Vitreous State Laboratory, The Catholic University of America, Washington, DC 20064

ABSTRACT

The effects of additions of the surfactant poly(acrylic acid) (PAA) on the rheological properties of simulated melter feeds for high-sodium-concentration waste vitrification was investigated experimentally. Significant changes in rheological properties were observed in certain feed types while little effect was found for others. In this application, these changes translate into increased achievable feed solids contents and glass production rates. It is believed that the pH of these feeds was an important distinguishing factor through its effects on establishing a repulsive inter-particle electrosteric force by surface modification of the suspended particles in the presence of PAA. Although only PAA was investigated in the present work, it is likely that similar effects are possible for feeds with different pHs through the use of other surfactants. This work illustrates the general principle that surfactant additions offer the potential for increasing the achievable solid content in melter feeds and, consequently, the achievable glass production rates.

INTRODUCTION

In many types of waste vitrification processes, the waste is combined with chemical additives to form an aqueous slurry, which upon vitrification, forms a glass of the desired composition and properties. The rheological properties of the intermediate aqueous melter feed slurry are of greater practical importance in the mixing and transport of the feed to the melter and also in the spreading of the slurry on the surface of the glass melt. These slurries typically exhibit a non-zero yield stress, strong shear-thinning, and a range of settling rates. These properties can be controlled to some extent by the selection of the chemical and physical characteristics of the glass-forming additives and by the solids content of the feed. In general, higher solids content lead to higher viscosity and higher yield stress, while lower solids contents lead to faster settling slurries. From the perspective of glass production rates, higher solids contents are generally preferred since these reduced the evaporative load on the melter. It is of interest, therefore, to identify methods for maintaining favorable rheological properties when the feed solids content is increased. The addition of relatively small amounts of organic or inorganic dispersing agents as rheology modifiers offers the potential to provide the necessary additional control of these properties.

The melter feeds investigated in the present work are of the general compositions of those that have been investigated for the vitrification of high-sodium low-activity nuclear wastes (LAW) at the Hanford site[1,2]. These high-sodium waste streams often contain significant amounts of sulfur, which limits the achievable waste loading in the glass product to range that depend on the sodium-to-sulfate ratio in the waste[1,2]. Accordingly, the melter feeds can be divided into three types: (i) low waste loading feeds with $2 - 7$ wt% Na_2O in the glass product; (ii) medium waste loading feeds with $8 - 15$ wt% Na_2O; and (iii) high waste loading feeds with $16 - 20$ wt% Na_2O. The water content in the final melter feed is determined by the sodium molarity in the starting waste, since only dry glass formers are added. However, the amount of the glass former additions to that waste varies inversely with the waste loading. Consequently, to achieve the same solids content in the final melter feed, the sodium molarity must be decreased as the waste loading is decreased. However, the practically achievable solids content and, therefore, the required starting sodium molarity in each case is determined by the rheological properties of the final melter feed.

The addition of relatively small amounts of dispersing agents might provide a means to improve the rheological properties of melter feeds such that the practically achievable feed solids content and, therefore, glass production rates, can be increased. For instance, it is known that long chain molecule organic surfactants under certain conditions can effectively cover the surface of suspended particles in slurries to form a repulsive electrosteric force between the modified particles that can prevent particles from agglomerating, thereby homogenizing and stabilizing the slurries[3,4]. One such compound, poly(acrylic acid) (PAA) is a commonly used surfactant in ceramic industry. It is known that PAA may become ionized and vary its chain shape in response to changes in pH level and ionic strength in slurries[3,4]. In this work, the effect of PAA on the rheological properties of representative melter feeds for high-sodium waste streams was investigated. The extent to which such additives can enhance the formation of homogenous slurries with high solid content and favorable rheological and settling properties was investigated for various melter feed types and concentrations of surfactant.

EXPERIMENTAL PROCEDURE

Three types of high-sodium low-activity waste were simulated based on published waste characterization information[5]. The three types are differentiated by their sodium oxide loadings in their respective product glasses[2], which are, in turn, determined by the ratios of sulfate to sodium in the original wastes: Type I,

II, and III correspond to sodium oxide loading in glass of 5.5 wt%, 14.4 wt%, and 18.5 wt%, respectively. Each waste simulant was prepared at medium and high sodium molarities that reflect the expected range of achievable dilutions for each waste type: 2.75 and 3.25 M Na for Type I; 6 and 7 M Na for Type II; and 8 and 11 M Na for Type III. The glass-forming chemicals (including sources of SiO_2, B_2O_3, Al_2O_3, Fe_2O_3, TiO_2, and ZrO_2) required to produce the requisite glass compositions were mixed into the simulated wastes to produce the melter feed samples. Appropriate amounts of the surfactant PAA (50% solution, molecular weight of 5000 g/mol, Polyscience, Warrington, PA) were added to achieve concentrations in the range of 0 – 1 wt% on a dry weight basis.

Densities of the melter feeds were measured by weighing a known volume of the material. An Accumet Research pH meter (AR15 or AR50) equipped with an automatic temperature compensation probe was used for pH measurement. A Haake RS75 rheometer was employed to determine the apparent viscosity of melter feeds by measuring the shear stress on the slurry at controlled shear rates that are increased stepwise from 0.01 to 200 s^{-1} with a delay of up to 30 seconds between steps. The yield stress was measured using the controlled-stress mode by identifying the discontinuity in fluid deformation versus torque on a shear-vane rotor; in general, this value differs from the value obtained by extrapolating the shear stress-shear rate curve to zero shear rate. These measurements were made after the sample was left undisturbed for at least 30 minutes. All of the above measurements were performed at a temperature of 25°C. A Microtrac S3000 particle size analyzer with an automated small-volume recirculator was employed for particle size distribution analysis at about 23°C using a 30-second ultrasonication prior to measurement and a particle refractive index of 1.54.

RESULTS

Figure 1 shows total solids content (TSC) in melter feeds versus sodium molarity in simulated waste for three typical melter feeds with sodium oxide loadings of 5.5 wt%, 14.4 wt%, and 18.5 wt%. In general, for all three types of feeds, higher TSC in melter feeds leads to unfavorable rheological properties, while lower TSC results in faster settling. Note the sensitivity of feed Type I to sodium molarity, which results from its low waste loading.

Table 1 lists the pH values and densities of the three types of melter feeds with various PAA additions. Type III has the highest pH, whereas Type I has the lowest pH values. This trend is a direct consequence of the waste loading: lower waste loading result in larger amounts of additives that tend to neutralize the alkaline simulated waste. In addition, the pH values are slightly lowered by both increasing the sodium molarity in waste (Type I and II) and by adding PAA (all three types of feeds).

Figures 2 – 4 plot apparent viscosity versus shear rate for the three types of feeds, at medium and high sodium molarities, with the PAA additions in the range of 0 – 1.0 wt%; the yield stresses of the melter feeds are listed in Table 1. It is apparent that the addition of surfactant PAA has a very large effect on the rheological properties of Type I feeds at high sodium molarity but little influence on Type II and Type III feeds. The effects of the PAA addition on Type I are dependent on the sodium molarity: both the apparent viscosity and yield stress increase with the PAA addition at 2.75 M Na but significantly decrease at 3.25 M Na. The rheological properties of Type I (3.25 M Na) are significantly improved through the PAA addition. For instance, at a low shear rate of 1 s^{-1}, the viscosity drops from 660 Poise without the PAA addition to 20 Poise with the addition of 1.0 wt% PAA; at the higher shear rate 100 s^{-1}, it drops from 18 Poise without PAA to 6 Poise at 1.0 wt% PAA. Both the yield stress and viscosity of the 3.25 M Na Type I feed with PAA are lower than those of the 2.75 M Na Type I feed.

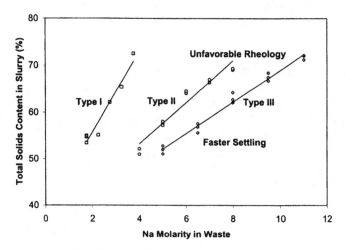

Figure 1. Total solids content versus Na molarity
for three types of melter feeds.

The significant changes in the rheological properties for the Type I feeds at 3.25 M Na suggests that the PAA addition has effected a surface modification of the suspended particles in these feeds; this is supported by the particle size analysis results, which are shown in Figures 5–7. A significant reduction in particle size with the PAA addition was observed in Type I at 3.25 M Na (Figure 6), but not at 2.75 M Na (Figure 5). Neither Type II nor Type III show significant shifts in particle size distribution with the PAA addition (Figure 7 shows results for Type III as an example), which is consistent with the observed effects on feed rheology. For the Type I feed at 2.75 M, there is actually evidence (Figure 5) that the PAA additions led to flocculation, in marked contrast to the behavior for the 3.25 M Type I feed.

DISCUSSION
    The properties of suspensions, such as the ones investigated in the present work, are strongly influenced by the nature of the double-layers that form between the surface of the particles and the bulk solution[6]. The pH of the solution

and the ionic strength both impact the size and gradient of the double layer and, thus, the "zeta-potential," which represents the electric potential at the outer boundary of the double layer. Measurements of the zeta-potential and its dependence on pH and ionic strength are, therefore, often useful in characterizing suspensions. Unfortunately, the very high solids contents and ionic strengths of the present slurries make such measurements very difficult. The effects of the attractive van der Waals force and the repulsive electrostatic force between like-charged particles, as captured in the DLVO theory[6], provides a basis for interpreting the effects of the inter-particle forces on slurry properties. However, in the presence of surfactant polymeric species (like PAA), an additional repulsive steric force arises from the steric interaction between particle surfaces that are covered with an adsorbed surfactant layer. Depending on the pH and ionic strength, this so-called electrosteric force can play an important role in preventing flocculation. Conversely, under other conditions, chain entanglement can actually enhance flocculation.

Table 1. pH, density and yield stress of melter feeds.

	Waste loading (wt% Na_2O)	Na Molarity	PAA addition (wt%)	pH	Density (g/cm³)	Yield stress (Pa)
Type I	5.5	2.75	0	8.3		37.3
			0.25	8.1	1.71	65.9
			0.5	8.0		73.0
		3.25	0	7.3		174.2
			0.5	7.2	1.79	35.2
			1	7.2		13.6
Type II	14.4	6	0	8.7		0.7
			0.5	8.6	1.70	0.5
			1	8.6		0.9
		7	0	8.4		3.3
			0.5	8.3	1.78	5.3
			1	8.3		4.6
Type III	18.5	8	0	12.3		0.4
			0.5	12.3	1.70	0.2
			1	12.2		0.3
		11	0	12.5		1.6
			0.5	12.3	1.86	1.3
			1	12.4		1.4

From Table 1, it can be seen that the pH level of the melter feed varies with waste loading and sodium concentration. The relatively low pH in Type I feed at 3.25 M Na may be a factor that favors particle surface modification in the presence of PAA in this feed since it is known that the role of PAA is highly pH dependent. In general, PAA tends to become ionized at high pH, with the portion

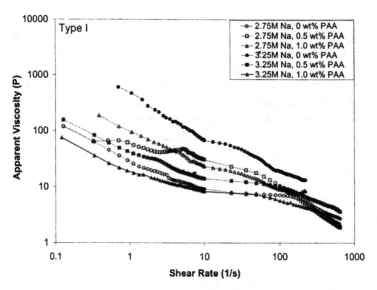

Figure 2. Apparent viscosity versus shear rate for Type I feeds at 2.75 M Na and 3.25 M Na with various PAA additions.

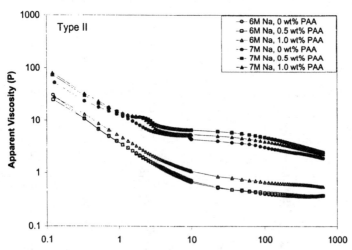

Figure 3. Apparent viscosity versus shear rate for Type II feeds at 6 M Na and 7 M Na with various PAA additions.

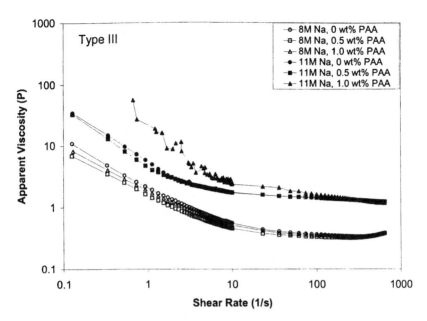

Figure 4. Apparent viscosity versus shear rate for Type III feeds at 8 M Na and 11 M Na with various PAA additions.

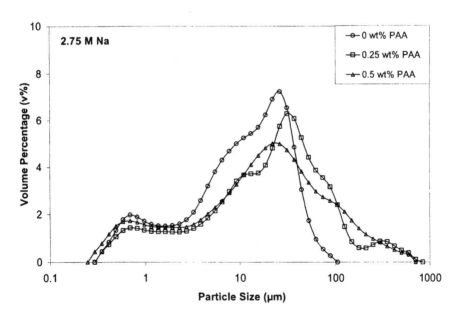

Figure 5. Particle size distributions for Type I feeds at 2.75 M Na with various PAA additions.

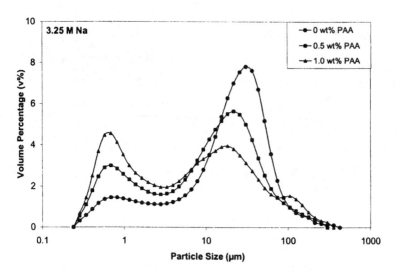

Figure 6. Particle size distributions for Type I feeds at 3.25 M Na with various PAA additions.

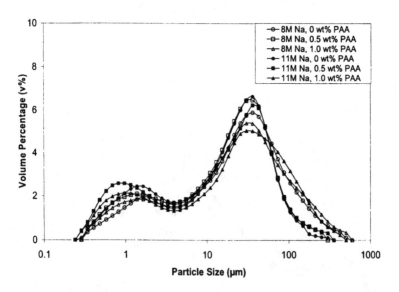

Figure 7. Particle size distributions for Type III feeds at 8 and 11 M Na with various PAA additions.

of the ionized PAA increasing with pH. It has been shown[3] that more than half of the added PAA molecules could have been ionized when the pH is above 7. The ionized PAA molecules are easily anchored to the positively charged sites as a part of double layer around the particles. In addition, the conformation of the PAA chain molecules is also affected by the pH level, with the PAA molecule changing from a coil-like shape at an extremely low pH to a fully stretched shape at an extremely high pH[3,4]; obviously, fully stretched PAA molecules on the surface of the particles would tend to increase the repulsive electrosteric force. It would be expected that the highly basic Type III feeds would promote the ionization of PAA molecules, which would occur to a lesser extent in Type I and Type II feeds.

The pH also strongly influences the number and sign of charged surface sites on a particle. When the pH is increased, more PAA molecules become ionized but fewer positively charged surface sites exist. The deficiency of positively charged sites on the particle surface may result in only partial coverage of PAA on the surface and an increased tendency for flocculation. The electrosteric effect can be expected to be strongest over a certain pH range when PAA molecules efficiently cover the surface of the particles; higher or lower pHs would then lead to increased flocculation. It should also be noted that the high ionic strength in the melter feeds can also affect the ionization and molecule chain shape of the PAA[4]. It would appear that with PAA additions, the combination of these factors was most favorable for Type I feed at 3.25 M Na. However, since different surfactants will exhibit these changes over different pH and ionic strength ranges, it is expected that similar changes could be brought about for other feed types by using other surfactants.

CONCLUSIONS

The addition of the surfactant PAA brought about significant changes in the rheological properties of Type I feeds at 3.25 M Na. The rheological properties were improved beyond those exhibited by the same feed without PAA additions at a concentration of 2.75 M Na, which represents more than an 18% increase in the achievable solids content. Such increases are likely to produce concomitant increases in glass production rates. The relatively low pH level in Type I at 3.25 M Na is considered to have an important role in establishing an inter-particle repulsive electrosteric force through particle surface modification on addition of PAA. These effects were much less evident in other feed types, probably due in part to their less favorable pHs. However, alternative surfactants that are better suited to these pH ranges may be capable of producing similar effects.

REFERENCES

[1]I.S. Muller, H. Gan, and I.L. Pegg, "Physical and Rheological Properties of Waste Simulants and Melter Feeds for RPP-WTP LAW Vitrification," Final Report, VSL-00R3520-1, Rev. 0, Vitreous State Laboratory, The Catholic University of America, Washington, D.C., 1/16/01.

[2]I.S. Muller, A.C. Buechele, and I.L. Pegg, "Glass Formulation and Testing with RPP-WTP LAW Simulants," Final Report, VSL-00R3560-2, Rev. 0, Vitreous State Laboratory, The Catholic University of America, Washington, D.C., 2/23/01.

[3]V.A. Hackley, "Colloidal Processing of Silicon Nitride with Poly(acrylic acid): I, Adsorption and Electrostatic Interactions", *J. Am. Ceram. Soc.*, **80** [9] 2315-25 (1997).

[4]J. A. Lewis, "Colloidal Processing of Ceramics", *J. Am. Ceram. Soc.*, **83** [10] 2341-59 (2000).

[5]R.A. Kirkbride et al., "Tank Farm Contractor Operation and Utilization Plan," CH2M Hill Hanford Group Inc., Richland, WA, HNF-SD-SP-012, Rev. 2, 4/19/00.

[6]P.W. Atkins, J. Depaula, "Physical Chemistry", 7th ed., W H Freeman & Co., pp. 520, 2001.

FREQUENCY MODULATED CONTINUOUS WAVE MONITORING OF REFRACTORY WALLS

B. Varghese[1], R. Zoughi[1], C. DeConink[1], M. Velez[2] and R. Moore[2]
[1]Applied Microwave Nondestructive Testing Laboratory (*amntl*)
Electrical and Computer Engineering Department
University of Missouri-Rolla
Rolla, Missouri 65409
zoughi@ece.umr.edu
(573)341-4656

[2]Ceramic Engineering Department
University of Missouri-Rolla
Rolla, Missouri 65409

ABSTRACT
    The refractory walls of glass tank furnaces degrade during their operational lifetime while premature furnace shutdowns may reduce productivity or even further damage the refractories due to temperature fluctuations. A systematic study using microwaves for wall thickness measurement under various conditions has been initiated with the adaptation of the Frequency-Modulated Continuous-Wave (FM-CW) Radar Technique. The system requires simple microwave hardware, which is designed to be portable allowing many measurements to be taken over the melt container surfaces yielding on-line information. The FM-CW technique can produce wall thickness data in real-time and basically requires a signal transmitter or horn which may also be made from high-temperature materials of appropriate dielectric constant. The horns could conceivably be permanently mounted at key locations in a glass tank melter or any industrial furnace, as the application of the technique can be expanded. This paper presents the design of such a microwave system as well as the results of some recent experiments for wall thickness measurement purposes.

INTRODUCTION

The glass melt furnace is at the heart of the glass industry. It is here that the raw materials are melted and fused to produce the crude molten glass. Alumina-Zirconia-Silica (AZS) is the most commonly used material for the side wall [1]. Regression of refractory linings of furnaces occurs due to a variety of mechanisms. This discussion will focus on the measurement of regression of refractories which are in direct contact with liquid corrodants. The rate of regression to a wall thickness which requires reline or extensive reconstruction vary widely. Some glasses are melted in the same lining for as long as 12 years. The operator of the furnace would like to know the rate of regression early in the life of a furnace, as there are many instances of "premature" wearing out of the lining. In spite of a near universal desire for a method to monitor the regression of the various linings enclosing liquid corrodants, there is no commercialized method for monitoring of the progressive thinning of refractory linings. There is a need for a reproducible and reliable method for measuring wall thickness that can be used while the furnace is operational. It is for this purpose that the potential of the FM-CW radar technique is being investigated.

BASICS OF FM-CW RADAR

Radars, in general, are capable of measuring several "attributes" of a "target" such as its velocity, distance or type. The conventional pulsed radar transmits a stream of pulses and determines the range of a target by monitoring the reflected pulses from the target. The Continuous-Wave (CW) radar on the other hand operates by sending a continuous electromagnetic wave. By sweeping the frequency of the transmitted signal, range tracking can be achieved. Frequency sweeping can be achieved by modulating a voltage-controlled oscillator with the triangular modulating waveform. The range detection ability of the FM-CW radar can be used to determine the thickness of a refractory wall. The block diagram for an FM-CW radar is shown in Figure 1.

In this technique a continuous signal is launched into the medium under consideration. If there is any discontinuity in the medium (which will be characterized by an impedance change), a reflected wave can be observed. The instantaneous frequency of the reflected wave is subsequently monitored. The reflected signal returns to the radar after a time delay, T, which is related to the thickness of the refractory wall. The time delay can be determined by knowing the difference between the transmitted frequency and the received frequency, the sweep frequency bandwidth (B) and the triangle-wave modulation rate (f_m). The difference between the transmitted frequency and the received frequency, known as the intermediate frequency (IF), is obtained by continuously mixing the transmitted and received signals together. The resulting IF output spectrum can be displayed on a spectrum analyzer. The measurement setup for determining the thickness of the refractory wall and a typical FM-CW radar output spectrum are shown in Figure 2 and Figure 3 respectively.

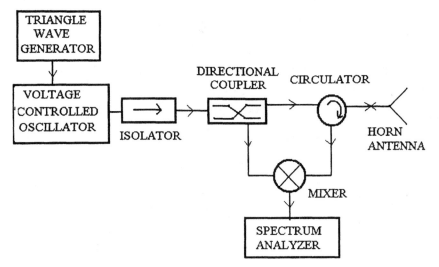

Figure 1: FM-CW radar block diagram.

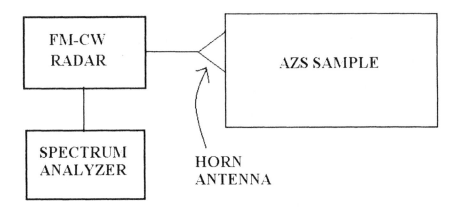

Figure 2: Test measurement setup.

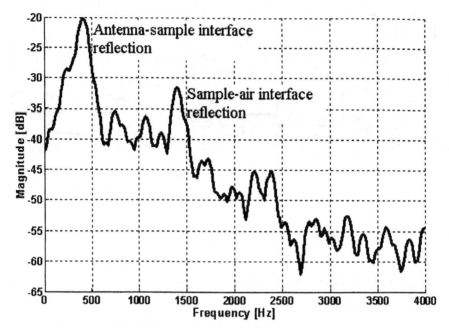

Figure 3: A typical FM-CW radar output spectrum.

Two frequency peaks can be seen in Figure 3. The first peak is due to the reflection at the antenna-sample interface while the second peak is due to the reflection at the sample-air interface. For determining the thickness of the wall, the difference in frequency between the antenna-sample interface, $F1$, and the sample-air interface, $F2$, must be considered. This difference in frequency is given by Δf_R. The thickness of the sample can be determined by the following expression [2]:

$$R = \frac{\Delta f_R v}{4 B f_m} \; (m) \tag{1}$$

where, v is the velocity of propagation given by:

$$v = \frac{3 \times 10^8}{\sqrt{\varepsilon_r}} \; (m/s) \tag{2}$$

and, ε_r is the dielectric constant of the refractory.

Environmental Issues and Waste Management Technologies IX

MEASUREMENTS AND RESULTS

Measurements were carried out on several AZS samples with different thicknesses. Before the thickness measurement can be carried out, the dielectric properties of the samples have to be determined. A simple way to determine an approximate value of the dielectric constant is to calibrate the system using samples of known thickness. In this way, the relative dielectric constant for the AZS samples was determined to be around 4.125 at X-band. As the samples used act like a dielectric waveguide, this procedure only gives an approximate value of the dielectric constant. This will not be the case in an actual furnace where the wall is continuous in the transverse directions.

For example for a 45 cm thick AZS sample, the output spectrum for $B = 900$ (MHz) and $f_m = 100$ (Hz) is shown in Figure 4.

$$\Delta f_R = 1560 - 410 (Hz)$$
$$\Delta f_R = 1150 (Hz)$$

also,

$$v = \frac{3 \times 10^8}{\sqrt{4.125}} (m/s)$$

$$v = 1.4777 \times 10^8 (m/s)$$

Now,

$$R = \frac{1150 \times 1.477 \times 10^8}{4 \times 900 \times 10^6 \times 100}$$

$$\Rightarrow R = 47.2 (cm)$$

Similarly, measurement for a 16 cm thick AZS sample is shown in Figure 5.

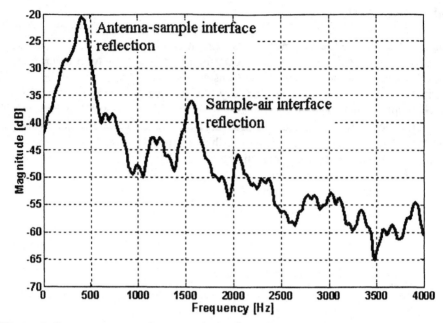

Figure 4: Output spectrum from the FM-CW radar for the 45cm AZS sample.
From Figure 4,

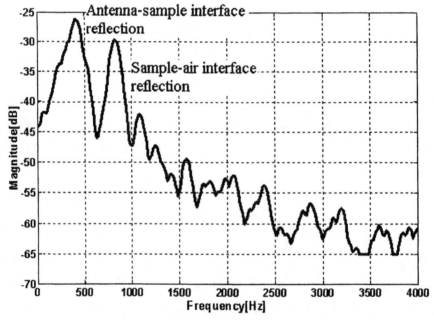

Figure 5: Output spectrum from the FM-CW radar for the 16 cm AZS sample.

From Figure 5, the measured thickness is 17.2 cm, whereas the actual thickness is 16 cm.

LIMITATIONS
The reflection at the antenna-wall interface will be prominent and increases the effective system noise level. Higher system noise reduces detection sensitivity. If the sample is too thin, then the reflection from the sample-air interface will be overlapped by the noise or cannot be properly deciphered. However, this limitation can be over come by increasing the bandwidth of the frequency sweep, B, or by increasing the modulation frequency f_m. When B is increased from 900 MHz to 1.1 MHz, the measurement results for the 8 cm AZS block results are shown in Figure 6.

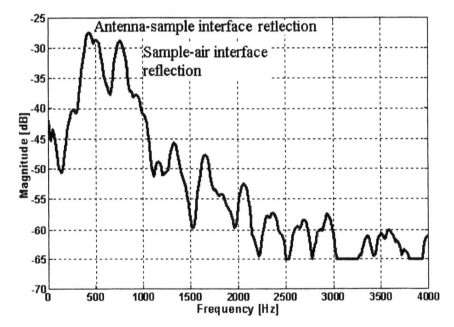

Figure 6: Output spectrum for the 8 cm thick AZS sample when the bandwidth is increased.

These measurements assume that there is only one type of refractory material being used to construct the wall, when actually this may not be the case. Different materials may have different dielectric properties and hence impedances. As a result at the interface of each type of refractory, there is a discontinuity, which will lead to a reflection. Thus, understanding of the construction of the furnace is required.

As measurements were made only at room temperature, it is not known at the moment if a temperature gradient will drastically change the dielectric properties of the refractory. The dielectric properties of the molten glass is also very critical. If the dielectric property of molten glass is similar to that of the wall, then there will be no impedance mismatch and hence no reflections at the wall-molten glass interface.

CONCLUSION AND SUMMARY

It has been shown that the FM-CW radar technique is a very effective and promising method for refractory wall thickness measurement. This technique provides good results with a high degree of accuracy. Incorporating a matched adapter into the design will provide maximum power transfer into the refractory wall. This will allow the reflected signals to be detected for refractory walls with reasonably high loss tangents. Since the design is rugged and portable, the system can be easily implemented in the industry.

REFERENCES

1. Douglas W. Freitag, *DOE/ORO 2076 Opportunities for Advanced Ceramics to Meet the Needs of the Industries of the Future, Chapter Six- Glass Industry,* U.S. Advanced Ceramics Association and Oak Ridge National Laboratory for the Office of Industrial Technologies Energy Efficiency and Renewable Energy U.S. Department of Energy, December 1998.

2. R. Zoughi, L.K. Wu and R.K. Moore, *SOURCECAT: A Very Fine Resolution Radar Scatterometer,* Microwave Journal, November 1985.

Combustion Control Experimentations at a Pilot Scale Glass Furnace

S. Keyvan, R. A. Rossow
Center for Artificial Intelligence in Engineering and Education
Mechanical and Aerospace Engineering Department
University of Missouri – Columbia
Columbia, MO 65211

M. Velez, W. L. Headrick, R. E. Moore
Ceramic Engineering Department
University of Missouri - Rolla
Rolla, MO 65409

C. Romero
Energy Research Center
Lehigh University
Bethlehem, PA 18015

ABSTRACT

In a multi-burner furnace, inefficient operation of individual burners could result in furnace operation at less than optimal fuel efficiency and elevated pollutant emissions. This paper presents various experimental explorations using a pilot scale glass furnace to investigate the impact of optimum combustion on flue gas emissions such as NOx. The glass furnace utilized is a 23-146 kW pilot scale furnace that can melt from about 45 kg to 900 kg of glass/day. Furnace design allows both air-gas and oxy-fuel combustion with different burner types and burner arrangements. The furnace is controlled through a LabView hardware and software control system.

Results from combustion control experimentation under ramp-up condition and various oxygen/fuel ratios from this pilot scale glass furnace are presented here. The oxygen/fuel ratio was varied from 1.8 to 2.4 with various combustion control experimentations in both step and ramp-up fashion. Using a spectrometer, spectral intensity data were collected over the ultraviolet/visible regions. The data was analyzed for specific radical chemiluminescence and the electromagnetic emission spectrum. Direct correlation and dynamic response was observed from the emission band from the hydroxyl flame radical, OH, to burner stoichiometry and flue gas NOx emissions. The results show a great promise for online combustion monitoring at the burner level for gas-fired glass furnace applications.

INTRODUCTION

For several years fiber-optic sensors [1-4] have been used to measure combustion temperatures, though many of these need to be placed in the combustion region to be effective. Elliot *et al* [5] and Cessou [6] have used non-intrusive methods for temperature measurement using lasers. Elliot *et al* add molecular seeds into the flame and Cessou employs a method involving laser excited OH radicals. Kang *et al* [7] listed several advantages of fiber-optic sensors including portability, small dimensions, and geometric versatility. Fiber optic sensor utilized here will have no impact on the combustion region as mentioned by Fisenko *et al* [8]. This sensor only needs a straight line of sight to the flame allowing it to collect data away from the combustion area.

The hydroxyl flame radical OH has a series of chemiluminescent peaks around 288 nm and another series around 310 nm. The specific peak around 310 nm was monitored for correlations to both burner stoichiometry and NO_x emissions. The wavelength of the emission line was monitored along with reference wavelengths above and below this point. The two reference wavelengths were averaged and subtracted from the OH peak to determine the actual OH value.

These tests were conducted at a pilot scale oxy-fuel fired glass furnace that utilized a 30 kW-145 kW furnace that is capable of melting from about 45 kg to 900 kg of glass per day. Furnace design allows both air-gas and oxy-fuel combustion with different burner types and burner arrangements. The furnace is controlled through a LabView hardware and software control system.

PROTOTYPE SENSOR

In order to place the fiber-optic assembly inside the harsh environment inside a conventional glass furnace a water-cooled probe casing was designed and built. A containment rod was used to hold the fiber-optic cable with housing for the collimating lens on the end. A larger water-cooled cylinder housed the containment rod with a flint glass window to provide additional protection to the lens. Minimal amounts of purge air are sent through the casing in order to prevent residue build-up on the lens. The disassembled probe is shown in Figure 1.

Two separate gratings process the information received from the probe requiring the initial cable to be bifurcated outside the probe. These gratings are housed in an Ocean Optics S2000 spectrometer which is a 2048-element linear CCD array spectrometer. The two gratings used include the primary grating that views the visible and near-IR region (400nm – 860nm) and the slave grating that measures the UV range (200 nm – 400 nm) intensities. The spectrometer is connected to a computer through an Ocean Optics ADC1000-USB interface allowing integration times from 5 ms to 60 s.

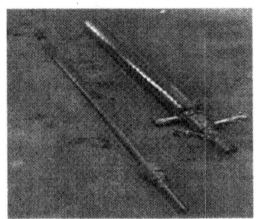

Figure 1: Disassembled Housing for Fiber-Optic Cable and Lens

EQUIPMENT SET-UP

The spectrometer was inserted into the pilot scale furnace such that the flame tip was head-on with the sensor. Standard hoses provide the water circulation used to cool the casing. A hose provides compressed air to prevent residue build-up on the lens with a built-in adjustment valve to allow the minimum amount of additional air to be added into the furnace. The set-up at the furnace can be seen in Figure 2. Figure 3 shows the set up for a CCD camera periscope to capture side view of the burner flame, as well as, a flue gas sampler at this furnace.

EXPERIMENTS

Experiments were conducted with changing burner stoichiometry by varying the Oxygen/Fuel (O/F) ratio. The O/F ratio is calculated by dividing the volumetric flows of O_2 by those of natural gas (NG). An O/F ratio of 2.1 is the nominal control value used at this furnace. Experiments were conducted at steady values of O/F ratios 1.8, 2.1, and 2.4 while changing NG flow rates. Additional experiments were conducted with varying O/F ratios. Another experiment was a step test were the O/F ratio was changed sharply from 1.8 to 2.4, allowed to settle for two minutes and then sharply dropped back to 1.8. Also, a ramp-up test was performed where the O/F ratio was gradually changed from 1.8 to 2.4 over a period of about one minute then sharply dropped back to 1.8.

Figure 2: Spectrometer Set-Up at Pilot Scale Glass Furnace

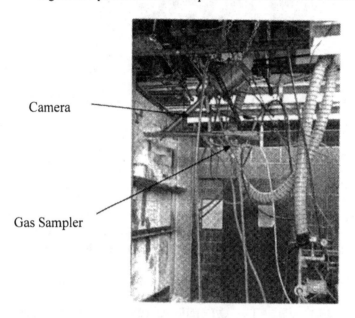

Figure3a: Camera (rear) and Flue Gas Sampler (front) Set-Up
at Pilot Scale Glass Furnace

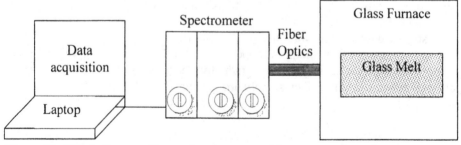

Figure 3b. Block diagram illustrating the spectral imaging and sensing system

Figure 4 shows sample side view flame images at three various O/F ratios in these experiments. A sample plot of spectral intensity as recorded by the spectrometer is shown in Figure 5. The dark line is the master grating that views the visible region and the lighter line is the slave grating that views the UV region.

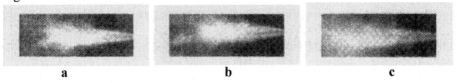

Figure 4: Flame images at (a) 2.4 O/F ratio, (b) 2.1 O/F ratio, c) 1.8 O/F ratio

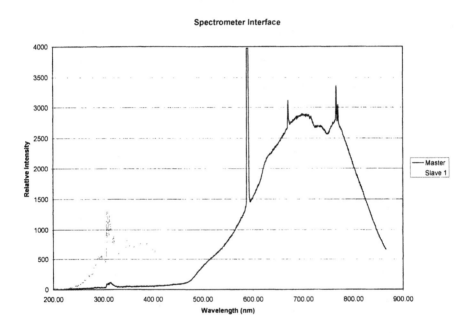

Figure 5: Sample Plot of UV and VI Spectrum

OH RESPONSE

As the O/F ratio changed, the emission spectrum in the UV range also changed as can be seen in Figure 6. Both the background intensity and the values of the emission spectrums changed as the O/F ratio increased. The background intensity increased as the O/F ratio increased (see Figure 6). Figure 7 shows the changes of the OH radical peak as the O/F ratio changes. The OH radical data is measured as a relative intensity by taking the peak intensity and subtracting the background intensity from it. The background intensity is found by averaging two wavelengths on either side of the peak that do not belong to the emission peak.

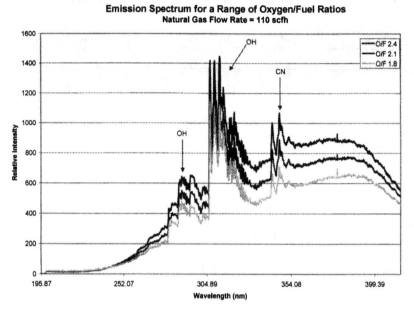

Figure 6: Emission Spectrum from 200 nm – 400 nm for Various O/F Ratios

As shown in Figure 7, the OH intensity increases as the O/F ratio increases for all three values of NG flow rates in this experiment. Figure 8 shows the relation between the NOx emission and O/F ratio for two set of natural gas (NG) flow rates. The dotted lower line represents the 2.27 m³/h natural gas flow rate. NOx emission is measured using a portable detector which measures the NOx content in the flue gas sample. NOx data in ppm represent a ten second data measurement. Even with the small size of the furnace and therefore large impact from small amount of air infiltration, the direct relation between O/F change and NOx emission is properly established. This relationship is also confirmed in separate experiments at commercial glass furnaces [9]. Comparison of Figure 7 and 8 clearly shows the variation of NOx with OH as O/F ratio changes from fuel rich to fuel lean condition.

Figures 9 and 10 show plot of OH peak intensity data as well as both the O/F ratio intended and measured values as function of time for ramp-up and step tests respectively. The OH radical follows the O/F ratio in the ramp-up test after applying a moving average process to the intensity. This process clearly shows the delay between the actual change and the measured response as can be seen in Figure 9.

The step test results are shown without any pre-processing of OH data. As shown in Figure 10, the OH radical follows the O/F ratio with a slight time delay and a few outlier data points.

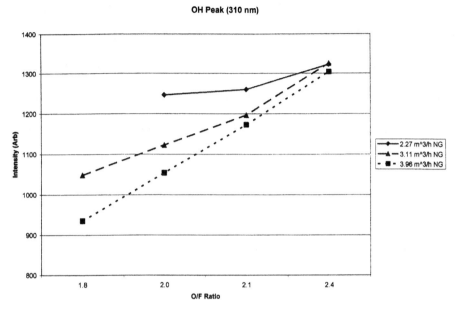

Figure 7: OH Peak Intensity Values

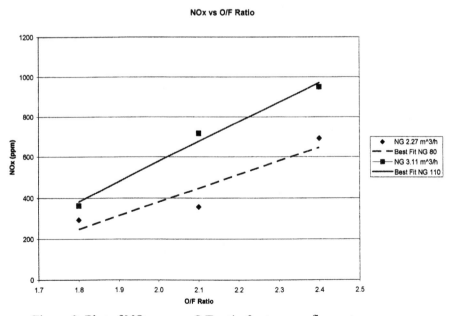

Figure 8: Plot of NOx versus O/F ratio for two gas flow rates.

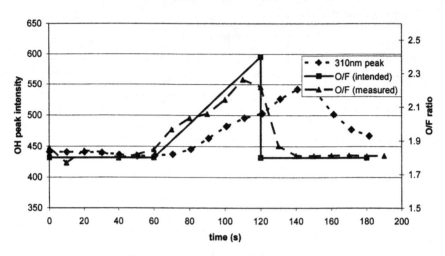

Figure 9: Ramp Test Results

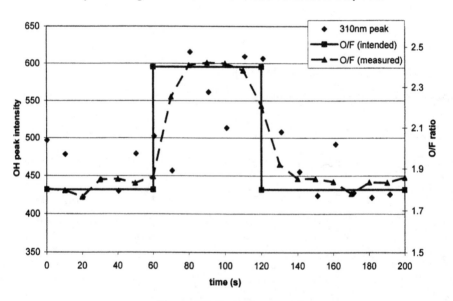

Figure 10: Step Test Results

CONCLUSIONS

The results from various combustion control experiments using a pilot scale glass furnace were presented in this paper. The objective was to obtain spectral data and flame images for oxy-fuel firing over a range of heat inputs and oxygen/fuel ratios. The heat rate to the furnace was varied from approximately 23 to 48 kW. The oxygen/fuel ratio was varied from 1.8 to 2.4. The purpose for conducting these experiments was to investigate the impact of O/F ratio on combustion radicals and products. The OH radical shows a very good response when O/F ratio is changed even under ramp-up condition. The results show the OH radical is certainly a good candidate for combustion behavior monitoring in oxy/fuel natural gas furnaces.

Furthermore, the results of NOx emissions as measured from the furnace flue gas sample show the expected variation with changes in O/F ratio even though the furnace size was small and hence sensitive to impact from air infiltration.

ACKNOWLEDGEMENTS

This work is supported by the Department of Energy under contract DE-FC07-00CH11032

REFERENCES

[1] Clausen S "Local measurement of gas temperature with an infrared fibre-optic probe" *Measurement Science Technology* **7** 888-896 (1996)

[2] Ewan B C R "A study of two optical fibre probe designs for use in high-temperature combustion gases" *Measurement Science Technology* **9** 1330-1335 (1998)

[3] Tong L, Shen Y, Ye L, Ding Z "A zirconia single-crystal fibre-optic sensor for contact measurement of temperatures above 2000 °C" *Measurement Science Technology* **10** 607-611 (1999)

[4] Grattan K T V, Zhang Z Y, Sun T, Shen Y, Tong L, Ding Z "Sapphire-ruby single-crystal fibre for application in high temperature optical fibre thermometers: studies at up to 1500 °C" *Measurement Science Technology* **12** 981-986 (2001)

[5] Elliot G S, Glumac N, Carter C D "Molecular filtered Rayleigh scattering applied to combustion" *Measurement Science Technology* **12** 452-456 (2001)

[6] Cessou A, Meier U, Stepowski D "Applications of planar laser induced fluorescence in turbulent reacting flows" *Measurement Science Technology* **11** 887-901 (2000)

[7] Kang H K, Bang H J, Hong C S, Kim C G 2002 "Simultaneous measurement of strain, temperature and vibration frequency using a fibre-optic sensor" *Measurement Science Technology* **13** 1191-1196 (2002)

[8] Fisenko A I, Ivashov SN 1999 "Determination of the true temperature of emitted radiation bodies from generalized Wien's displacement law" *Journal of Physics D: Applied Physics* **32** 2882-2885 (1999)

[9] http://www.oit.doe.gov/sens_cont/meetings_annual_0602.shtml

Waste Vitrification Programs

COMPLETION OF THE VITRIFICATION CAMPAIGN AT THE WEST VALLEY DEMONSTRATION PROJECT

R. A. Palmer and H. M. Houston
West Valley Nuclear Services Co.
10282 Rock Springs Road
West Valley, NY 14171-9799

A. J. Misercola
U. S. Department of Energy
West Valley Demonstration Project
10282 Rock Springs Road
West Valley, NY 14171-9799

ABSTRACT
    The West Valley Demonstration Project (WVDP) operated a slurry-fed ceramic melter to vitrify high-level nuclear waste (HLW) for the U.S. Department of Energy (DOE) from June 1996 until September 2002. A total of 68 batches of HLW were mixed with glass-forming chemicals to produce 275 canister of glass. The WVDP melter was put into service in October 1995 and radioactive waste was first added in June 1996. The final canister was poured in August 2002 and the melter emptied (using the evacuated canister method) and shut down in September 2002.

    Data collected and recorded during the vitrification campaign includes: waste form chemistry, radiochemistry and durability; canister fill-height; and the results of smear testing to detect contamination on a canister. Production Records for the canisters of vitrified HLW will summarize the data generated by the processes designed by the WVDP to comply with these Waste Acceptance Product Specifications (WAPS). This paper will review some of the data collected from the WVDP vitrification campaign and discuss some of the activities necessary for melter shutdown and melter disassembly.

INTRODUCTION AND BACKGROUND
    The West Valley Demonstration Project (WVDP or the Project) is a High-Level Waste (HLW) solidification and radiological cleanup project being conducted by the United States Department of Energy (DOE) at the site of the only commercial nuclear fuel reprocessing facility to have operated in the United States. As mandated by a Federal Law enacted in 1980, Public Law 96-386 (or the Act), one of the DOE's primary responsibilities at the 220-acre WVDP site in West Valley, New York is to safely solidify liquid HLW into a durable solid suitable for transport and disposal at an approved Federal repository.

    The HLW requiring solidification was generated from 1966, when the site was operated as a commercial reprocessing facility by Nuclear Fuel Services (NFS), until 1972 when NFS suspended reprocessing operations at the site. Most of the HLW requiring solidification resulted from neutralization of waste produced by using the plutonium uranium extraction (PUREX) process to extract useable product (plutonium and uranium) from spent nuclear fuel. A lesser amount of the HLW requiring solidification was produced by using the thorium extraction (THOREX) process to extract useable product from one core of mixed uranium-thorium fuel that was reprocessed at the site.

The overall method developed for processing HLW involved preparing the HLW for processing by separating (removing) the majority of the radioactive species (mainly Cs-137) from the waste liquid, combining the resulting HLW with the waste sludge, and processing the HLW mixture into the approved glass waste form. After removing HLW from the waste liquid, the resulting low-level waste (LLW) liquid also was processed into an approved waste form, cement encapsulated LLW. The basic overall method and systems used to process liquid HLW into approved HLW and LLW forms are illustrated in Fig. 1.

HLW Processing Flow Sheet

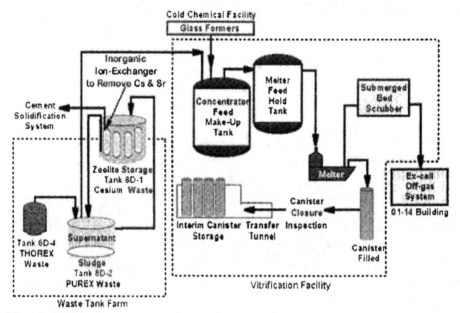

Fig. 1 HLW Process

Full-scale testing of the vitrification system design was accomplished through the Functional and Checkout Testing of Systems (FACTS) Program. Conducted from 1984 until 1989, this program provided the opportunity to evaluate process system, subsystem and component performance, confirm the HLW glass qualification approach and use the Vitrification System to produce high quality glass on a production schedule.

Thirty seven different system tests were performed and approximately 150,000 kilograms of glass were made during the FACTS program using non-radioactive isotopes in lieu of radioactive species to produce a waste glass as close as practical to the projected HLW. Following the final FACTS run, much of the equipment was disassembled for examination of test components and conversion for radioactive service, including the melter. A number of components that performed well during FACTS were reassembled for reuse, including the tanks used to prepare and feed slurry the melter, the Concentrator Feed Makeup Tank (CFMT) and Melter Feed Hold Tank (MFHT), as well as the facility used to prepare the glass former recipe to be blended with the HLW slurry, the Cold Chemical System (CCS).

HLW solidification was carried out in two phases. Phase I involved transferring the HLW slurry mixture from HLW tank 8D-2 to the CFMT where it was blended with batches of cold chemicals (including glass formers), concentrated, transferred to MFHT, continuously fed to the melter, and poured into stainless steel canisters to produce the final HLW glass form. During Phase I, which was conducted from July 1996 until June 1998, 211 canisters were made, for a production total of 436,546 kilograms of HLW glass. Following the same basic processing sequence, Phase II was conducted from July 1998 until August 2002, resulting in the production of 64 canisters, for a production total of an additional 132,756 kilograms of HLW glass.

The basic process used to produce canisters of HLW glass began by combining the radioactive species captured in the zeolite media during HLW pre-treatment with the HLW sludge, blending in recycled waste liquids resulting from off-gas treatment, and transferring the homogenous HLW slurry from tank 8D-2 to the CFMT, where the first step of HLW processing was initiated. This processing step involved taking slurry samples for chemical and radiochemical analysis to determine the exact glass-making recipe needed to process the slurry into HLW glass. While slurry samples were analyzed, the batch of slurry in the CFMT was concentrated to remove excess water. After the batch recipe was determined and a pre-mix of chemicals prepared, the pre-mix was transferred into the CFMT, mixed with the concentrated slurry and sampled again to ensure target glass composition. This feed preparation cycle was the most critical stage in HLW glass production because the time it took to prepare a feed batch had to be less than the time it took to transfer a full tank of feed from the MFHT to the melter, about 180 hours. The greatest portion of feed preparation time involved slurry analysis. Modifications made to improve batch preparation cycle time during both testing and early radioactive operations proved to be critical to the success of Phase I operations.

The next step in HLW processing involved transferring feed to the melter, allowing water to evaporate, salts to decompose, and remaining solids to calcine. Inside the melter, calcined wastes and glass-formers melted and fused into a glass pool where they homogenized. During melter operation, homogenized molten glass was periodically airlifted into a stainless steel canister held in position under the melter by the canister turntable -- a four-position, four-canister device that provides one position for filling, one position for canister removal and replacement, and two positions for filled canisters to cool.

After being allowed to cool, the filled canister was moved from the canister turntable to the weld station, where a stainless steel lid was welded onto the canister.

From the weld station, the welded canister was moved to the canister decontamination station, where the surface of the canister was decontaminated by chemical etching. Decontamination solution used was recirculated back into the melter feed. The final step in canister production involved moving decontaminated canisters from the Vit Cell to the High-Level Waste Interim Storage Facility (HLWISF) which is equipped with racks to hold the HLW canisters.

During the glass-melting process, steam, volatile elements evaporating from the glass pool, and feed particles entrained in the process off-gas were vented to the off-gas treatment system. The first component of this system, the Submerged Bed Scrubber (SBS), was used to quench the off-gas and remove particulate from it by scrubbing it through a submerged bed of ceramic spheres. Quenched off-gases were then drawn through a mist eliminator and high-efficiency mist eliminator (HEME) to remove mist and fine particulate. Scrubbed and treated off-gas was then heated and passed through a high-efficiency particulate air (HEPA) filter to remove particulate. Essentially free of radiological pollution at this point, the treated off-gas stream was directed through an underground trench to another building where final stages of HEPA filtering and oxides of nitrogen (NOx) abatement were completed before venting the treated off-gas through the Main Plant stack.

VITRIFICATION OPERATING EXPERIENCE

Systems and system hardware used in HLW processing and glass-filled canister production at the WVDP functioned beyond design expectations. Several aspects of system operation required close attention as canisters of HLW glass were produced. Melter operation and related functions affecting glass production played a major role. Specific factors that influenced system performance and experienced gained from managing events related to these factors during radioactive operations are described in the following discussions of melter operation and HLW processing that took place at the WVDP during HLW glass production.

The melter is designed to discharge molten glass by pulling homogenized bulk glass from the main melter cavity through an under flow drain (as illustrated in the cut-away view of the melter and turntable shown in Fig. 2). Glass is directed from the under flow drain up a sloped channel that rises to a trough. The high point of the channel is the melter overflow. Normally the melter is operated at a glass level of one to two inches below overflow, using the airlift to raise glass up to the trough where it then flows by gravity to the canister. The cavity holding the trough is independently heated to maintain glass temperature during pouring. Glass discharge from the melter is a recognized point of potential failure.

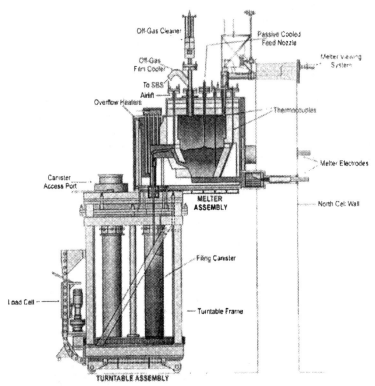

Fig.2 HLW Melter and Turntable

Trough wear, erosion, and corrosion caused by glass flow has often been observed in both commercial and waste melters. Beyond these factors, glass pouring itself can produce dripping and thin pour streams that solidify into glass structures. Formation of such structures can functionally disable the discharge chamber. For these reasons, the final design of the HLW melter included two separated glass discharge chambers, each one equipped with removable silicon carbide heaters. These heaters were used to keep the trough and pour spout hot enough to maintain glass viscosity. The heaters that were installed are remotely replaceable units that can be used in either discharge chamber. During melter operation, only one pour spout was heated while the second chamber was maintained unheated, preventing unwanted glass pours from the "cold" discharge chamber.

Experience with silicon carbide heater assemblies during FACTS operation indicated a service lifetime from six to 18 months, with an expected average service life of nine months. The first assembly failed after about 16 months of service. As a failed heater assembly represents a large, heavy block of highly contaminated waste, efforts were made to formulate an approach to increase their service life. The approach developed by WVNSCO engineers to extend service life was focused on reducing stress placed on the heaters by temperature transients. This was accomplished by providing back up power to the heater supplies and changing threshold limits on the control system to limit temperature swings. Effective reduction of stress on the heaters extended the service of the second set of heaters to

4 ½ years, after which the internal resistance in the silicon carbide heater bars used in heater assemblies increased until power supplies could not provide enough energy to heat the assembly to required temperatures.

A failure associated with operation of the discharge chamber occurred when the glass exit port plugged with glass. As previously described, problems with pouring such as the formation of drips can cause fibers to build up as glass is being poured. The drips form when melter pressure fluctuates. Damage to the pour tip also can cause pour problems by dividing the glass streams, each one acting like a drip. Although a direct cause of the exit port plug has not been identified, it may have been caused by the formation of drips. An examination of the discharge chamber is needed to determine the exact cause of the plug. (Figure 3 is a photograph of the failed trough taken after the melter was shut down.) As a result, the secondary discharge chamber had to be brought on-line with a fresh heater assembly near the end of radioactive operations. As part of the process of bringing the discharge chamber into service, an airlift tube was installed in the riser after heating the discharge chamber. Platinum was selected for the air lift tube (pipe) because it is impervious to the glass. The expectation was that it would last for the remaining life of the melter, as well as being thin-walled and flexible enough to bend as it slid down the sloped riser channel. However, flexibility proved to be detrimental in getting the tube down the riser as it bent and folded on itself (accordion style), instead of extending down the riser. Although the airlift was made operational, the design insertion depth was never obtained.

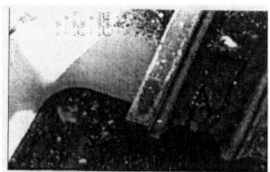

Figure 3. Discharge trough in WVDP melter

Heat up of the east discharge chamber was performed within a prescribed schedule. A slow ramp to temperature was used to limit stresses from different rates of thermal expansion within the refractory and the metal dam. Even with the slow ramp (heat up), cracking in the refractory caused air to leak between the discharge chamber and the main melter cavity. Air flow in the west discharge chamber was controlled by an orifice in the connecting jumper on top of the melter. This orifice was effectively bypassed in the east discharge chamber. Air apparently traveled past the airlift and through the seal between the lid and main cavity. This event indicates that additional work on discharge chamber seal design to reduce air flow is warranted.

Instrumentation and process lines enter the melter through openings (penetrations) in the melter lid. Mechanical design of lid penetrations is somewhat complex because linings used in the penetrations have to be durable enough to withstand remote installation of various services and the corrosive effects of melter off-gas. Original designs used heavy Inconel 690 tubes (pipes) welded into the lid with fragile, low density, castable refractory placed behind the liners to provide lid insulation. Corrosion of these liners during FACTS testing caused them to fail within three years and damage the lid refractory.

As a result, alumina liners were incorporated into the final design of the melter. Thermal stress on these liners during melter start-up caused them to crack, making it necessary to remove them and replace them with Inconel liners. The replacement liners included channels to draw air through the liners for cooling and flushing of corrosive vapors from the penetration region. As no damage was reported when periodic liner inspections were performed during the first two years of melter operation, subsequent inspections were limited to examination of liners that had to be removed to perform other maintenance tasks. Liner inspections performed after six years of melter operation showed only minor visible damage.

There is considerable variation in off-gas flow rates and the amount of particulate produced by an operating slurry-fed ceramic melter. During the early phases of vitrification program development, peak variability was measured at six times normal off-gas flow. This value was used as a design input for subsequent designs and behaved adequately as melter testing progressed. The final design for the WVDP melter used externally supplied air injection to make up the difference between nominal and peak off-gas flow rates. Steady melter pressure was maintained during radioactive operations by controlling it with an air injection control loop that served to dampen (filter) pressure signals. This made it possible to slow down responses to rapid fluctuations in melter pressure. Slower responses limited the possibility for over reaction to short term pressure spikes, providing for better control of melter pressure.

Figure 4. MFHT Agitator

Excess particulate exiting the melter off-gas system was managed by using a film cooler mounted near the off-gas penetration in the melter lid. Acting like a curtain of air, this device works by intercepting particulate in the off gas stream. Set in the melter off gas penetration, the cooler includes a series of louvers that cool and distribute air as it leaves the melter plenum and moves up the off-gas line. The final design for the film cooler is based on a series of designs developed at the WVDP and the Savannah River site to address plugging experienced during vitrification process development. This design uses a cleaner (brush) to periodically remove particulate by moving the brush from its resting position from above the film cooler down into the melter cavity and back up again. Used during early melter operation, brush operation diminished and was then discontinued because particulate build-up in the off-gas line was no longer observed. A plug did form in the off-gas jumper during radioactive operations, but not near the film cooler. The plug occurred at the bend in the jumper where flow is diverted downward. Plug formation was indicated soon after radioactive operations began by an observable rise in jumper differential pressure. Additional plug formation was mitigated by adding a water purge to the injection/control air run through the jumper. Although periodic water purges cleared away most of the particulate build-up, build-ups continued to occur such that water purges were far less effective at melter-shutdown than they were when first used.

The homogeneous slurry mixture needed to make HLW glass was prepared and fed to melter using two slurry tanks, the Concentrator Feed Make-up Tank (CFMT) and the Melter Feed Hold Tank (MFHT). Homogeneity of slurry mixtures held in each tank was maintained by using commercial agitators that had been modified for remote service. Although the cell was not equipped to perform preventive maintenance on these agitators, it was possible to add oil to the agitator gear box using in-cell equipment. Radiation resistant lubricants were specified for periodic additions of oil. Generally one frame size larger than standard motors, the agitator motors were specified with a 1.15 service factor but were not radiation resistant.

Both agitators failed after five years of radioactive service. Failure mode was identical for both agitators. Failure occurred within the flexible coupling, located between the motor and the gearbox, the metal grid holding the two mating hubs together (see Figure 4). The agitator manufacturer did indicate that grid failure occurs if the coupling is not greased periodically. When the motor housing was remotely removed from the coupling, a broken grid and hardened grease was found. As no method was available to replace the grid or the coupling, the entire agitator assembly had to be replaced. Blade wear was noticed on all blades after the agitators were removed from each tank. MFHT agitator failure occurred first because it was used on a full 24 hour cycle to keep feed slurry suspended during melter feed operations. Its blades showed the greatest amount of wear (see Figure 4, inset). Although the CFMT was used steadily, it was shut off when new batches of glass forming chemicals were being prepared.

A number of preparations had to be made so that the replacement agitators could be moved into the Vit Cell, including construction of a tipping frame used to fit the agitator assemblies through the shield door during transfer cart operation. Replacement was accomplished by bolting the agitator onto the tipping frame, placing the tipping frame onto transfer cart, and moving the transfer cart into the Vit Cell. Once inside the Vit Cell, the frame was tipped upright, allowing lubricating oil to be pumped into the agitator gear box before installation.

Electrical connections for each agitator were run through a rigid jumper with PUREX connectors. During agitator replacement, flexible jumpers were substituted for the original rigid jumpers because they are difficult to fit. Flexible jumpers also proved to be problematic because it was difficult to orient the jumper connection with the agitator. Flexible jumpers had to be made more rigid before completing assembly.

At the beginning of radioactive operations, four different techniques were developed to ensure that each canister produced was no less than 85% full. Weight measurement and level detection were the main techniques used to confirm canister filling. Mass balance calculations based on feed depletion and glass production, and glass pour viewing also were used to verify canister filling. Of all techniques used, level detection proved to be the most effective. Canister weight measurement proved to be the least effective. Canister level detection was accomplished using the Infrared Level Detection System (ILDS).

Canister weight measurement was derived from data generated by one of four load cells under each pour position on the canister turntable. Electronic summing of each load cells produced final canister weight. As canisters were rolled into position, movement on to and off of the weighing platform damaged the load cell assemblies, resulting in the need to make significant adjustments to achieve credible output. When it became apparent that the ILDS produced highly reliable results, canister weight measurement as a fill detection technique was abandoned.

The ILDS uses an infrared sensitive imaging radiometer (camera) mounted inside a shielded enclosure to detect areas of high temperature representing freshly poured, hot glass. Layering of glass from separate glass pours makes it possible to determine fill levels by detecting temperature differences. Level detection using the ILDS was so precise that it was possible to consistently produce canisters that were 90% full. The shielded enclosure that houses the ILDS radiometer is designed to extend the life of the unit, which is about one year in continuous service. An internal (Sterling) cooler was used to maintain the detection unit at cryogenic temperatures. The ILDS units used during radioactive operations lasted up to two years with intermittent service until the cooler failed. Shielding of other radiation sensitive components was sufficient to prevent radiation alone from causing ILDS failure.

Several efforts were made during pre-operational testing to enhance rates of glass production. One such effort made use of a technique known as "power skewing." This involves adjusting the amount of heat generated by each of the melter's three power circuits to change internal glass (natural) circulation rates, thus increasing heat transfer and glass production. Use of the technique was abandoned after various combinations of settings tested yielded insignificant results. However, measurable increases in glass production were noted near the end of the six-year period of melter operation, independent of any active effort. This was most apparent during the last year of melter operation when glass convection patterns and temperature profiles from the two control thermocouple sets were found to be substantially different from earlier operations. Normally, temperatures toward the bottom of glass pool are the lowest, increasing steadily until the cold cap is reached. In the last year of melter operation, temperature readings from the middle of the glass pool were lowest, with higher temperatures being recorded both above and below the middle of the glass pool. This new temperature gradient suggests a circulation pattern with heating from below. Changes in glass pool temperatures and circulation patterns resulted in glass production rates increasing from 35 to 40 kg of glass per hour to 45 to 50 kg of glass per hour.

Changes in heat distribution also were noted toward the end of melter operation. Such changes in heat distribution in the glass pool are believed to be attributable, in part, to formation of a conductive sludge layer that built up over time near the bottom of the melter cavity. Modeling during formation of the sludge layer showed that localized hot spots moved and were more intense as the depth of the sludge layer increased. The first noticeable change in melter operation was temperature instability. All of the glass thermocouples in one of the thermowells changed rapidly from 50 to 75 °C, either as an increase or decrease in temperature. Rapid fluctuations (temperature excursions) caused interruptions in melter power control. They also affected all parts of melter operation, making the entire process more difficult to control.

One of the core requirements for designing the facility was to incorporate it as a new system into existing structures and facilities. From a practical standpoint, this limited the amount of space available for constructing the cell. Although the final system design proved to be highly successful in supporting six years of radioactive operation, it restricted the ability to maintain and manage expended equipment and material inside the Vit Cell. As a result, several provisions had to be made to address the amount of expended equipment that was accumulating. These provisions included adding a maintenance station to perform simple maintenance tasks and developing a sorting and size reduction program to manage the amount of expended material resulting from routine in-cell operations more effectively. Considerable resources were dedicated to maintenance and housekeeping during radioactive operations.

MELTER SHUT-DOWN
The final batch of glass made in the campaign contained no waste from tank 8D-2 but was made up of residual radioactive material flushed from the vitrification system. The inventory of radioactivity contained within the entire primary processing system diminished with each flushing cycle. Eventually, a point of diminishing returns was reached, signaling the end of flushing effectiveness. At this time, the composition of residual molten material remaining in the melter (the primary system component used in glass production) consisted of a small quantity of radioactive material and large quantities of glass former materials needed to produce borosilicate waste-glass.

The last HLW canister was poured on August 14, 2002. At that time, the level of glass in the melter was the minimum required for routine pouring. The melter was placed in an idling mode while the Evacuated Canister System (ECS) was deployed to remove this radioactively dilute, residual molten material from the melter before system operations were brought to a formal end. Figure 5 shows the evacuated canister hovering in the vitrification cell after draining about half of the molten glass from the melter.

Figure 5. Evacuated Canister

The ECS consists of a stainless steel canister of the same size and dimensions as a standard HLW canister that is equipped with a special L-shaped snorkel assembly made of 304L stainless steel. Both the canister and snorkel assembly fit into a stainless steel cradle that allows the entire canister assembly to be balanced over the melter as molten glass is drawn out by a vacuum applied to the canister.

Specific features of the evacuated canister design include:

- Increased the thickness of the cylindrical portion of the canister as well as its top and bottom heads to maintain structural integrity during glass transfer;

- Reinforced ribs at the top and bottom heads of the canister to eliminate the possibility of buckling at high temperatures under a vacuum;

- A dam at the canister neck to prevent backflow during removal of the canister snorkel (lid) assembly; and

- A nickel-coated O-ring to buffer sealing surface anomalies and improve the seal between the canister and snorkel (lid) assembly.

Two of these canisters were prepared to empty the melter. A vacuum was drawn on the canisters and monitored for leaks for several weeks prior to use of the system. Once the rest of the facility was readied for final melter shut-down, the canisters were brought into the vitrification cell. The temperature in the melter was raised to about 1250°C to attempt to keep the glass molten for a period long enough to remove it from the melter.

A plug and a nozzle insert through which the snorkel would be inserted were removed from the top of the melter. The unit was then moved into position over the melter and held there as the melter was powered down. At this point, the snorkel was lowered into the melter and briefly held in position to allow the explosively-bonded plug at the snorkel tip to melt. Rapid heating of the unit was visible within three minutes, indicating the start of glass flow. The snorkel was then lowered to within one inch of the melter's top flange and held in position until observable change in dynamometer readings and unit appearance indicated the cessation of glass flow. After this was confirmed by checking dynamometer readings, the snorkel was lifted about 30.4 cm (1 ft) and held for several minutes to allow residual glass on the snorkel to drip back into the melter before recording the readings from the dynoamometers, removing the snorkel from the melter, replacing the temporary cover, and returning the filled unit to the rack. With the first unit filled and in place on the rack, the sequence was then repeated using the second unit to complete the glass-removal process.

The entire process was accomplished within one hour. After the fill-sequence was completed, each unit was lifted off the support-rack and set into an upright position near the melter to cool for a two-day period while the support-rack was loaded back onto the transfer cart and moved out of the cell. The final stage of evacuated canister operation involved removing the snorkels from the canisters and then processing them following standard system procedures for measuring, sampling, weighing, welding, and decontaminating canisters.

WASTE FORM PROPERTIES

Each batch was tailor-made to the federal repository's Waste Acceptance Product Specifications (WAPS)[1] according to the procedures prescribed by the Waste Form Compliance Plan (WCP)[2] and Waste Form Qualification Report (WQR)[3].

WAPS Specification 1.1 requires the reporting of the chemical composition of the glass waste form. (WVDP strategies for compliance with all of the specifications can be found in References 2 and 3.) All oxides present greater than 0.5 weight percent must be reported. The target glass composition and the range of oxide components are reported in Table I. Approximately 10% of the canisters of glass have been sampled and analyzed for those 15 oxides. (Data in the table are from the chemical analysis of shard samples taken from 27 canisters over the course of the vitrification campaign.)

Specification 1.2 is the Radionuclide Inventory Specification. Again, approximately 10% of the canisters have been analyzed and an average will be reported for the population. All radionuclides that have half-lives longer than 10 years and that are, or will be, present in concentrations greater than 0.05 percent of the total radioactive inventory, indexed to the years 2015 and 3115 must be reported. To meet this specification, only Cs-137 and Sr-90 will actually be measured on the waste form itself (from shard samples taken from the top of the canister). Other radionuclides will be calculated based on the known ratio to these more predominant species. Table II shows the reportable radionuclides for each year required. Figure 6 shows the Cs-137 and Sr-90 concentrations measured in the glass shards taken from 27 production canisters over the six-year vitrification campaign.

Table I. Range of Oxides in WVDP Glass

Oxide	Target Composition (wt%)	Measured Minimum (wt%)	Measured Maximum (wt%)
Al_2O_3	6.00	5.6	7.1
B_2O_3	12.89	11.2	14.8
CaO	0.48	0.21	0.6
Fe_2O_3	12.02	10.7	13.5
K_2O	5.00	4.1	5.3
Li_2O	3.71	3.3	4.2
MgO	0.89	0.7	1.3
MnO	0.82	0.7	0.9
Na_2O	8.00	7.1	8.6
P_2O_3	1.20	1.0	1.4
SiO_2	40.98	39.5	48.4
ThO_2	3.56	0.1	3.6
TiO_2	0.80	0.7	0.9
UO_3	0.63	0.1	0.8
ZrO_2	1.32	1.2	1.4

Table II. Reportable Radionuclides

Radionuclide	1996	2015	3115
Sr-90	YES	YES	NO
Cs-137	YES	YES	NO
C-14	NO	NO	NO
Ni-59	NO	NO	YES
Co-60	NO	NO	NO
Ni-63	YES	YES	YES
Se-79	NO	NO	NO
Nb-93m	NO	NO	YES
Zr-93	NO	NO	YES
Tc-99	NO	NO	YES
Pd-107	NO	NO	NO
Sn-126	NO	NO	YES
Cs-134	NO	NO	NO
Cs-135	NO	NO	YES
Sm-151	YES	YES	YES
Eu-154	NO	NO	NO
Ac-227	NO	NO	YES
Pa-231	NO	NO	YES
Th-232	NO	NO	YES
U-233	NO	NO	YES
U-234	NO	NO	YES
U-235	NO	NO	NO
U-236	NO	NO	NO
U-238	NO	NO	NO
Np-236	NO	NO	NO
Np-237	NO	NO	YES

Radionuclide	1996	2015	3115
Pu-236	NO	NO	NO
Pu-238	**YES**	**YES**	**YES**
Pu-239	**YES**	**YES**	**YES**
Pu-240	**YES**	**YES**	**YES**
Pu-241	**YES**	**YES**	NO
Pu-242	NO	NO	NO
Am-241	**YES**	**YES**	**YES**
Am-242m	NO	NO	**YES**
Am-243	NO	NO	**YES**
Cm-242	**YES**	NO	**YES**
Cm-243	NO	NO	NO
Cm-244	**YES**	**YES**	NO

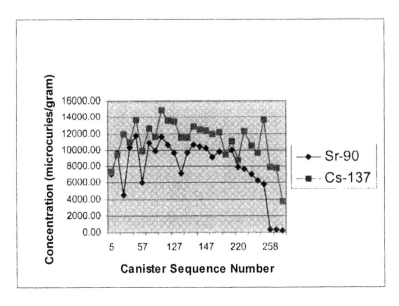

Figure 6. Sr-90 and Cs-137 as measured in the glass shards

Specification 1.3 is, essentially, the waste form durability specification. The glass must be shown to be more durable than a standard glass (known at the "EA glass") as measured by the Product Consistency Test (PCT). To show compliance with this specification, the WVDP predicts PCT results based on the chemical analysis of the production glass samples described for Specifications 1.1 and 1.2, and compares these predictions to measured EA glass data. The PCT predictions are based on a DOE-accepted regression model correlating measured PCT results to glass composition. The predicted normalized PCT releases for boron, lithium, and sodium are compared to measurements from the benchmark EA glass to demonstrate compliance with this specification. (See Table III.)

Table III. Product Consistency Test Results

Glass	Cation Release, mg/L		
	Boron	Sodium	Lithium
Benchmark EA glass	16,780	13,274	9,036
Production Glass Average (from Shard Analysis)	769	648	785

Specification 1.6 describes the International Atomic Energy Agency (IAEA) safeguards regarding the uranium and plutonium content of each canister. The total amount (in grams) as well as concentration (in grams per cubic meter) of ten of the isotopes of these elements will be reported along with the radionuclide content as described in Specification 1.2. Additionally, Specification 3.14 requires the reporting of the concentration of plutonium in each canistered waste form. The WVDP plans to comply with this specification using shards removed from the top of canistered glass (used to meet Specification 1.1), measuring the quantity of Sr-90 in the shard, and relating this value to the quantity of plutonium in the canistered waste form using scaling factors from the waste characterization program. This plutonium value will then be divided by the quantity of glass in the canistered waste form to generate the plutonium concentration value. This calculated plutonium concentration will be listed in the Production Records. Typical results of these calculations for these two requirements are reported in Table IV.

Table IV. Uranium and Plutonium Content for Canister WV-135

Radionuclide	Canister Content (g)	Canister Concentration (g/m³)	Ratio by Weight of Total Element (%)
U-233	4.59E+00	--	0.05
U-234	3.44E+00	--	0.04
U-235	1.67E-01	--	0.00
U-236	1.62E+01	--	0.17
U-238	9.61E+03	--	99.75
U Total	9.63E+03	--	100.00
Pu-236	5.67E-04	--	0.00
Pu-238	2.24E+00	--	1.08
Pu-239	1.67E+02	--	80.77
Pu-240	3.21E+01	--	15.50
Pu-241	3.42E+00	--	1.65
Pu-242	2.07E+00	--	1.00
Pu Total	2.07E+02	275.62	100.00

Specification 3.6 requires that the canisters be at least 80 percent full. As discussed above, the ILDS was used to ensure that the WVDP canisters would be at least 85 percent full. The actual average production value of all 275 canisters is 90.50 percent full. The official method for determining the fill height of each production canister is a measuring device which physically probes the height of the glass in several places after the canister has cooled and been removed from the loading turntable.

SUMMARY

Production of the vitrified high-level waste at the West Valley Demonstration Project is now complete. The Vitrification Facility has now been shut down and the HLW canisters are in temporary storage on site. The Production Records will be completed during the summer of 2003 and will be available for review by the regulators.

REFERENCES

[1]U. S. Department of Energy, Office of Environmental Restoration and Waste Management, "Waste Acceptance Product Specifications for Vitrified High-Level Waste Forms," U.S. Department of Energy Report, EM-WAPS, Rev. 2, (December 1996).

[2]West Valley Demonstration Project, "Waste Form Compliance Plan (WCP) WVDP-185", West Valley Nuclear Services Company (1996).

[3]West Valley Demonstration Project, "Waste Form Qualification Report (WQR) WVDP-186", West Valley Nuclear Services Company (1996).

REVIEW OF THE FRENCH VITRIFICATION PROGRAM

R. Do Quang
COGEMA
2 rue Paul Dautier
BP 4
78141 Velizy-Villacoublay Cedex, France

P. Mougnard
COGEMA
50440 Beaumont La Hague Cedex, France

C. Ladirat
CEA Marcoule
BP 171
30207 Bagnols sur Cèze, France

A. Prod'homme
SGN
1 rue des Hérons
78182 St Quentin Yvelines Cedex, France

ABSTRACT

The vitrification of high-level liquid waste produced from nuclear fuel reprocessing has been carried out industrially for over 20 years by COGEMA, with two main objectives : containment of the long lived fission products and reduction of the final volume of waste.

Research performed by the French Atomic Energy Commission (CEA) in the 1950's led to the selection of borosilicate glass as the most suitable containment matrix for waste from spent nuclear fuel and to the commissioning of the Marcoule Vitrification Facility (AVM) in 1978. In this plant, vitrified waste is obtained by first evaporating and calcining the nitric acid feed solution-containing fission products. The calcine is then fed together with glass frit into an induction-heated metal melter to have a glass throughput of 15 kg/h. Based on the industrial experience gained in the Marcoule Vitrification Facility, the process was implemented at an even larger scale in the late 1980's in the R7 and T7 facilities of the La Hague reprocessing plant. Both facilities are equipped with three processing lines having each a glass production capacity of 25 kg/h.

Consistent and long-term R&D programs associated to industrial feed back from operation have enabled continuous improvement of the process. For instance for R7/T7, the average melter lifetime now exceeds the design basis value by more than a factor of two (5000 hours instead of 2000 hours) and since 1996, melter are equipped with mechanical stirring in order to deal with a higher noble metals content and to increase the capacity of the vitrification lines.

So far, COGEMA's AVM (at Marcoule), R7 and T7 facilities (at la Hague) have produced more than 12400 high-level glass canisters, representing more than 4900 tons of glass and $163\ 10^6$ TBq. More than a technical success, in-line vitrification of HLW produced by operating reprocessing plants has become a commercial reality that led, in 1995, to the first return of glass canisters to COGEMA customers. Moreover AVM is now playing an important role by vitrifying HLW solutions coming from the decommissioning of COGEMA UP1 reprocessing facility.

To go a step further in the vitrification in order to obtain greater operating flexibility, to increase plant availability and to open new fields for high waste loading matrix, CEA and COGEMA have developed the Cold Crucible Induction Melter vitrification technology, where the heat is transferred directly to the melt with no impact on the melter itself. This technology is particularly well suited for the vitrification of La Hague highly concentrated and corrosive solutions coming from uranium/molybdenum fuel reprocessing, where the standard hot melter currently used in R7 will be remotely replaced by a cold crucible.

In this paper, the vitrification process currently operated in the COGEMA facilities will be described as well as the major milestones of French industrial HLW vitrification.

INTRODUCTION

Vitrification of high-level liquid waste is the internationally recognized standard to both minimize the impact to the environment resulting from waste disposal and the volume of conditioned waste. Many countries such as the USA, France, the United Kingdom, Germany/Belgium, Japan, Russia, have vitrified high level waste and several more countries are studying application of the vitrification technology

The first work on vitrification of radioactive waste began in France in 1957 at the Saclay nuclear center with early the selection by CEA (French Atomic Energy Commission) of .

- Borosilicate glass as the most suitable containment matrix for waste from spent nuclear fuel.

  Due to their amorphous structure, glasses appeared to be a category of materials capable of incorporating most of the fission product oxides in their vitrous network. By the mid 60s, borosilicate glasses were selected for the vitrification of HLW solutions as the best compromise in terms of containment (leach resistance, thermal stability, resistance to irradiation), technological feasibility, and cost (via the volume reduction factor). Today, borosilicate glasses have become a worldwide standard and have been chosen for nearly all vitrification processes of HLW solutions.

  The first high-level radioactive glass blocks, weighing 3 kg each and containing some 111 TBq, were fabricated at the Marcoule industrial site in 1965, in graphite crucible.

- Induction-heated vitrification technology : the obvious advantage of this solution is the simplicity of the joule heating of a metallic melter by using electric inductors and the fact that the heating system is outside the metallic melter (melting pot) and thus
  - Independent of the melting pot
  - In sensitive to glass melting (no wear, no corrosion, no shorting)
  - Not directly contaminated by High Active Level glass
  - Easy to start (even if the metallic melter is full or empty), to stop, to maintain or to replace.

  Induction technology development was conducted in parallel with glass formulation studies

The way from preliminary R&D choices to large scale industrial implementation has been and is always the result of combined approach which involves close links between research, engineering and operating team, and also judicious build up of results and experience.

As a consequence, COGEMA[*], CEA[**†] and SGN[***‡] have integrated a unique experience in the vitrification field of high-level waste comming from reprocessing activities through :

---

[*] Industrial Operator

[**] The French Atomic Energy Commission : COGEMA's R&D and R&T provider

[***] COGEMA's Engineering

- The design and operation of three industrial vitrification facilities with high records of safety, reliability and product quality;
- The design of various glass formulations including those used in the AVM, R7 and T7 facilities which, together, have produced up to now more than 12,400 glass canisters (corresponding to more than 160 10[6] TBq immobilized in 4900 tons of glass) ;
- Continuous efforts to improve at the same time the technology (from hot to cold crucible) and the associated matrix formulations, with constant emphasis on quality and volume reduction.

FROM R&D TO INDUSTRIAL VITRIFICATION FACILITY

The piver pilot facility

The first industrial-scale prototype unit in the world intended for vitrification of concentrated fission product solutions began operating in 1968 at Marcoule, in southern France.

The heart of the process was a tapered cylindrical Inconel vessel 35 cm in diameter and 3 m high, heated by a series (7) of superimposed inductors designed to supply power as necessary according to the level reached by the molten glass inside. It was a four-step batch process capable of vitrifying 200 liters of fission product solution in 100 kg of glass with a 25% loading factor.

The first step consisted in evaporation of a mixture of fission products in solution and glass frit in suspension in water, during which the pot heating level was adjusted as the vessel was filled.

In the second calcination-melting step, the nitrates in solution were decomposed and glass was formed. This was achieved by heating the pot from about 300°C to 1200°C by raising the induction voltage.

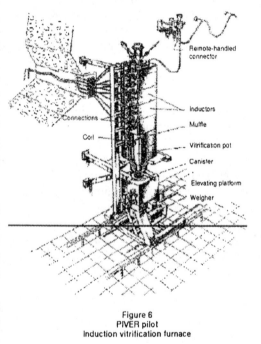

Figure 6
PIVER pilot
Induction vitrification furnace

The glass was then refined for 2 to 3 hours at 1200°C.

In the final step, the glass was poured into a canister simply by induction heating of a cold glass plug in the metal pouring nozzle.

The entire melter assembly was already designed to allow dismantling by means of an overhead crane and a few special tools. Thermocouples were installed on the vessel wall and inside the pot. The melting pot was insulated by a muffle; The muffle was also designed to protect the inductors from outflow of molten glass in the event of a pot failure. The melting pot and the muffle were designed as consumable items. The melter itself, consisting of the concrete-lined inductor blocks, was never replaced.

The PIVER pilot vitrification unit operated from 1968 to 1970, during which time it converted 25 m³ of HLW into 12 metric tons of glass poured into 164 canisters containing a total of $1,6 \times 10^5$ TBq. The facility was restarted in 1972 to vitrify very high-level solutions obtained by reprocessing fuel from the PHENIX fast breeder reactor, and produced another 500 kg of glass containing 4×10^5 TBq, representing a specific activity of 0,07 TBq per gram of glass- a record level at that time. PIVER was decommissioned in the early 1990s.

Industrial French vitrification design

Learning from the PIVER vitrification experience, the basic principles leading to the choice and design of the French industrial two-step vitrification process with hot induction metallic melter [1] are :

- The separation of the functions (calcinations/melting), to have simpler and more compact equipment and to limit the size of the melter, allowing complete in-cell assembly and disassembly with moderate size overhead cranes, master-slave manipulators and remote controlled tools.
- Easy remote maintenance of the process equipment with optimization of solid wastes generated during operation

Thus, in the two step process, the nitric acid solution containing the concentrated fission products solution coming from reprocessing operation is sent to a rotary calciner which assume evaporating, drying and calcining functions. Aluminum nitrate is added to the feed prior to calcination to avoid sticking in the calciner (melting of NaNO3). Sugar is also added to the feed prior to calcination to reduce some of the nitrates and to limit ruthenium volatility. At the outlet of the calciner, the calcine falls directly into the melting pot along with the glass frit which is fed separately. The melting pot is fed continuously but is batch poured. The melting pot is made of base nickel alloys; the glass in the melter is heated to a temperature of 1100°C and is fully oxidized.

Off-gas treatment comprises a hot wet scrubber with tilted baffles, a water vapor condenser, an absorption column, a washing column, a ruthenium filter and three HEPA filters. The most active gas washing solutions are recycled from the wet scrubber to the calciner.

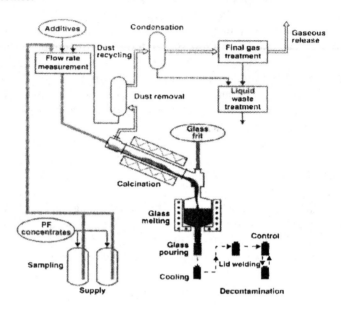

French industrial two step Vitrification process

FRENCH INDUSTRIAL VITRIFICATION FACILITIES

AVM : The World's First Industrial HLW Vitrification Facility

Process and technology developments performed in the Marcoule pilot facility led to the commissioning of the AVM (Marcoule Vitrification Facility) which was the first vitrification facility to operate in-line with a reprocessing plant. The AVM started active operation in June 1978.

AVM design

This facility was designed to meet the needs of the Marcoule UP1 reprocessing plant and has a nominal glass throughput capacity of 15 kg/h. Initially, the backlog of stored waste was vitrified. Then, the facility began in-line processing of the HLW solutions resulting from the reprocessing of GCR fuels and research reactor fuels at the UP1 plant.

The evaporative capacity of the calciner is about 40 l/h. The melter is fed continuously and is batch poured. The glass in the melter is heated up to a temperature of 1100°C. A glass batch is poured into a canister every 8 to 12 hours depending on the composition of the HLW solutions being treated. The canister, which has a volume of 150 liters, is filled with three batches of 125 kg each.

The melter of the AVM facility is made of inconel 601. It is cylindrical in shape with a diameter of 0.35 m and an overall height of 1,7 m. The melter is heated by induction. The temperature is controlled during the different operating stages by means of the power delivered through four inductor coils.

The AVM is a compact facility built around the hot vitrification cell containing almost all of the process equipment, including the fission products solution feeding system, the rotary calciner, the hot melter, the pouring station, the glass canister welding equipment, as well as the primary components of the off-gas treatment system (scrubber and condenser).

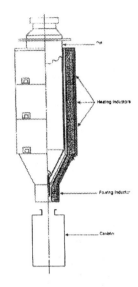

AVM Vitrification Melter

Moreover, solid wastes such as worn components are dismantled in the same cell and packaged into containers of the same size as the glass canisters. The interim storage is leaning with the vitrification cell.

All mechanical and process equipment subject to wear are designed for remote assembly and disassembly by means of a 20 kN overhead crane and master-slave manipulators. Replacements are introduced by the overhead crane from its hoist park outside the crane. Special remotely operated jumpers have been designed for radioactive fluid connections.

AVM Records of Operation

The AVM facility's records of operation by the end of march 2003 are given in Table I.

Table I : AVM records of operation

Weight of glass produced since start-up (metric tons)	1110
Number of canisters produced since start-up	2960
Total activity immobilized from start-up (10^6 TBq)	16,7

With an availability of about 70 % for only one processing line, AVM has first demonstrated the industrial maturity of the two-step continuous vitrification process.

During these 25 years of operation, no major difficulties leading to long-term shutdowns have been encountered. Nevertheless, all the important maintenance operations planned for the design stage of the facility have been performed : the metallic melter is periodically replaced after more than 7000 hour of operation; the calciner tube, its driving motor, the inductor coils, the scrubber and the condenser have been replaced. All these exceptional in-cell maintenance operations have proved the efficiency of remote maintenance when taken into account in the design of the process equipment as well as the facility's layout.

AVM robustness in 'non-standard' operating conditions

The UP1 plant stopped its reprocessing activity in 1999 and has been entering in a phase of rinsing and decommissioning. During these operations, the AVM facility plays an important role as it vitrifies all the HLW solutions produced by the decontamination of the UP1 equipment and as it continues to produce specified AVM glass.

The current strategy applied by COGEMA is to vitrify these decontamination effluents by diluting with in tank fission product solutions, previously stored before the end of UP1 reprocessing activities, and to continue like this to produce specified AVM glass.

The sensitivity of this approach lies in a sharp batch management of the flow to be vitrified, asking for a close cooperation between teams involved into UP1 decommissioning and AVM operators. Preparation of effluents to be vitrified is the key phase to succeed in respecting glass specifications (if needed the composition of the frit may be slightly adjusted) all the more as many nuclides are incorporated as close as possible of their limit so that the volume of waste generated by decommissioning operations can be limited.

The robustness and flexibility of the two-step continuous process as implemented at AVM facility has been demonstrated, AVM being operated successfully since 1999 in these 'non-standard' conditions (i.e. the physical properties of the treated solution, especially oxide contents, are different from the ones for which equipment have been initially designed and operated). One direct consequence of the dilution, for example, is to increase by a factor of two the time necessary to produce a glass canister. Thanks to the high availability of equipment, the COGEMA team operating the AVM facility is taking up the challenge. Till now, objectives in terms of schedule as well as in terms of number of canister produced have been reached

The AVM facility will continue operations until 2005.

R7/T7 : COGEMA's Modern Commercial HLW Vitrification Facilities

Based on the industrial experience gained in the Marcoule Vitrification Facility, the AVM vitrification process was implemented at larger scale in the late 1980's in the R7 and T7 facilities in order to operate it in line with the UP2 and UP3 reprocessing plants. Both vitrification facilities are equipped with three vitrification lines having each a nominal glass production capacity of 25 kg/h.

R7/T7 glass Formulation

The glass formulation was adapted to commercial Light Water Reactor fission products solutions, including alkaline liquid waste concentrates as well as platinoid-rich clarification fines. R7/T7 glass formulation was designed to hold, a maximum of 18.5 wt % of radioactive waste oxides (fission products, actinides, noble metals and Zr fines), or equivalently an overall maximum waste-loading ratio of 28 %. This limit was set to avoid excessive heating of the glass during interim storage. The glass product has a high activity (predominantly 137Cs, 90Sr) and significant amounts of noble metals (3 wt% max.). The maximum $\beta\gamma$ activity at the time of vitrification is of 28150 TBq per canister (each canister receiving about 400 kg of glass). The maximum contact dose rate of the canister at the time of production can be greater than 10^5 rad/h

In addition, since the glass withstands a high heat load due to its radioactive content, it has been specifically designed to avoid devitrification during interim storage. The maximum amount of crystals after severe heat treatment (with respect to nucleation and growth) does not exceed 5% in the presence of noble metals. This devitrification doesn't have any impact on glass durability [2].

Industrial glass samples coming from R7 and T7 facilities have been characterized [2]. Satisfactory quality of the glass has been demonstrated; glasses were homogeneous with no undissolved feed and their characteristics were in full agreement with the expected values.

The R7/T7 formulation is known worldwide to have an outstanding durability, especially in the long term. Normalized releases using a powder test very similar to the 7-day PCT are less than 1/10 of the US acceptability criteria.

Table II : From AVM To R7/T7 reference glasses
(FP: fission products, Act. : Actinides, MP : Metallic particles)

wt %	SiO_2	B_2O_3	Al_2O_3	Na_2O	Fe_2O_3	NiO	Cr_2O_3	FP	Act.	MP
AVM	40,0	16,6	10,0	16,5	0,9	0,2	0,4	3,2	0,4	0,7
R7/T7	45,1	13,9	4,9	10	2,9	0,4	0,5	10,4	2,7	1,6

The R7/T7 design

The R7 and T7 facilities were designed on the basis of the industrial experience acquired in the AVM facility. The AVM vitrification process was implemented at a larger scale in order to operate the R7 and T7 facilities in-line with the UP2 and UP3 reprocessing plants. Each facility was designed to have a glass throughput of 50 kg/h. A three line design was selected, each line being capable of producing 25 kg/h of glass. With two lines in service and one line on stand-by, each of the vitrification facilities could meet the production requirements and allow sufficient flexibility of operation.

To achieve the required capacity on each of the vitrification lines, the calciner was scaled-up (by a factor of about 2). The shape of the hot induction melter was also modified. Its base was made oval-shaped (long axis 1m; short axis 0,35 m; total height 1,4 m; weight around 400 kg) to limit the resistance to heat transfer from the walls of the melter to the feed materials falling at the top of the glass melt; is made of base nickel alloys.

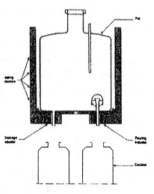

The glass is heated up to a temperature of 1100°C and is batch poured every 8 to 12 hours. Each canister is filled with two batches of 200 kg.

R7/T7 Induction melter

The off gas treatment (the wet scrubber, the condenser and the NOx column) has been also redesigned in order to be compliant with the waste treatment capacity of a vitrification line.

The decontamination factor after the wet scrubbers is greater than 10^6 and the total decontamination factor for off-gas treatment is of about 10^{10}.

Maintenance operations are fully integrated into the process design and method of operations, which is of utmost importance to minimize downtime and to increase availability for production. Thus, the process equipment of each vitrification line are located in a separate cell, while pouring and cooling cells are common to the three lines. All of these cells are equipped with cranes, master slaves manipulators and shielded windows for remote maintenance. They are associated with parking cells allowing crane maintenance as well as introduction of new replacement equipment. Pieces of equipment considered to be the least reliable are designed to be modular, so that their main sub-components are relatively compact and easy to replace remotely. The volume of secondary waste generated by maintenance operations is thus minimized.

R7/T7 record of operation

a. R7 Start-Up

R7 entered active service in June 1989 and began to treat the backlog of the HLW solutions that had been accumulated since the start of the first La Hague plant, UP2. These solutions included no clarification fines and no alkaline liquid waste concentrates, but represented an important volume (about 1200 m^3), nearly saturating the storage capacities for HLW solutions. Thus, the challenge for the R7 start-up was to rapidly reach nominal capacity.

From this point of view, the start-up was a complete success even though the operators had to cope with two significant problems :

- the melter lifetime was lower than the design basis value (2000 hours) thus requiring frequent replacements;
- the containment at the bellows assembly located between the pouring nozzle and the canister was not sufficient during the pouring stage thus leading to an increase of the activity in the cell ventilation filters.

In spite of these problems, fission products stock resorption and production goals were met thanks to the ease of maintenance.

b. T7 Start Up

The T7 facility is dedicated to the treatment of the HLW solutions produced by UP3 plant. It entered in active service in July 1992, 3 years after its twin facility R7.

The design of T7 took into account the feedback from R7 operations :

- implementation of a new connecting device between the pouring nozzle and the canister, associated with an improved off-gas system ;
- addition of a washable in-cell pre-filtering device on the ventilation line of the main hot cells (vitrification, pouring, dismantling) in order to protect HEPA filters from contamination ;
- modification of the cranes to improve the reliability of sub-components and to reduce maintenance;
- improvement of the canister decontamination device

Moreover, T7 operators had the unique opportunity to train on R7 before T7 active start-up.

As a consequence, T7 was able to quickly reach its production goals, and the improvements mentioned previously proved to be beneficial in terms of reduction of operating costs, reduction of the volume of waste, reduction of doses to personnel, and availability. For instance, the use of washable metallic pre-filters led to a reduction, by a factor of about 10, of the number of HEPA filter replacements.

c. R7 Upgrading

At the beginning of 1994, all of the backlog of fission product solutions from the UP2 plant had been vitrified in the R7 facility. Since HLW solutions produced in 1994 and 1995 would require at least one year of cooling time in order to let their radioactivity decay prior to vitrification, it was decided to interrupt operations and to upgrade the R7 facility to the same level as the T7 facility, by implementing all of the major improvements that T7 had benefited from before its start-up in 1992.

The main objective during the project was to minimize the doses to the personnel and the careful training on inactive mock-ups was an essential element to achieve this goal. The two others goals were to minimize the volume of waste and to comply with the deadlines in order not to interfere with the production schedule of the facility.

The work was conducted in two steps. During the first step, from February to June 1994, only one line was stopped, while the other vitrification lines continued to be operated. During the second step, from July 1994 to March 1995, all the lines were stopped.

All of the goals set at the start of the project were achieved : the doses to the personnel were 10 % lower than those estimated, waste volumes generated by the modifications as well as costs were very close to those projected, and R7 resumed operation in March 1995, 10 days earlier than planned.

d. Major On-Line Developments

Since the start of operations, the R7 and T7 facilities have demonstrated the industrial maturity of the French two step vitrification process. Nevertheless, COGEMA and CEA have been continuously improving its performance through consistent and long term R&D programs. All of the developments have been aimed at maintaining safety at its highest, improving availability, reducing secondary waste generated by operations, and minimizing operating costs.

The plant's layout and maintenance concepts, already described previously, have played an important role in meeting these different objectives. The fact that each facility has three lines in parallel has also enabled testing of most major developments before deployment on all the vitrification lines.

One of the major on-line developments undertaken has been the work performed to extend the melter's lifetime.

At the start of operations in the R7 facility, the lifetime of the oval-shaped metallic melter was less than the design basis value (2000 hours) due to the combined effects of thermal, and mechanical stresses as well as corrosion in the gaseous phase.

An important R&D program was launched. Comprehensive studies were performed in order to better understand the electromagnetic, thermal and mechanical behavior of the melter at the different stages of operation as well as the corrosion mechanisms at play. In particular, the power transfer from the inductor coils to the melter wall was analyzed in detail.

These studies helped to determine which species were responsible for corrosion. They also showed that the thermal power released in the melter's wall and therefore the temperature gradients in the melter could lead to high levels of stress as well as condensation of certain corrosive species. As a result of these studies, the design of the melter and the method of controlling the temperatures were modified.

These changes led to a sharp increase in the lifetime of the melter. At present, after 14 years of operation, the lifetime of the melter widely exceeds the design basis values with an average melter lifetime of 5000 hours corresponding to more than 150 glass canisters produced with a single melter.

Today R7/T7 use one melting pot per year and per vitrification line, that having a great impact on process downtime and secondary wastes downtime. Moreover, less than one week is necessary to stop a vitrification line, to change the melter and to restart the vitrification operation. The melting pot replacement is now based on preventive maintenance to change equipment or sub-equipment before the failure in order to minimize the number of component or sub-component to be changed and the level of component's contamination.

Another major development was the implementation, in 1996, of mechanical stirring in the melter in order to increase the noble metals content in the feed and at the same time maintain the throughput capacity of the vitrification lines.

The waste processed in the R7 and T7 vitrification facilities has a high content in noble metals which tends to lead to the precipitation of RuO_2 and metallic aggregates. The maximum specified values for the noble metals content in the R7/T7 glass is 3 %.

Initially, the melters were only equipped with bubble stirring devices which were efficient in limiting settling phenomena for moderately high values of the noble metals content. Mechanical stirring proved to be successful in achieving the objectives since the noble metals content in the glass was increased from an initial 1.6 wt. % to 3 wt. %, without any pouring problems or any metal accumulation in the bottom of the melter.

In fact, mechanical stirring also participate to melter life time performance with better temperature melter homogeneity.

2003 marks the 25th anniversary since the start of HLW vitrification in COGEMA's commercial reprocessing plants. Over this period of time, outstanding records of operation have been established by the R7 and T7 facilities. These are given in Table III below.

Table III : R7/T7 records of operation

Number of canisters produced since start-up	9480
Weight of glass produced since start-up (metric tons)	3792
Total activity immobilized since start-up (10^6 TBq)	146,2

R7 and T7 are mature vitrification facilities which vitrify in line fission products and fines coming from UP2 and UP3 reprocessing plants.

To improve flexibility in term of waste to be treated, in volume of glass to be produced by developing new high waste loading glass matrix, in glass throughput, to increase availability, and to reduce still more the secondary waste generated during operation, COGEMA has developed with the CEA and its engineering subsidiary SGN the Cold Crucible Induction Melter (CCIM) technology.

FROM HOT INDUCTION MELTER TO COLD CRUCIBLE INDUCTION MELTER

Basic Principles of the Cold Crucible Induction Melter

The cold crucible is a water-cooled induction melter in which the glass frit and the waste are melted by direct high frequency induction. The cooling of the equipment produces a solidified glass layer which acts as a protection against corrosion along the melter inner wall. Since the heat is transferred directly to the melt, and since the CCIM is water-cooled, there is practically no upper bound on the operating temperature with no impact on the melter itself. CCIM can then process melts that are either too corrosive, too viscous or too refractory for standard metallic or ceramic crucible melters. Some silicotitanate melts, for instance, were processed at temperatures of about 1500°C.

In addition , mechanical stirring of the melt, directly derived from those used in the presently operating La Hague facilities, guarantees homogeneity of temperature and composition and enables high throughputs.

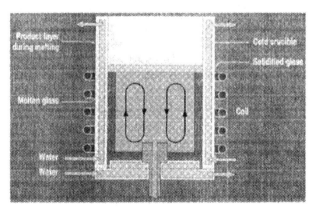

Cold Crucible Melter Technology

This technology developed by CEA and COGEMA since the 1980s, benefits from 25 years of HLW French industrial vitrification experience and ensures a compact design, a long equipment service life, and extensive flexibility in dealing with different types of waste.

It combines several favourable features :

- high temperatures in the melt ,
- short residence time in the melter,
- mechanical stirring of the melt,
- bottom pouring.

In addition to HLW conditioning, a vitrification process based on the CCIM can also be used for other types of radioactive (low level, intermediate level, transuranic) or mixed waste with varied chemical and physical waste forms including liquids, slurries, sludges, solids and mixtures of these physicochemical forms.

Indeed, the introduction of this technology has opened new perspectives in term of waste formulation and enabled the CEA to design a totally new range of matrices for very varied applications [3].

The UMo Project

COGEMA is committed to condition some hard-to-process legacy solutions, coming from the reprocessing in the seventies of spent Mo-Sn-Al fuel (used in gas cooled reactor), which are still stored in La Hague [4]. The main features of these High Level Liquid Wastes called « UMo solutions » are to have high molybdenum and phosphorus contents (about 90 g/l in MoO_3 and 15 g/l in P_2O_5), and to be less radioactive than the current fission products coming from actual reprocessing activities.

Such high amounts of molybdenum cannot be accommodated :

- in the present R7T7 glass formulation, because its solubility is limited to about 4 wt % in oxides (At higher values a water-soluble segregated yellow molybdate phase may appear),
- with standard vitrification technologies due to its unacceptable corrosive power at high temperature.

That is why the basic data associated to the UMo project were the following :

- development of a specific new glass formulation with a maximal MoO_3 waste loading (higher than 10 wt %) in order to minimize the UMo vitrification campaign duration (a 5 year-period corresponding to the production of less than 800 glass canisters) and to reduce the final waste storage volume,
- implementation of a CCIM in one of the three R7 vitrification line with conservation of the 2 step-vitrification process design, and glass throughput (25kg/h),
- use of the glass canister design used in R7/T7 to optimise the waste standardization management from the vitrification line to the interim storage.

A specific high temperature glass-ceramic formulation (a calcium and zirconium-enriched alumino-boro-silicate matrix) able to incorporate a nominal value of 12 % of MoO_3 and elaborated at 1250°C has been developed by the CEA and has been qualified through lab and pilot testing.

Table IV : UMo glass-ceramic composition

wt %	SiO_2	B_2O_3	Al_2O_3	Na_2O	P_2O_5	MoO_3	ZnO	ZrO_2	CaO
UMo	36,0	13,0	6,2	8,8	3,7	12	5,6	7,1	5,7

The process and the associated technologies have been qualified in parallel on a full-scale pilot prototype to ensure maximum representativeness of the test conditions and to allow accurate determination of directly applicable process control parameters. The CCIM developed in this purpose, is a compact one-piece unit of 950 kg, with 650 mm in diameter, and is equipped with cooled retractable stirring device.

On the full-scale pilot the following process features were investigated :

- the calcining conditions for UMo solutions were adjusted according to their particular composition (high Mo and P concentrations);

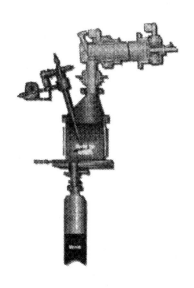

UMo's CCIM

- the vitrification parameters including, stirring parameters were optimised to ensure glass thermal homogeneity while maintaining a cold cap layer on the surface of the melt : despite the high temperature of the molten glass (1250°C), this layer limited the quantity of volatilized material;
- the process control parameters were determined for normal operation and for transient conditions (water ingress, replacement of an instrumentation probe, etc.);
- the decontamination factors were determined for the equipment items and compliant with La Hague standard;
- the wall temperature reached by the glass canister on pouring allowed the use of R7T7-type glass canisters.

Engineering studies (basic and details design, nuclearization, safety assessment) have been performed in parallel in order to replace remotely a hot induction melter by a CCIM in one of the R7 vitrification cell and to operate it. It constitute a challenge for engineering as all modifications :

- have to be performed remotely (the surrounding dose rate in the vitrification cell is about 50 Gy/h),
- must also be compatible with actual layout of the cell to minimize the modifications of the existing equipment (calciner, off-gas treatment system, glass pouring system).
- must be as short as possible to minimize the down-time of the production mean.

The UMo project is now engaged in a phase of exchange with safety authorities, in order to get the approval on the glass specification.

The vitrification of UMo solution is, in accordance with La Hague vitrification plan management scheduled in 2008.

CONCLUSION

2003 marks in France 25 years of industrial operation of HA vitrification facilities, demonstrating the success of the French vitrification program.

From Piver to T7, feedback from hot operations and the long-term R&D programs conducted jointly with the CEA have helped to continuously improve the vitrification process in all of its aspects (glass formulation, process, associated technologies, operations and maintenance).

AVM, the first HLW industrial vitrification facilities in operation in the world is now playing an important role by vitrifying decontamination effluent coming from the rinsing and decommissioning of the UP1 reprocessing plant.

The R7 and T7 vitrification facilities, in-line with COGEMA's two major commercial La Hague reprocessing plants, have had outstanding records of operation, not only from the standpoint of total glass production and plant availability but also with respect to safety (doses to personnel), remote in-cell maintainability, and secondary waste generated.

The next major milestone in the evolution of the vitrification process will be the deployment of the Cold Crucible Induction Melter technology in the R7 facility. The technology will be applied to the vitrification of highly concentrated and corrosive solutions produced by Uranium/Molybdenum fuel reprocessing which would not be possible with standard technologies.

From a more general point of view, introduction of high temperature technology open new field of operation [5] to deal with various waste compositions and new perspectives in term of waste formulation.

References

1. R. Do-Quang , JL. Desvaux, P. Mougnard, A. Jouan, C. Ladirat - " COGEMA experience in operating and dismantling HLW melter", ACERS 2002, St Louis

2. P. Cheron, P. Chevalier, R. Do-Quang, M. Senoo, T. Banba, C. Fillet, N. Jacquetfrancillon and al - "Examination and characterization of an active glass sample produced by COGEMA", MRS proceedings Vol.353-1995

3. V. Petitjean, C.Fillet, R. Boen, C. Veyer, T. Flament - "Development of Vitrification process and glass formulation for nuclear waste conditioning", Waste Management 2002, Tucson

4. R. Do-Quang, V. Petitjean, F. Hollebecque, O. Pinet, T.Flament, A.Prod'homme - "Vitrification of HLW produced by uranium/molybdenium fuel reprocessing in COGEMA's Cold Crucible Melter", Waste Management 2003, Tucson

5. R. Do Quang, A. Jensen, A. Prod'homme, R. Fatoux, J. Lacombe - " Integrated pilot plant for a large Cold Crucible Induction Melter", Waste Management 2002, Tucson

EXAMINATION OF DWPF MELTER MATERIALS AFTER 8 YEARS OF SERVICE

Daniel C. Iverson, Kenneth J. Imrich, Dennis F. Bickford, James T. Gee, Charles F. Jenkins and Frank M. Heckendorn
Westinghouse Savannah River Company
Aiken, SC 29808

ABSTRACT

The first Defense Waste Processing Facility high level radioactive waste glass melter was successfully operated for eight years. Recent failure of melter heaters and decrease in glass production necessitated its removal. Prior to removing the melter from the facility, a remote in situ visual inspection of the refractory and Inconel™ 690 components was performed. The vapor space and glass contact refractory blocks were in excellent condition, showing little evidence of spalling or corrosion. Inconel 690 top head components and lid heaters in the vapor space were also in good condition, considering the service. Upper electrodes experienced significant deflection, which probably resulted from extended operation in excess of 1150 °C. Condition of the melter components examined during the remote visual inspection is summarized in this paper.

BACKGROUND

Operating History

Approximately 136 million liters (36 million gallons) of highly radioactive liquid waste from the production of nuclear materials at the United States Department of Energy's Savannah River Site (SRS) are presently stored in large underground carbon steel tanks. SRS is currently in the process of tank farm closure and is dispositioning the inventory of high level waste. One of the technologies developed at SRS for immobilizing the waste to allow for controlled decay of long-lived radionuclides is vitrification. The process begins with the transfer of liquid high level waste sludge to the Defense Waste Processing Facility (DWPF). Where the chemistry is adjusted by addition of nitric and formic acids. Borosilicate frit is then added and the slurry is concentrated until a solids content of 45 to 50 wt% is attained. The slurry is finally fed to a melter where it is melted. The resultant high level waste glass is poured into stainless steel canisters, which are subsequently sealed.

DWPF's first glass melter began operation in May 1994 and was shut down for replacement in November 2002. It operated continuously for over eight years, including 2 years of non-radioactive cold chemical operations followed by 6 years of radioactive waste processing. Over 1400 canisters, 2.4 X 10^6 kg (5.2 million pounds) of glass, have been successfully poured. This represents about 27 percent of the total glass to be produced in this facility.

The DWPF Melter (Figure 1) is a refractory lined cylindrical vessel. Heat is provided via Joule heating of the glass and by resistance heaters located in the melter plenum, riser, pour spout, and drain valve. Internal dimensions are approximately 1.83 m (6 ft) diameter by 2.2 m (7 ft) high. Nominal glass depth is 86 cm (34 in). Glass contact material is Monofrax™ K-3, a fused-cast chromia-alumina refractory. Vapor

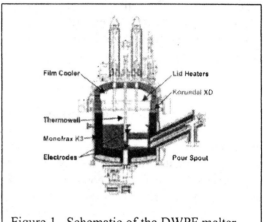

Figure 1. Schematic of the DWPF melter.

space refractory is Korundal™ XD. Metal components in contact with the glass and vapor space are Inconel™ 690. These include thermowells, a bubbler, off gas film coolers, lid heaters, borescopes, and electrodes.

The melter was operated at a glass temperature of 1050 to 1200 °C. Lid heaters were maintained at 975 °C, which kept the vapor space at 500 to 900 °C during feeding and idle conditions, respectively. An air purge of 204 kg/hr (450 lb/hr) was maintained into the plenum to promote combustion of organics in the feed. Glass production rates varied from 36 to 91 kg/hr (80 to 200 lbs/hr). Nominal high level waste melter feed composition is shown in Table 1.

Prior Inspections

Following completion of cold chemical runs in 1996, an extensive visual inspection of melter top head and off gas components was performed. The inspection focused on vapor space and molten glass attack. No significant degradation was observed, except for the chloride salt attack and oxidation of the borescope (Ref. 1) and minor oxidation

Figure 2. Film cooler suspended by canyon crane. Inlet (bottom) shows minor attack.

of the tip of the off gas film cooler (Figure 2).

During early radioactive operation, glass pouring problems led to remote visual inspection of the melter pour spout. This was accomplished by the use of a high temperature remotely operated borescope. Examination revealed significant material loss on the glass contact side of the pour spout (Figure 3). This figure

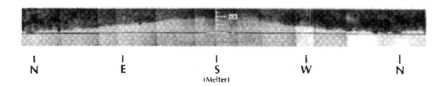

N	E	S	W	N
		(Melter)		

Figure 3. Panoramic view of the Inconel 690 pour spout showing 6.4 mm (0.25 in) metal loss on the glass disengagement point (S)

represents a 360 degree panoramic view of the pour spout bore, showing the extent of material loss, estimated to be 6.4 mm (0.25 in) at the glass disengagement point at that time. The need to reestablish desired glass flow characteristics led to the development of a remotely deployed replaceable pour spout insert. Subsequent metallurgical examination of a degraded insert established a nominal corrosion rate 4.78 mm/yr (0.188 in/yr) and the mechanism for the observed attack (Ref. 2).

Table I. Nominal Melter Feed Composition

Analyses		Analyses	
Total Solids	48.40 wt%	Calcium	0.998 wt%
Calcined Solids	43.64 wt%	Chromium	0.065 wt%
Density	1.395 g/mL	Copper	0.022 wt%
pH	6.8	Iron	8.815 wt%
TOC	8401 ppm	Potassium	0.136 wt%
Formate	26792 ppm	Lithium	1.492 wt%
Chloride	< 1,050 ppm	Magnesium	1.419 wt%
Fluoride	< 1,050 ppm	Manganese	1.172 wt%
Nitrate	12391 ppm	Sodium	8.010 wt%
Nitrite	< 1,050 ppm	Nickel	0.438 wt%
Phosphate	< 1,050 ppm	Silicon	22.911 wt%
Sulfate	< 1,052 ppm	Titanium	0.036 wt%
Aluminum	2.454 wt%	Uranium	2.999 wt%
Boron	2.325 wt%	Zirconium	0.073 wt%

The melter center thermowell was removed in May 2000 and visually inspected. It experienced slight melt line attack, but otherwise appeared to be in good condition. The Inconel 690 dip tube bubbler melt level detection device was removed in April 2002 after approximately seven years of service. It originally had a 7.62 cm (3 in) outer diameter with a 1.27 cm (0.5 in) diameter center bored hole through which argon was bubbled at 0.014 to 0.028 m^3/hr (0.5 to 1 scfh). Remote video inspection revealed that through wall attack had occurred on one side over approximately 76.2 cm (30 in) of the submerged length ending just below the melt line (Figure 4). This relatively high rate of material loss can be attributed to the "scrubbing" effect of continuous bubbling. A portion of the tip, approximately 15 cm (6 in), was completely lost.

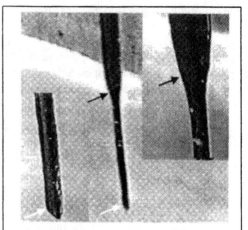

Figure 4. Photograph of bubbler showing necking at melt line and through wall attack between arrows.

Near the end of melter life the zone, 1 drain valve heater failed. This strip heater was fabricated from Inconel 690. The component could not be inspected after failure, so the reason for failure is unknown.

Melter End Of Life Inspection

Following melter shut down, an inspection was performed using a remotely deployed video camera suspended 12.2 m (40 ft) below the facility remote operated crane (Figure 5). This camera, called Mini-Sputnik, is a remotely controlled, video-based viewing system developed at SRS for inspecting tanks, vessels, and other difficult to access areas in the DWPF Canyon. The system was designed to be deployed in a vessel through a 6.4 cm (2.5 in) opening to a depth of 1.83 m (6 ft) using the canyon crane. An operator in the crane control room controls all system functions

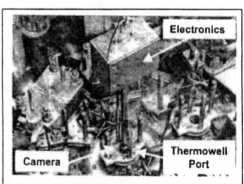

Figure 5. Mini-Sputnik being lowered through the center thermowell port on the melter.

including lights, pan-tilt-zoom and focus. Because the system is NTSC video based, resolution was limited.

Refractory Materials

The glass contact refractory, Monofrax™ K-3, was in excellent condition. Minimal spalling was observed in several localized regions around the melt line. It was significantly less than that observed in pilot scale melters. Refractory attack was expected to be greatest in the vicinity of the melt line, but only minimal corrosive attack was observed (Figure 6 and 7). Corners of the refractory blocks exhibited some rounding; however, the joints were still square indicating that no significant shifting of the refractory blocks had occurred.

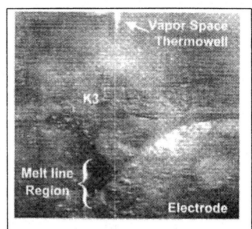

Figure 6. Photo showing condition of K3 blocks near melt line.

Gross material loss was low. This qualitative observation was based on relative distance between the back face of the upper electrodes and the refractory wall. This represents a refractory corrosion rate significantly lower than that predicted by the initial pilot scale work. Melter design life was 2 years based on the pilot scale work and a corrosion allowance of 4 inches of refractory. The present results are better than those seen in the Large Slurry Fed Melter and Scale Glass Melter at SRS (Ref. 3).

The vapor space refractory, Korundal™ XD, was in excellent condition (Figure 7). No missing pieces or shifted blocks were seen. No signs of chipping or spalling were observed around the top head penetrations or dome heaters. The entire refractory surface was glazed.

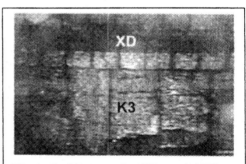

Figure 7. Photograph of Korundal XD refractory in the vapor space.

Inconel 690 Components

The lower electrodes were not visible due to residual glass in the melter. Upper electrodes were completely visible and exhibited rounding of corners and edges below the melt line and an obvious downward deflection (Figure 8). The

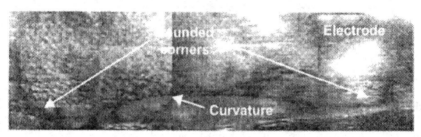

Figure 8. Upper electrode, Top photo: Upper portion showing downward deflection. Bottom photo: Composite showing curvature and rounding of corners on the bottom portion.

deflection was probably a result of creep. The faces of the electrodes exhibited an irregular surface morphology; however, it was not clear whether it was due to localized material loss or build-up of deposits.

The upper electrodes were originally 101.6 cm (40 in) wide by 38.1 cm (15 in) high by 7.62 cm (3 in) thick. The 215.5 kg (475 lb) electrode was supported by a centrally located 10.2 cm (4 in) diameter bus and was mounted flush against the refractory wall. Downward deflection of the ends was significant and was symmetric around the bus (Figure 9). Initially it was thought that the observed shape was due to material

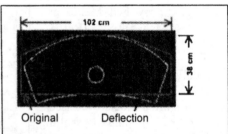

Figure 9. Schematic of the upper electrode showing the original shape and the observed deflection.

loss resulting from corrosion. A more detailed examination revealed that the shape was due to deflection rather than corrosion. This is based on the following:

1) Normal corrosion mechanisms cannot explain the concave shape of the bottom of the electrode.

2) The upper portion shows curvature, yet the edges appear sharp suggesting they were not significantly corroded.

3) The overall appearance suggests uniform symmetric slumping about the axis of the electrode bus. The appearance is similar to that expected under high creep rate conditions.

Existing high temperature creep data (Ref. 4, 5) show that at DWPF operating temperatures significant slumping is not expected. However, these data were generated using materials that had not been exposed to melter environments or at extended times (years). A change in mechanical properties due to long term elevated temperature exposure may explain the observed deflection. This supposition is based on an earlier metallurgical examination of the Inconel 690 drain valve exposed to 1150 °C glass for 7 years in the Integrated DWPF Melter System (IDMS). An extensive change in morphology was noted, including internal void formation, extremely large grains, and precipitates on grain boundaries through the entire microstructure (Figure 10). Electrode temperature may also have been a factor. This is based on modeling, which shows that higher

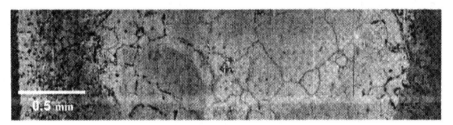

Figure 10. Photomicrograph showing the microstructure of an Inconel 690 through-wall section of the IDMS drain valve.

temperatures can be expected at the electrode face than in the bulk glass. Another factor is that the DWPF melter was at times operated up to 50 °C higher than the 1150 °C design temperature. (The melting point of Inconel 690 is 1345 °C.)

Off Gas Film Coolers

Both the primary and backup off gas film coolers were in good condition (Figure 11). Heavy deposits were observed on the tips that penetrated into the plenum but the contour of the lower lip was still visible. A small separation in a weld along the lower edge of the backup film cooler was observed. In the IDMS, a one tenth scale pilot melter, the film cooler was replaced after approximately 6 years of operation because several inches of material from the lower portion were lost.

Figure 11. Primary (A) and backup film coolers (B). Weld separation on the lower lip of backup film cooler is evident.

Thermowells

The vapor space thermowell was in excellent condition, with no visual evidence of degradation even at the refractory/plenum interface (Figure 12). The side thermowell, which was partially immersed in the melt pool, experienced necking and loss of material as a result of exposure to the molten glass. This thermowell was still functioning after 6 years of service. This component was replaced prior to the beginning of radioactive operations.

Figure 12. Vapor space (A) and side (B) thermowells. Molten glass attack in the immersion zone did not affect operability of side thermowell after six years of service.

Lid Heaters

The melter lid heaters (vapor space heaters) consist of eight resistance heated Inconel 690 tubes. They are 8.3 cm (3-1/4 in) OD by 1.27 cm (1/2 in) wall by approximately 1.65 m long (Figure 13). One of the heaters had failed (stopped passing current) prior to melter shutdown but the cause of failure was not evident during the inspection. Aside from that failure, the condition of the heaters was better than expected. They showed only minor deflection (upward in some cases) and no evidence of significant corrosive attack (Figure 13). This is supported by the operational history, which did not indicate a significant change in heater resistance over time. Slight upward bowing of the tubes may be attributed to differential thermal expansion between the refractory wall and the outer supporting cooled melter shell. Significant feedstock deposits were noted at the ends of the heaters. However, no evidence of hot spots or other damage was associated with the buildup. It is felt that the overall good condition of the heaters can be attributed to careful control of heater temperature during operations.

CONCLUSIONS

Based on the results of the remote visual inspection of the DWPF melter following eight years of operations it is concluded that:

- Overall, the melter was in better condition than expected based on earlier pilot melter studies.

- Loss of refractory due to spalling and corrosion was minimal.

- Corrosion of Inconel 690 components was no more severe than predicted by pilot scale melter and laboratory experience.

- Lid heaters showed no significant degradation or deflection.

- It could not be determined why one of the four lid heater circuits had failed.

- Dip tube bubbler showed significant degradation, which was attributed to "scrubbing" effect.

- Significant deflection of the upper electrodes was observed. This may be attributed to changes in material properties and operating at temperatures approaching the melting point of the material, 1345 °C. Risk of damage to Inconel 690 components is increased with increasing temperature above 1150°C.

ACKNOWLEDGMENTS

The authors would like to express appreciation to the DWPF remote crane operators for their support and skilled crane operation. This document was prepared in connection with work done under Contract No. DE-AC09-96SR18500 with the U. S. Department of Energy.

REFERENCES

[1]K.J. Imrich, Degradation of a N06690 Borescope in a Radioactive Waste/Glass Melter System, Heat-Resistant Materials II, ASM International, 1995.

[2]K.J. Imrich and D.C. Iverson, Metallurgical Evaluation of an Inconel 690 Insert from a Radioactive Waste Glass Melter Pour Spout, Environmental Issues and Waste Management Technologies, Ceramic Transactions, Vol. 107, American Ceramic Society, 2000.

[3]D.C. Iverson and D.F. Bickford, Evaluation of Materials Performance in a Large-Scale Glass Melter after Two Years of Vitrifying Simulated SRP Defense Waste, Materials Research Society Symposium Proceedings, Vol. 44, 1984.

[4]V. Venkatesh and H.J. Rack, Influence of Microstructural Instabilities on Elevated Temperature Creep Deformation of Inconel 690, Materials Science and Technology, Vol. 15, 1999.

[5]V. Isoard, Thermo-Mechanical Stability of Inconel 690 at Elevated Temperatures, M.S. Thesis, Clemson University, 1995.

TESTING TO DEMONSTRATE REGULATORY COMPLIANCE OF GLASS WASTE FORMS FOR IMMOBILIZATION OF RADIOACTIVE WASTES AT THE HANFORD SITE

David B. Blumenkranz, P.E.
Washington Group International
2435 Stevens Center Place,
Mail Stop: MS4-C1
Richland, WA 99352

Sam Kelly
Washington Group International
2435 Stevens Center Place
Mail Stop: MS1-B
Richland, WA 99352

David J. Swanberg, P.E.
Science Applications International
Corporation
250 Port of Benton Blvd
Richland, WA 99352

Chris A. Musick
Bechtel National Incorporated
2435 Stevens Center Place
Mail Stop: MS1-B
Richland, WA 99352

ABSTRACT

The United States Department of Energy's (DOE's) Hanford Site is the current home of approximately 50 million gallons of mixed (radioactive and hazardous) waste stored in underground tanks. The waste is designated as a listed waste under the Resource Conservation and Recovery Act of 1976 (RCRA), and is subject to land disposal restrictions (LDR) in accordance with state and federal regulations.

The waste will be separated into high-level and low-activity fractions, and vitrified into immobilized high-level waste (IHLW) and immobilized low-activity waste (ILAW), respectively. The IHLW must be delisted to facilitate disposal in a geologic repository. Approval of a LDR treatability variance is needed to facilitate any land disposal pathway for both the IHLW and the ILAW. To support development of an upfront delisting petition for the IHLW, and an upfront petition for a LDR treatability variance for the IHLW and ILAW, the WTP is conducting extensive research. The data requirements to support development of the petitions were evaluated using the United States

Environmental Protection Agency (EPA) *Guidance for the Data Quality Objectives Process* [1]. The DQO Process identified the research requirements for glass sampling and analysis [2]. The IHLW and ILAW must comply with the requirements of the LDR program for the constituents regulated through those regulations. Additionally, for IHLW, the treated waste must no longer meet the criteria that caused it to be listed. The research conducted to provide data that would support such requirements.

The following specific research and technology development (R&T) activities are considered important for the petitions:

- Demonstrating the relationship between glass composition and performance as a waste form; and
- Demonstrating the means for controlling glass composition;

Separate lists of constituents of potential concern (COPCs) were established for the delisting and LDR petitions. The COPC selection process was used to identify both inorganic and organic constituents of potential concern, however, given that organic compounds will be destroyed and/or volatilized during the vitrification process, there is no requirement for organics destruction testing. As a result, planning was focused on testing to demonstrate the immobilization of inorganic COPCs.

From the DQO process [2], action limits were derived to represent numerical performance standards for comparing test results to determine if treated waste met the requirements for delisting and a LDR treatability variance. The action limits for delisting are based on the EPA-required Delisting Risk Assessment Software (DRAS) [3]. The action limits for LDR are based on minimizing threats to human health and the environment posed by land disposal of the waste, and allow the project to propose alternate treatment standards. Data will be evaluated against the DRAS levels and the LDR Universal Treatment Standards (UTS) in 40 CFR 268.48.

The R&T strategy chosen is focused on crucible-scale testing of simulated IHLW and ILAW. Various glasses will be formulated based upon a glass composition region that includes projected compositions that will result from the vitrification of Hanford tank wastes. Data collection activities must be conducted under the appropriate regulatory quality control procedures to ensure that the results meet compliance needs. The SW-846 Method 1311 "Toxicity Characteristic Leachate Procedure" (TCLP) will be applied to ascertain glass leach resistance [3]. In order to focus research activities on those COPCs that would be most likely to exceed the regulatory criteria, the approach to testing is based on two groups of COPCs:

- COPCs with a low likelihood of reaching action limit
- COPCs with a higher likelihood of reaching action limit

The R&T program was designed in response to a DQO that incorporates three different but increasingly complex statistical designs, depending on the TCLP response of the constituent of interest **Error! Reference source not found.**. The test design requires the collection of a sufficient number of samples to compute a 90 % upper confidence interval (UCI) for the mean TCLP leachate concentration for comparison with applicable action limits. The use of an UCI for this comparison enables data users to assess the impacts of error and uncertainty associated with the data, and facilitates decisions within error tolerances that were chosen during the DQO process.

A small number of glasses will be "spiked" with inorganic COPCs (oxides of hazardous metals) at levels higher than expected in plant operation in order to determine whether COPCs have the potential to reach the action limit. In those cases where experience suggests (or initial data indicate) that there is a potential for a constituent to leach from the glass waste form at or above action limits, testing will progress to involve simultaneously analyzing multiple glasses with multiple spiking levels of those COPCs. Eventually, data collected may be used to identify constituents of concern, and support the development of TCLP response-glass composition models to be used in the glass formulation process.

A significant aspect of the development strategy is to gather sufficient data to demonstrate, upfront, that glass waste forms covering a wide range of compositions will effectively immobilize the hazardous constituents (COPCs). Once demonstrated, mathematical models relating glass composition to TCLP response can be used during production to predict the TCLP performance of the glass product based on analysis of the waste feed to the vitrification process. If necessary, feed additives may be adjusted to yield a glass product with the desired TCLP leach resistance properties. This strategy obviates any need to obtain samples of the glass product and conduct TCLP testing during production to demonstrate compliance, as the immobilized waste forms may be highly radioactive. Obtaining and testing samples of the treated waste represents a significant risk to workers, unnecessary waste generation, and added cost.

INTRODUCTION

The United States Department of Energy (DOE) has assigned the Hanford Tank Waste Treatment and Immobilization Plant (WTP) with responsibility for treating the high-level radioactive wastes currently stored in Hanford Site waste tanks in order to prepare them for disposal. The waste materials contain Resource Conservation and Recovery Act of 1976 (RCRA)-regulated hazardous wastes, both listed wastes identified in 40 CFR 261.31 through 40 CFR 261.33, and constituents that are subject to the land disposal restrictions (LDR) contained in 40 CFR 268 and the Washington Administrative Code (WAC) 173-303-140. The RCRA regulated materials also are subject to management under the analogous provisions of the Washington State Dangerous Waste Regulations found in WAC 173-303.

The waste will be separated into high-level and low-activity fractions, and vitrified into immobilized high-level waste (IHLW) and immobilized low-activity waste (ILAW), respectively. The IHLW must be delisted to facilitate disposal in a geologic repository. The federal high-level waste repository will not be permitted under federal RCRA or corresponding State regulations, and will not accept hazardous wastes for disposal. The EPA and the State of Washington, however, have established procedures for delisting wastes that no longer meet the criteria that caused them to be listed as hazardous.

In addition to delisting the IHLW for disposal at a deep geologic repository, the treated waste forms must comply with the requirements of the land disposal restrictions found in 40 CFR 268 and WAC 173-303-140. ILAW and IHLW from the WTP must either meet the performance-based treatment standards for the LDR-regulated constituents of concern (COCs), or be treated according to specified LDR treatment technologies. Currently, the LDR treatment standard for waste codes D002 and D004 through D011 for high-level radioactive wastes is a specified technology of HLVIT (high-level vitrification). The treatment standards for other waste codes in the Hanford tank wastes are Universal Treatment Standard (UTS) concentration-based standards appearing in 40 CFR 268.48. That is, no specified technology treatment standard has been established for high-level wastes containing waste codes other than D002 and D004 through D011.

Delisting is normally conditional upon analysis of the treated process materials demonstrating that the treated materials comply with specified performance standards. Likewise, compliance with the LDR requirements can be demonstrated by sampling the treated glass product to show that it meets applicable UTS criteria for the regulated constituents in the waste. This approach raises concerns, however, due to radiological exposure of sampling and analytical personnel, and the practical logistics associated with sampling and analysis of the glass product. EPA has stated in 55 FR 22627 that, "...the Agency believes that the potential hazards associated with exposure to radioactivity during analysis of this high-level mixed waste preclude setting a concentration based treatment standard". Accordingly, WTP has determined that the more appropriate method for complying with delisting requirements will be to establish that the treatment process removes the hazardous constituents from the waste or immobilizes them within the waste sufficiently to allow the IHLW to be delisted. With regard to LDRs, the WTP has determined that, because of the radiological and chemical exposures associated with the sampling and analysis required to demonstrate compliance with existing numerical standards, a performance-based standard applicable to Hanford tank wastes is inappropriate. Therefore, WTP has decided that a petition for a treatability variance from the existing performance-based standards is the preferred method for demonstrating compliance with LDR requirements for IHLW and ILAW. Compliance with applicable regulations will be achieved through development of an upfront delisting petition for the IHLW, and an upfront petition for a LDR treatability variance for the IHLW and ILAW, supported with appropriate research data.

DATA QUALITY OBJECTIVES

Unlike waste streams vitrified at other DOE facilities, the Hanford tank waste composition varies widely, and as a result, for an upfront petition to be successful, the vitrification process must be shown to be robust enough to adequately remove or immobilize hazardous constituents. For IHLW, this means collecting data to demonstrate that the treated waste no longer meets the criteria that caused it to be listed, generally through analysis of the treated waste forms and their simulants. Additionally, IHLW and ILAW must comply with the requirements of the LDR program for the constituents regulated through those regulations.

In light of the complexity of the situation, and the cost and importance of successfully petitioning for delisting and for a treatability variance, the data requirements to support development of the petitions were evaluated using the United States Environmental Protection Agency's (EPA) *Guidance for the Data Quality Objectives Process* [1]. The DQO Process identified the research requirements for glass sampling and analysis, as well as providing the appropriate regulatory forum for negotiation of research scope. This is of particular importance since there is no precedence for vitrifying a waste stream with such high composition variation.

Problem and Decision Statements

The problem statement sets the context for the DQO process. EPA's DQO guidelines recommend the development of a concise description of the problem to be investigated and the resulting decisions which must be addressed through the evaluation of the data to be gathered. The fundamental problem dealt with during the DQO was to ensure that the data collection approach would provide data that are adequate to support petition development and acceptance. For delisting, the intent is to prepare a petition for an upfront, conditional delisting based on a technology demonstration showing that the vitrification process removes or stabilizes the constituents of concern such that the health-based delisting levels are met. The goal for achieving compliance with the land disposal regulations is to receive approval of a petition for a treatability variance that establishes vitrification as the specified method of treatment for Hanford tank wastes for all waste codes and all hazardous constituents. Data collection and research activities were designed with these goals in mind.

Both Federal and State agencies will be petitioned. The decisions whether or not IHLW qualifies for delisting, and whether or not Hanford waste vitrification qualifies for a LDR treatability variance, must be clearly supported by the data. To this end, available information and existing data were discussed with regulatory agencies and DOE officials. Results of Hanford tank waste characterization, vitrification and testing of actual and simulated Hanford tank wastes (including toxicity characteristic leaching procedure (TCLP) results), and glass composition control strategies were discussed at length. Demonstrating the relationship between glass composition and glass performance (TCLP response), and demonstrating the means for controlling glass composition, were identified as the specific research goals, fundamental to up-front petition development.

Identifying Constituents of Potential Concern

Separate lists of constituents of potential concern (COPCs) were established for the delisting and LDR petitions. A large realm of regulated constituents were considered for inclusion as COPCs. Logic for the exclusion of certain constituents as COPCs was based primarily on identifying compounds that were never used at Hanford, and those that are unstable in the oxidizing, caustic, and high radiation tank environment. The COPC selection process was used to identify both inorganic and organic constituents of potential concern, however, given that organic compounds will be destroyed and/or volatilized during the vitrification process, there is no requirement for organics destruction testing. Evidence of organics destruction was presented to regulatory agencies in the course of DQO discussions, and adequately demonstrated the volatilization and/or destruction of organic compounds during vitrification. As a result, planning was focused on testing to demonstrate the immobilization of inorganic constituents. Delisting assessment will include testing for antimony, arsenic, barium, beryllium, cadmium, chromium, copper, lead, mercury, nickel, selenium, silver, thallium, vanadium, and zinc. Since many of these constituents are already covered by the HLVIT treatment standard, only antimony, beryllium, nickel, and thallium require extensive testing to support a LDR treatability variance.

Identifying Applicable Action Limits

From the DQO process [2], action limits were derived to represent numerical performance standards for comparing test results to determine if treated waste met the requirements for delisting and a LDR treatability variance. The EPA has recently adopted a new model for delisting entitled Delisting Risk Assessment Software (DRAS) [3]. DRAS is used to calculate health-based delisting limits for hazardous constituents that could leach from the waste under the assumption that the waste is disposed of in a solid waste landfill, and that a remote user consumes groundwater contaminated from leachate. The action limits for LDR are based on minimizing threats to human health and the environment posed by land disposal of the waste. Data will be evaluated against the DRAS levels and the UTS appearing in 40 CFR 268.40.

Scale of Testing, Applicable Test Methods and Quality Control

The research strategy chosen will focus on crucible-scale testing of simulated IHLW and ILAW. This is appropriate since, as previously discussed, the radiological and chemical exposures associated with the sampling and analysis of actual waste pose unwarranted hazards to research personnel. Simulated materials offer flexibility in that precise adjustments to composition can be made. Various glasses will be formulated based upon a glass composition region that includes projected compositions that will result from the production scale vitrification of initial Hanford tank wastes.

The SW-846 Method 1311 "Toxicity Characteristic Leaching Procedure" will be applied to ascertain glass leach resistance (TCLP response). Both DRAS limits and the UTS for inorganic COPCs are based upon TCLP results, so a comparison to regulatory thresholds is straightforward. However, data collection activities must be conducted under the appropriate regulatory quality control procedures to ensure that the results meet compliance needs. Quality control requirements of SW-846 do apply, and can be burdensome to analytical facilities that do not routinely perform work according to SW-846 analytical protocols.

Statistical Optimization of the Research Scope

In order to show that the vitrification process removes or stabilizes constituents such that health-based delisting levels are met, and that vitrification is suitable for treatment for LDR constituents, the TCLP response must be measured for glasses that sufficiently represent the variety of tank waste to be processed. The TCLP results of these glasses will be compared to action limits to support decisions regarding delisting and LDR. The TCLP results are subject to sample handling and analysis error, and are really only a representation of the possible TCLP response for a glass. Since the true mean TCLP leachate concentration for a COPC cannot be known, decisions are made based on sample means (averages), which are used to represent the true mean TCLP leachate concentration. Sample data represent incomplete information (because not every possible sample was analyzed) and have uncertainty associated with them that results from various sources of error. Therefore, it is possible to make an incorrect decision when using sample data. To minimize this potential, a statistical analysis was performed to determine the optimal number of samples and appropriate quantities (sample results) for comparison against the action limit(s).

The decision error of concern is incorrectly deciding that the true mean TCLP leachate concentration is less than the action limit, when in fact it is greater than the action limit. Making this error would result in applying for delisting or a treatability variance when in fact it is not appropriate to do so. The 90 % upper-confidence interval (UCI) of the true mean was chosen as the appropriate quantity for comparison against the action limit. If the 90 % UCI estimate of the mean TCLP leachate concentration is less than the action limit, then there is sufficient evidence to conclude that the true mean TCLP leachate concentration is below the action limit, and delisting and/or a treatability variance is warranted.

RESEARCH PROGRAM

In order to focus research activities on those COPCs that will most likely exceed the regulatory criteria, the approach to testing is based on the two groups of COPCs:

- COPCs with a low likelihood of reaching the action limit
- COPCs with a higher likelihood of reaching the action limit

The likelihood of COPCs leaching from glass waste forms in the TCLP has been shown, using research quality "proof-of-principle" data, to depend on two factors, 1) the concentration of the COPC of interest in the glass, and 2) the characteristic TCLP leach resistance of the glass. The latter has in-turn been shown to be related to the composition of the glass, primarily in terms of the dozen or so major oxide constituents that make up greater than 90 percent by weight (wt%) of the glass on an oxide basis. This phenomenon has been observed as a result of years of development of borosilicate glass compositions for the immobilization of radioactive wastes.

The research program to support the regulatory petitions was designed to incorporate three different but increasingly complex statistical designs, depending on the anticipated difference between a COPC's TCLP response and it's action limit. The statistical basis for the three-tiered approach to inorganics testing is described below. Stated succinctly, the three test cases are:

- Case 1: Experimental TCLP leachate results are analyzed for each glass composition individually and for a single COPC spiking level.
- Case 2: Experimental TCLP leachate results are analyzed for each glass composition individually, with multiple COPC spiking levels.
- Case 3: Experimental TCLP leachate results are analyzed for multiple glass compositions simultaneously at multiple COPC spiking levels.

The test design requires the collection of a sufficient number of samples to compute a UCI for the mean TCLP leachate concentration. The UCI is in-turn used for comparison with action limit derived from the DRAS model or UTS, as appropriate. The use of an UCI for this comparison enables data users to assess the impacts of error and uncertainty associated with the data, and facilitates decisions within error tolerances that were chosen during the DQO process. Each of the glass testing phases for inorganic COPCs (Cases 1, 2, and 3) requires an increasingly sophisticated statistical model to analyze the data that results from that testing phase. The third case is the most complex. The second is a simplification of the third case, and the first case is a further simplification of the second case. The term "spiking" used above denotes increased concentration of the COPC of interest, relative to the contribution of other constituents or glass formers that make up the composition of the glass being tested.

A small number of glasses will be "spiked" with inorganic COPCs (oxides of hazardous metals) in order to determine whether those constituents can affect glass quality from a regulatory perspective and to identify possible upper limits of COPC concentrations in glass (Case 1 and 2 testing scenarios). The results from this testing may support the further development of glass formulation models if needed, and possibly the identification of needed feed limits and glass former requirements. In those cases where experience suggests (or initial data indicate) that there is a potential for a constituent to leach from glass above action limits, testing will progress to involve simultaneously analyzing multiple glasses with multiple spiking levels of inorganic COPCs (Case 3 testing scenario). Eventually, data collected will be used to identify constituents of concern, and support the development of any needed TCLP response models to be used in the glass formulation process.

A significant aspect of the petition development and corresponding research strategy is the generation of sufficient data to demonstrate, upfront, that glass waste forms covering a wide range of compositions will effectively immobilize the hazardous constituents (COPCs). The use of a tiered approach to testing (Case 1, 2, and 3 test scenarios) provides a cost effective means of providing data sufficient for the development of regulatory petitions, and also facilitates the development of sufficient data for those COPCs which are expected to be present in high enough concentrations to have an adverse impact on glass quality from a regulatory perspective.

Test Implementation

The Case 1 and 2 Experimental Approach: The Case 1 experimental approach is warranted in those cases where data for the target constituents, when processed at the maximum reasonable COPC concentration, results in constituent leachate concentrations that are below the action limits. The Case 2 testing approach is similar, except that sequentially increased COPC concentrations are tested to gain additional information about the effect on leachate concentration.

A glass composition for testing TCLP response has been developed and is being utilized to show that inorganics with low leachability will be sufficiently immobilized. Previous data [5] have indicated that leach properties of RCRA constituents can be predicted based upon normalized boron release. Therefore, glasses intentionally designed to have a high normalized boron release under TCLP conditions are being tested along with nominal glass compositions, in simulants representing both IHLW and ILAW. The amount of each COPC added depends on the anticipated concentration in waste and the action limit. Generally, COPCs are added in concentrations that equate to the higher of; 100 times their corresponding action limits, or the maximum expected concentration in the glass based on their estimated concentration in the waste. The maximum estimated concentrations are derived from maximum concentrations specified in the WTP contract (glass or waste feed), Hanford tank farm inventory records and process flowsheet model results, or WTP dynamic flowsheet model results.

Constituents that are being tested in simulated IHLW include arsenic, barium, beryllium, chromium, copper, lead, mercury, nickel, selenium, silver, vanadium, and zinc. The Case 1 testing scenario for IHLW includes a "bounding" glass (estimated normalized boron release of roughly 340 mg/l) spiked and analyzed in triplicate, and a single nominal IHLW glass composition (estimated normalized boron release of 26 mg/l), spiked and analyzed for comparison to the bounding glass. In the case of antimony and thallium in IHLW, available, but limited data indicate the potential for these constituents to leach under TCLP conditions at levels that could approach or exceed action limits. Antimony and thallium are therefore being tested according to the Case 2 testing scenario. The concentration of these constituents is being varied over a range of four different concentrations while holding the remaining glass component concentrations essentially constant. The range of concentrations has been selected such that the resulting TCLP release data are expected to span the action level and provide useful data for statistical analysis of the results.

For ILAW, antimony, beryllium, nickel, and thallium are the COPCs of interest. Three nominal ILAW compositions have been developed, spiked, and tested (six

---

* The term "bounding" used here is to describe a glass composition that has a very high predicted normalized boron release, and represents a worst case formulation from a TCLP response perspective, conservatively bounding the experimental glass composition region.

replicate samples each). The use of replicate sampling is required in order to estimate uncertainty of the results, verify the correct number of samples were taken (by statistical analysis), and to compute the UCI. Preliminary data on ILAW Case 1 test results have been evaluated. None of the results, nor the corresponding UCIs exceed the UTS. In fact, all are below the action level by an order-of-magnitude or greater. The preliminary data indicate that ILAW will meet the requirements for land disposal and that vitrification is an appropriate method of treatment for Hanford tank wastes for all applicable waste codes and hazardous constituents. The data confirm that testing under a Case 2 or Case 3 test scenario is not warranted for ILAW.

The Case 3 Experimental Approach: The approach to testing inorganic COPCs with the potential to reach the action limit involves making and testing multiple glass compositions with varied concentrations of COPCs. To formulate glass compositions which are meaningful, an assessment of available data was preformed to determine which COPCs are most likely to leach from glass under TCLP conditions at levels which could approach or exceed the action limits. Waste characterization data, WTP dynamic flowsheet model results, previous glass TCLP results, and the action limit concentration were all considered in determination of the maximum likely concentrations of these COPCs. From this assessment, cadmium, and thallium were identified as the most likely COPCs to be problematic. Accordingly, these constituents were treated as individual variables in the matrix design of glasses to be tested. Antimony and selenium were also systematically varied in the matrix as the next most problematic COPCs, and chromium, zinc, and nickel were varied in the matrix because of their impacts on other properties of interest. The remaining COPCs (arsenic, barium, beryllium, copper, lead, silver, and vanadium) were grouped into a single matrix variable. The resulting test matrix has 19 variables, which include all the COPCs of interest, plus other Hanford tank waste feed components (non-RCRA metals) and glass former contributions, representing 41 different elements in all, with 24 of those at varying concentrations. Overall, 92 glass compositions are being tested. The data will be used to construct TCLP response models for those COPCs whose leachate concentrations have the potential to approach the applicable action level.

CONCLUSION

Once demonstrated, mathematical models relating glass composition to TCLP response may be used during production to predict the TCLP performance of the glass product based on composition analysis of the feed to the vitrification process. If necessary, feed additives may be adjusted to yield a glass product with the desired TCLP leach resistance properties. Statistical analysis of the TCLP model will be used to determine how well the model fits the data and to quantify uncertainties associated with the model predictions. Uncertainties in sampling and analyzing the composition of the feed to the vitrification process will also be accounted for in assessing the overall uncertainty in the quality (leach resistance) of the glass product. This strategy obviates any need to obtain samples of the glass product and conduct TCLP testing during production to demonstrate compliance.

The TCLP response models will be based on knowledge of glass behavior under TCLP conditions and statistical analysis of data. Uncertainties in the model predictions will be quantified in order to show with high confidence during production that the waste forms will meet delisting and LDR requirements. The data derived from the research program described above will support petition development. The data will be presented in petitions demonstrating that the vitrification process removes or stabilizes the constituents of concern such that the health-based delisting levels are met, and a treatability variance for HLVIT is warranted.

ACKNOWLEDGMENTS

The authors would like to thank the many individuals whose efforts and dedication were instrumental in the production and success of this project. Special thanks go to Environmental Quality Management for their role as facilitator and co-authors of the DQO, Battelle's Pacific Northwest National Laboratory for their statistical support, Vitreous State Laboratory of the Catholic University of America for their role in describing glass chemistry, formulation, and properties. Thanks also go to Mr. Steven Weil, Bechtel-SAIC, LLC., for his technical and regulatory support.

The success of this project would not have been possible without the cooperation and support of the Department of Energy's Office of River Protection, the Environmental Protection Agency, and the Washington State Department of Ecology.

REFERENCES

[1] EPA, *Guidance for the Data Quality Objectives Process*, EPA QA/G-4. EPA/600/R-96/055, US Environmental Protection Agency, Washington, D.C., USA, 2000

[2] D. B. Blumenkranz and J. C. Cook, *Data Quality Objectives Process in Support of LDR/Delisting at the WTP*, 24590-WTP-RPT-ENV-01-012, Rev. 2, Bechtel National Inc., Richland, Washington, March 26, 2003

[3] EPA, *User's Guide for the US EPA Region 6 Delisting Risk Assessment Software (DRAS) version 2.0*, EPA906-D-98-001. US Environmental Protection Agency, Washington, D.C., USA.

[4] EPA, *Test Methods for Evaluation Solid Waste Physical/Chemical Methods*, SW-846, 3rd edition, as amended by Updates I (July 1992), IIA (August, 1993), IIB (January 1995), and III., U.S. Environmental Protection Agency, Washington, D.C., USA, 1997

[5] H. Gan H, and I. L. Pegg, "Effect of Glass Composition on the Leaching Behavior of HLW Glasses Under TCLP Conditions," *Proceedings of the International Symposium on Environmental Issues and Waste Management Technologies in the Ceramic and Nuclear Industries VII*, Ceramic Transactions, Vol. 132, pp. 335-344, Edited by G. L. Smith, S. K. Sundaram and D. R. Spearing, The American Ceramic Society, Westerville, Ohio, 2002

COLD CRUCIBLE INDUCTION-HEATED MELTER TEST RESULTS WITH SURROGATE DOE HIGH-LEVEL WASTES

C.C. Herman, D.F. Bickford,
    and D.K. Peeler
Savannah River Technology Center
P.O. Box 808
Aiken, South Carolina 29808

R. Goles and J.D. Vienna
Pacific Northwest National Laboratory
P.O. Box 999 / MS K6-24
Richland, Washington 99352

A. Aloy
V.G. Khlopin Radium Institute
2nd Murinskiy Ave., 28
St. Petersburg, 194021, Russia

S. Stefanovsky
SIA Radon
7th Rostovskii per. 2/14
Moscow 119121 Russia

D. Gombert and J. Richardson
Idaho National Engineering and Environmental Laboratory
P.O. Box 1625
Idaho Falls, Idaho 83402

ABSTRACT

The Department of Energy (DOE) is evaluating Cold Crucible Induction-Heated Melter (CCIM) technology for the immobilization of High Level Waste (HLW). HLW contains a varied array of chemical compounds, a number of which can limit the waste loading or cause operating difficulties with melt rate or equipment corrosion. Therefore, the DOE could benefit through the use of advanced melters that allow for increased waste throughput and longer life. The CCIM has the potential to increase waste loading and throughput through increased melt temperature and to increase melter life by minimizing melter material corrosion. However, using a technology that has not been extensively tested in the United States or on DOE wastes involves significant risks that need to be addressed and successfully resolved before applicability is considered in DOE waste treatment facilities. Some of the risks include: service life of melter materials; ability to accommodate electrically conductive noble metal fission products; power requirements and control stability (with slurry feeding and secondary phases); ability to meet production rate goals with liquid feed; ability to increase waste loading and adequately treat offgas emissions. To try to address these issues, DOE contracted two Russian Institutes, Khlopin Radium Institute and SIA Radon, to perform testing with DOE high level wastes simulants. The results of the initial testing with Idaho National Engineering and Environmental Laboratory (INEEL) and Hanford HLW surrogates will be presented.

BACKGROUND

Existing DOE waste melter systems have limited operating temperatures and melt rate capacities, are difficult to construct and dispose of, and are susceptible to power loss from electrical shorting by noble metal fission products. The CCIM is purported to have several advantages over joule-heated melters (JHM). The CCIM is constructed of water cooled metal segments arranged to form a crucible. This crucible contains the feed or melt and is protected by the slag layer that forms along the crucible walls upon heating of the feed. Heating of the glass is accomplished using an inductive coil that surrounds the crucible. When energized the coil produces a field that efficiently penetrates both the water-cooled crucible wall and the nonconductive, corrosion-resistant slag layer formed on the crucible's inner surfaces. The design concept relies upon inductive coupling of the water-cooled high-frequency electrical coil with the glass, causing eddy-currents that produce heat and mixing. [1] Molten glass can be discharged from the CCIM through a water-cooled freeze-valve pouring device that can be installed as one of the segments for side discharge or on the bottom to facilitate bottom pouring. Some of the potential advantages of the CCIM technology include:

- High resistance to corrosion - materials of construction are protected by a corrosion-resistant slag; thus, the system has the potential for longer life and to require less maintenance.
- Higher operating temperatures – the elimination of refractory and electrodes allow higher operating temperature glass formulations to be used which has the potential to increase waste loading and throughput and enhance solubility of troublesome components, but this must be weighed against the potential for increased volatility.
- Reduced melter foot print – equivalent or higher productivity in a smaller package reduces hot-cell space requirements and decommissioning waste while lowering construction costs.
- Reduced melter hold-up – higher projected CCIM melt volume turnover rates minimizes crystal growth and secondary phase accumulations. In addition, a CCIM equipped with a bottom freeze valve can be completely drained, resulting in a reduced radiation field for maintenance and decommissioning, a minimization of secondary wastes, and lessen the impact of failure. This also lowers susceptibility to noble metals accumulation.

To assess the validity of the potential advantages, DOE testing was initiated as a collaborative effort between the Savannah River Technology Center (SRTC), the INEEL, the Pacific Northwest National Laboratory (PNNL), the Khlopin

Radium Institute in St. Petersburg, and the SIA Radon facility in Moscow. The specific objectives of the testing program were as follows:

- Determine CCIM readiness for DOE production applications by demonstrating the technology on liquid/slurry feeds and assessing the potential operating and safety risks of implementation;
- Define the engineering parameters that must be addressed to establish: scale-dependent induction frequency and uniform heating conditions, electrical and thermal stability, power requirements, and engineering design details; and
- Support the DOE accelerated clean-up mission.

The test program initially focused on bench-scale tests to provide data for preliminary mass balance calculations (volatility) and electrical design. After successful testing at the smaller scale, larger lab and engineering-scale tests were then planned and performed to verify integrated design concepts, provide temperature profiles, establish mass and power balance, and generate offgas emission characteristics.

The Khlopin Radium Institute is a Russian MINATOM institute and performs research and development activities with nuclear materials. Khlopin is actively involved in radioactive waste stabilization. They use a small, bench-scale (86 mm diameter) CCIM unit to perform feasibility testing for a given waste or glass composition. As part of the DOE testing, a larger 3-liter CCIM unit (155 mm diameter) was fabricated. The intent of this unit was to demonstrate continuous processing on a slightly larger-scale. Figure 1a shows a picture of the Khlopin 86 mm CCIM utilized in DOE testing.

SIA Radon was founded in 1961 to provide low and intermediate waste processing research and industrial waste processing services for the Moscow region. The industrial waste processed is primarily from medical facilities and industry, but does include some low level waste from power plants. Radon is currently operating three CCIM units for the stabilization of low-level radioactive waste. Radon operates several sized CCIM units to evaluate applicability of the technology. A new 216 mm diameter crucible was manufactured for the DOE testing and was integrated into their system containing a feed and offgas system. A larger diameter (~500 mm) crucible is also being fabricated to support DOE engineering-scale testing. The Radon facility provides the opportunity for integrated testing on a larger scale than that used in the Khlopin testing. Figure 1b shows the 216 mm Radon unit used for DOE testing.

DOE WASTE STREAMS EVALUATED

While the CCIM advantages may apply to many DOE waste streams, the applicability to the individual DOE sites may be more limited due to existing facility constraints and planned treatment paths. When the program was initiated, the INEEL, the Savannah River Site (SRS), and the Hanford Reservation were interested in the technology.

INEEL had planned to build a vitrification plant to treat their wastes. In fact, the projected size requirements for a CCIM supported possible installation in an

Figure 1a. Khlopin CCIM melter   Figure 1b. Radon CCIM Melter

existing facility and also provided a potential throughput increase that would help meet processing commitments. As a result, the INEEL Sodium Bearing Waste (SBW) was selected as a stream for CCIM testing. However, as the INEEL SBW stream was being tested, DOE eliminated vitrification as a treatment path, so all SBW-CCIM testing activities were terminated. The information gained, however, should be considered applicable to waste streams with similar components.

INEEL SBW is a high nitrate feed with high concentrations of aluminum and sulfate. The surrogate was based on recipes used in DOE joule-heated melter testing. Table I provides the target composition of the major oxides. Cesium is not shown in the table, but was added at a spiked amount to assess volatility.

Table I. INEEL SBW surrogate composition (wt%)

Oxide/Species	Target	Oxide/ Species	Target
Al_2O_3	27.58	MnO	0.816
B_2O_3	0.349	Na_2O	52.00
CaO	2.157	Cl	0.925
Fe_2O_3	1.415	F	0.533
K_2O	7.544	SO_3	4.556
MgO	0.3956		

At the SRS, CCIM advantages include the possibility to minimize noble metal accumulation and the potential to increase waste loading to achieve accelerated mission goals. Compared to the existing Defense Waste Processing Facility (DWPF) melter, a CCIM has less fabrication material and potentially less fabrication time and should produce a reduced glass inventory upon disposal due to easier glass removal. These factors also make management of and recovery from a failed melter easier for all DOE sites.

Similar melter and waste processing advantages may be obtained for the treatment of HLW at Hanford, but retrofitting of the melter into the existing design appears to be more problematic than for SRS. Nonetheless, Hanford's C-102/AY-106 stream was selected as a second simulant stream to represent a high alkaline waste for both Hanford and SRS. The troublesome components in this waste are similar to those present in SRS wastes. At the time of the selection, the best available information on the surrogate was based on recipes developed by Catholic University of America - Vitreous State Laboratory[2]. This surrogate had also been used in DOE joule-heated melter testing[3].

The C-106/AY-102 simulant represents a blend of tanks that will be processed during Hanford HLW vitrification efforts. The predominant cations are iron and aluminum. The composition presented in Table II takes into account sludge washing that will be performed as part of the HLW pretreatment process and the projected products from the low-activity waste pretreatment processes.

Table II – High alkaline waste (C-106/AY-102) surrogate composition (wt%)

Oxide/Species	Target	Oxide/Species	Target
Al_2O_3	19.60	Na_2O	15.21
CaO	1.30	NiO	0.28
Cs_2O	0.19	SiO_2	10.71
Fe_2O_3	21.33	SrO	13.37
MnO	8.07	ZrO_2	0.21
CO_3	7.644	NO_3	0.939
NO_2	0.002	TOC	1.150

MELTER TESTING AND RESULTS

INEEL SBW Testing

The initial testing at Khlopin and Radon utilized the INEEL SBW surrogate in their smaller-scale units. The main objectives were to demonstrate the viability of slurry feeding and to produce a processable and durable glass. Only limited testing was performed on the INEEL SBW surrogate; therefore, optimization of the treatment of the slurry feed was not achieved. DOE recommended a glass composition of 20 wt% Waste Loading (WL) containing glass formers of B_2O_3, CaO, Fe_2O_3, Li_2O, MgO, Na_2O, SiO_2, V_2O_5 and ZrO_2.[4] Khlopin slightly adjusted the DOE recommended glass composition by increasing the alkali to silica ratio to accommodate CCIM processing and tested 20 wt% WL and 30 wt% WL. Radon tested the 20 wt% WL glass composition recommended by the DOE.

Samples of the glasses produced at both facilities were transmitted to DOE for characterization. Khlopin transmitted a sample from each waste loading, while Radon transmitted samples from four of the ten containers produced. All samples were characterized for chemical composition, crystallinity, and durability using the Product Consistency Test (PCT).

The glass samples were dissolved using lithium metaborate and sodium peroxide fusions and analyzed using Inductively Coupled Plasma – Atomic Emission Spectroscopy (ICP-AES). The chemical composition results for the major oxides are given in Table III, along with the target compositions. The Radon results present an average of the four samples analyzed.

Table III. Chemical composition of the INEEL SBW glass (wt%)

Oxide/ Species	Khlopin 30 wt% WL target	Khlopin 30 wt% WL actual	Khlopin 20 wt% WL target	Khlopin 20 wt% WL actual	Radon 20 wt% WL target	Radon 20 wt% WL actual
Al_2O_3	8.32	8.67	5.52	5.88	5.61	9.11
B_2O_3	4.56	3.31	4.96	1.37	4.99	6.31
CaO	4.55	3.46	4.41	3.50	4.52	4.91
Cs_2O	1.00	0.96	0.45	0.17	N/A	N/A
Fe_2O_3	1.55	1.43	1.49	1.52	1.53	1.47
K_2O	1.27	0.916	1.50	0.862	1.53	1.66
Li_2O	2.20	2.12	4.85	4.63	2.49	3.03
MgO	1.64	1.58	1.46	1.56	1.50	1.54
Na_2O	17.44	17.7	13.8	12.8	12.3	13.7
SiO_2	49.94	56.1	53.9	59.9	55.3	51.1
V_2O_5	3.58	3.60	3.87	4.00	3.90	4.00
ZrO_2	1.79	1.80	1.94	1.88	1.99	1.82
SO_3	1.37	0.170	0.904	0.134	0.96	0.79

In general, both of the Khlopin glasses were relatively close to their target composition. A more detailed assessment indicated slightly lower B_2O_3, CaO, K_2O and higher SiO_2 concentrations than targeted. Slightly higher volatility for the boron and alkali oxides was seen for the 20 wt% WL Khlopin glass. Cs retention was ~96% in the 30 wt% WL run, while the 20 wt% WL test had ~62% Cs volatility. Some problems with melt viscosity were seen in the 20 wt% WL run, and the melt temperature was raised from 1150 to 1250°C, which may have caused the additional volatilization. A salt layer was not observed in either test possibly indicating high sulfur volatility.

The Radon glass compositions were also fairly close to the target composition, although analyses indicated slightly lower SiO_2 and higher Al_2O_3, B_2O_3, Li_2O, and Na_2O concentrations. Sulfur retention was ~82%. A melt temperature of 1350°C was used during the runs, but volatility of boron and alkali oxides did not appear to be an issue. Cs was not added to the Radon glass. Although not shown, the glass compositions from container to container were fairly similar with Al_2O_3 increasing and B_2O_3 and SiO_2 decreasing as the run progressed.

The PCT, ASTM C-1285-97, was performed on the glasses. Table IV gives the normalized releases for the glasses, along with the accepted values for the

HLW Environmental Assessment (EA) benchmark glass[5]. The Radon results represent the average result for the four containers, but no real differences were seen between the four Radon samples. Measured durability was an order of magnitude better than the HLW EA glass.

Table IV. Average PCT normalized release for INEEL SBW glasses

Glass ID	B (g/L)	Li (g/L)	Na(g/L)	Si (g/L)
Khlopin 30 wt% WL	0.443	0.706	0.844	0.299
Khlopin 20 wt% WL	0.846	1.310	1.146	0.455
Average Radon Glass	0.304	0.562	0.508	0.218
EA HLW Benchmark[5]	16.695	9.565	13.346	3.922

A sample of each of the glasses was submitted for X-Ray Diffraction (XRD) to determine if any crystals were present. Both Khlopin glasses were reported to be amorphous by XRD. The three glasses from the Radon early pours contained quartz. Although no quantitative measurements were performed, the amounts seemed to decrease as more glass was poured. The last container sample contained no quartz, but had a cubic phase that was suspected to be a metal phase.

Alkaline (Hanford C-106/AY-102) Surrogate Testing

Testing with a DOE simulant of high alkaline waste was performed to determine the suitability of the CCIM technology. A glass formulation with a 70 wt% WL was recommended for both Russian facilities. The frit/glass formers included B_2O_3, K_2O, Li_2O, and SiO_2.[6]

Khlopin Scoping Test: Khlopin completed an initial lab-scale test to enable them to determine the suitability of the glass formulation for their system. Slurry feeding was not performed in this initial testing. Once the feed materials were reacted and the desired properties were observed, the melter was shut-down and the glass monolith was allowed to cool. Some problems with crystallization and viscosity were experienced during the testing, which were attributed to the different operating parameters necessary for a CCIM versus a JHM.

The glass monolith was sectioned for characterization by Khlopin and DOE and contained two visually distinct regions. The "Interior" was black and shiny, while the "Exterior" was more crystalline. XRD analyses determined that both samples contained magnetite and nepheline. This was consistent with DOE findings for the composition when subjected to the DWPF canister centerline cooling (ccc) temperature regime.[7] The ccc temperature regime cools the glass at a much slower rate, which occurs to the CCIM monolith upon cooling. In addition, a temperature extreme exists at the crucible wall compared to the melt pool throughout melting due to the design.

The PCT was performed, and results were an order of magnitude better than the HLW EA benchmark glass.[5] The Khlopin glasses had normalized B and Si releases (0.21 and 0.14 g/L) similar to and normalized Li and Na releases (0.38 and 0.52 g/L) slightly higher than DOE glasses of similar composition. Even though the "Exterior" glass visually appeared to contain a much greater amount of crystals, its durability was comparable to the "Interior" sample.

Khlopin and Radon Test: After the initial scoping test, the glass formulation for Khlopin testing was revised to maintain acceptable molten glass physical and electrical characteristics for the laboratory-scale CCIM with excessive cooling-rate properties. The relative proportion of the glass formers was changed slightly and the use of K_2O was eliminated, but the target waste loading was maintained. The Radon testing used the original DOE target glass formulation.

The Khlopin test used slurry feeding and pouring over an 8-hour period. The actual demonstration consisted of two separate batch-processing campaigns separated by a 30-minute fining period and a glass pour using a side discharge. Melt temperature was ~1350°C.

Since the Radon melter was a larger scale and was equipped with integrated feed and pouring systems, a continuous 55-hour processing campaign was demonstrated. Twenty-four pours were conducted producing 130 kg of glass from 458 kg of feed. Waste processing at Radon was conducted in a quasi-continuous, batch-wise manner. The slurry was quickly transferred to the melter, allowed to dry, calcine, and melt until only 30% cold-cap coverage remained, and then more feed was transferred to the melter. The cold-cap was allowed to completely melt and a fining period was employed to homogenize the melt before the pour was initiated. Pouring was initiated once sufficient glass had accumulated and was accomplished by raising the water-cooled plug from the bottom freeze valve seat. The nominal melt temperature for this run was 1260°C. Although too small to reliably predict full-scale performance, the Radon CCIM achieved an average specific glass production rate (1.6 metric tonnes/day/meter^2) that comfortably exceeded the cold-lid, unagitated JHM reference design value of 0.5 MT/d/m^2.[3]

Samples were once again transmitted to DOE for chemical composition, crystalline content, and durability characterization. For Khlopin, samples were obtained from the homogenized glass remaining in the crucible and poured from the crucible. Radon transmitted samples from the 16th and 18th containers.

The analytical results for the major oxides in the glass samples are given in Table V, along with the target compositions. B volatility was not a problem for any of the glasses. K_2O was not reported for the Khlopin glasses since it was not added as a glass former. For the Khlopin glasses, MnO and SiO_2 were higher than targeted, while Na_2O was lower than targeted. Glass compositional analyses on the Radon glasses indicated that a glass batching error occurred before the pouring of the 16th container. The glass had a low Fe and high Mn content, so it was quite different from the target composition. In addition to the obvious error, Al_2O_3, B_2O_3, and K_2O were higher than targeted, and Na_2O was lower than targeted. Glass compositional analyses indicated that Cs volatility was ~18-50%. Sr volatility was ~22-29% for the Khlopin glass and ~7-12% for the Radon glass. Khlopin offgas data supported the Cs volatility, but not the Sr volatility indicating partitioning of <5% for Sr.

Table V. Chemical composition of high alkaline glasses (wt%)

Oxide	Khopin target	Khlopin crucible glass	Khlopin poured glass	Radon target	Radon container #16 glass	Radon container #18 glass
Al_2O_3	14.8	14.1	13.6	14.8	16.8	16.5
B_2O_3	4.00	4.97	4.41	3.00	3.84	3.86
CaO	0.980	0.743	0.816	0.980	1.09	1.08
Cs_2O	0.140	0.115	0.076	0.140	0.074	0.069
Fe_2O_3	16.1	15.1	15.2	16.1	1.06	1.07
K_2O	N/A	N/A	N/A	0.500	0.753	0.866
Li_2O	2.00	1.93	1.85	0.501	0.500	0.466
MnO	6.09	6.95	7.55	6.09	20.3	20.6
Na_2O	11.5	8.80	9.95	11.5	10.6	9.84
SiO_2	32.1	37.9	36.6	34.1	33.5	34.5
SrO	10.1	7.19	7.85	10.1	9.43	8.93
ZrO_2	0.161	0.157	0.157	0.161	0.085	0.176

XRD of the Khlopin glasses indicated 9 to 17 wt% crystal content depending on the sample and was primarily spinel with the remainder being nepheline. The spinel appeared to be uniformly distributed throughout the glass. The glass extracted from the inside of the crucible after cooling contained an increased presence of nepheline and spinel crystals. XRD of the Radon glass indicated crystalline content of >47% in container #16 and >35% in container #18. Both were dominated by nepheline and carnegieite (a high temperature form of nepheline). Spinel comprised a smaller fraction of the crystalline species in the glass and was manganese/oxygen rich. Scanning Electron Microscopy of the Radon glasses also indicated major amounts of nepheline and spinel and identified the presence of elemental silver inclusions.

The PCT results are reported in Table VI. Although slightly different durability was seen for the Khlopin glasses, both were an order of magnitude better than the EA benchmark glass[5]. The PCT was performed on two samples from both Radon containers and the average results are reported. Even though both containers contained glass containing significant amounts of crystallinity and were considered out of specification from a composition perspective, these glasses were an order of magnitude better than the EA glass[5].

Table VI. Average PCT normalized release for high alkaline glass

Glass ID	B (g/L)	Li (g/L)	Na (g/L)	Si (g/L)
Khlopin Crucible Glass	0.309	0.283	0.237	0.067
Khlopin Poured Glass	0.188	0.225	0.422	0.157
Radon Container #16	0.744	0.765	0.393	0.123
Radon Container #18	0.513	0.453	0.406	0.119

CONCLUSIONS

Lab- and research-scale CCIM testing with a simulant of the INEEL SBW and an alkaline surrogate representative of Hanford and SRS waste showed promising results. Slurry feeding, which had not been widely demonstrated and presented a concern for DOE waste vitrification in the CCIM, was proven viable. Batch pouring through a bottom and side drain configuration was also successfully demonstrated. Acceptable glass durability was also demonstrated. Glass characterization indicated that not all glasses met the target compositions and some crystalline species were present; however, durability and processability were not significantly affected. Preliminary data on cesium and strontium retention were obtained and represent a concern with the higher melt temperatures. Since the intent of vitrification is to stabilize the radionuclides, this is an issue that needs to be addressed during future testing. An integrated engineering-scale test is planned in the summer of 2003 at the Radon facility. This demonstration will provide further information on volatility, power requirements, wear/corrosion, and processing rates for DOE wastes.

REFERENCES

[1] I. Sobolev, et.al., "Cold Crucible Vitrification of Radioactive Waste", Waste Management 97, Tucson, Arizona, March 2 – 6, 1997.

[2] W.K. Kot, et.al., "Physical and Rheological Properties of Waste Simulants and Melter Feeds for RPP-WTP HLW Vitrification", VSL-00R2520-1, Vitreous State Laboratory, The Catholic University of American, Washington DC (2000).

[3] R.W. Goles, et.al., "Test Summary Report Vitrification Demonstration of an Optimized Hanford C-106/AY-102 Waste-Glass Formulation", PNNL-14063, Pacific Northwest National Laboratory, Richland, Washington (2002).

[4] J.D. Vienna, et.al., "Glass Formulation Development for INEEL Sodium-Bearing Waste", PNNL-14050, Pacific Northwest National Laboratory, Richland, Washington (2002).

[5] C.M. Jantzen, et.al., "Characterization of the Defense Waste Processing Facility (DWPF) Environmental Assessment (EA) Reference Material", WSRC-TR-92-346, Revision 1, Westinghouse Savannah River Company, Aiken, South Carolina (1993).

[6] D.K. Peeler, et.al., "Development of High Waste Loading Glasses for Advanced Melter Technologies", WSRC-TR-2002-00426, Westinghouse Savannah River Company, Aiken, South Carolina (2002).

[7] S.L. Marra and C.M. Jantzen "Characterization of Projected DWPF Glasses Heat Treated to Simulate Canister Centerline Cooling", WSRC-TR-92-142, Westinghouse Savannah River Company, Aiken, South Carolina (1992).

CRUCIBLE-SCALE VITRIFICATION STUDIES WITH HANFORD TANK AZ-102 HIGH SULFATE-CONTAINING LOW ACTIVITY WASTE

C.L. Crawford, R.F. Schumacher, D.M. Ferrara and N.E. Bibler
Savannah River Technology Center
773-41A
Aiken, SC 29808

ABSTRACT

A proof-of-technology demonstration for the Hanford River Protection Project (RPP) Waste Treatment Plant (WTP) was performed by the Savannah River Technology Center (SRTC) at the Savannah River Site (SRS). As part of this demonstration, treated AZ-102 Low-Activity Waste (LAW) supernatant was vitrified using a crucible-scale furnace. Samples from the low-activity AZ-102 glass waste form were characterized for metals and radionuclides. The glass was also leach-tested using the ASTM Product Consistency Test (PCT) protocol and USEPA Toxicity Characteristic Leaching Procedure (TCLP). These tests used the AZ-102 glass formulation LAWB88 that targeted AZ-102 waste loading at 5 wt% Na_2O. The purpose of this report is to document the characterization and leach testing of this AZ-102 glass that was both initially rapidly cooled and later remelted and canister centerline cooled. Previous crucible-scale vitrification testing with pretreated AZ-102 at SRTC had resulted in a product LAW glass with crystalline surface material. That previous testing used a former glass formulation (LAWB53). Thus, another significant goal of this present work was to investigate the influence of reformulating the glass recipe on crystalline formation in the product glass. The glass produced in this present study using glass formulation (LAWB88) was first rapidly quenched and analyzed. Latter testing involved remelting of the AZ-102 glass followed by simulated canister centerline cooling.

Glass samples were dissolved using an acid dissolution method and using a peroxide fusion dissolution method with an acid strike. The resulting solutions were analyzed to determine the concentration of metals and radionuclides in the glass. The sum of metal oxides from both acid dissolution and peroxide fusion dissolution of the AZ-102 LAW glasses indicated all major constituents (those at or above 0.5 weight percent) were determined. Results were typically within ten percent of the target for the AZ-102 glasses and for the standard Low-Activity Reference (LRM) glass. Measured densities for the AZ-102 glass were ~ 2.6 g/cc and the measured density of the LRM glass was ~ 2.5 g/cc compared to a reference value of 2.51 g/cc. Radionuclides measured in the AZ-102 glass were close to target values calculated from measured radionuclide specific activities in the treated feed. The activities for Cs-137, Sr-90, and Tc-99 were all below the specified WTP contract upper limits for LAW glass. The transuranics measured in the AZ-102 glass were also well below the upper limit of 100 nCi/g for LAW glass. PCT results for both the quenched and canister-cooled AZ-102 glass indicate that normalized releases for B, Si, and Na are all below the BNI-WTP contract upper limits for normalized release of 2 g/m^2. The PCT results for the standard LRM glass also matched previous data for leach testing of that glass. Results for the Toxicity Characteristic Leaching Procedure (TCLP) test on the AZ-102 glass indicate all hazardous metals are below regulatory limits. No crystalline phase was observed in the product AZ-102 LAW quenched or canister-cooled glasses.

INTRODUCTION

The tasks addressed in this report are part of continuation work on a proof-of-technology demonstration performed by the SRTC for Bechtel National, Inc. (BNI). In the initial demonstration, a sample of AZ-102 high-level radioactive waste was treated to remove suspended solids and most radionuclides. The resulting LAW radioactive supernate was concentrated, mixed with glass-forming minerals, and vitrified in a platinum crucible. The initial glass product contained crystals on the bottom glass surface and was not analyzed.[1] A simulant was developed at SRTC that produced the same crystalline surface when vitrified. This simulant was subsequently used by researchers at the Vitreous State Laboratory (VSL) at the Catholic University of America (CUA) to investigate the crystalline formation and to develop a reformulated glass recipe for the AZ-102 LAW supernatant. The AZ-102 glass was reformulated and a second batch of AZ-102 LAW radioactive glass was produced and analyzed at SRTC. The scope of the task described in this report was to provide the reformulated, rapidly cooled glass waste form characterization data, to report durability testing (waste form leachability) of the glass per the ASTM C1285-97 PCT and to report TCLP results. Further testing of the original rapid-cooled glass involved remelting and canister centerline cooling, followed by repeated PCT and crystalline analysis.

EXPERIMENTAL

Testing associated with the task described in this report included preparation and analysis of the AZ-102 glass waste form samples. During the vitrification task, an AZ-102 low-activity glass waste form and a nonradioactive AZ-102 simulant glass waste form were produced in platinum/gold crucibles in a custom-designed DelTech Model DE-29-TL-610 Top Loading Laboratory Furnace capable of 1200°C with programmable set point temperature control. Glass sample preparation at SRTC generated solid samples that were crushed to particles with diameters of less than 0.9 centimeter, which is the maximum particle size for the TCLP. These glass particles were used prepare ground glass powders (100 to 200 mesh) for leachability testing per the PCT and for crystalline analysis by X-ray Diffraction (XRD) and Scanning Electron Microscopy (SEM). Ground glass powders were also produced (-200 mesh) for dissolution and analysis per ASTM C1463-00 dissolution methods (acid dissolution method and the peroxide fusion method).

AZ-102 GLASS FORMULATION

Vitrification of the pretreated AZ-102 supernatant used a reformulation (LAWB88) of the original formulation (LAWB53) that produced pyroxene crystalline surface material.[1] Comparison of the two different formulations (LAWB53 vs. LAWB88) is shown in Table I. Table I shows oxide concentrations in glass for the glass former additive components. Both formulations targeted sodium at 5 wt% as oxide in the glass. Comparison of the two glass formulations indicates an increase in B, Ca, Si, and Zn in going from the original to the reformulated glass recipe. Conversely, Fe, Li, Mg were decreased in the latter formulation relative to the original. Also, the glass former rutile (TiO_2) was not used in the reformulation. The AZ-102 simulant described in Ref. 1 was used by VSL to develop the reformulation recipe for the glass. Scanning Electron Microscopy images pertaining to the VSL investigations using both the original AZ-102 formulation (LAWB53) and the latter AZ-102 'reformulation' (LAWB88) are presented in Figures 1-2. Figure 1 shows a cross-section from the bottom of the original LAWB53 formulation glass with platinum-nucleated pyroxene crystals. Figures 2 shows the absence of any crystalline phases when the reformulated LAWB88 glass was produced and canister-cooled. No crystals were observed also in glasses that were heat-treated at 700°C.

CRUCIBLE-SCALE VITRIFICATION

Crucible-scale vitrifications on the actual radioactive AZ-102 feeds using the reformulated LAWB88 were performed at SRTC using 95%Pt/5%Au crucibles that were located inside of the quartz glass off gas system inside of the Deltech furnace. The off gas system was employed to trap off gas from the vitrification process in order to capture the products for return to Hanford as residue. Vitrification tests involved loading the 600-mL crucibles with the prescribed amount of treated AZ-102 supernatant and glass former blend. This mixture (representing the 'melter feed slurry') was briefly mixed by hand with a stir rod before placing into the furnace. Due to expansion of the static melts during the vitrification process (evaporation, calcining, and melting), the heating sequence involved first slow heating to evaporate off the liquid in the slurry, followed by calcining, and then melting at 1150°C. Initial tests produced melted glass samples that were allowed to rapidly cool by turning off the furnace power supply. Later tests involved remelting of the glass followed by a 62-hour prototypical cooling curve to reach 400°C.

CONCENTRATION OF METAL ANALYTES AND GLASS DENSITY

Samples of the rapid-cooled glasses were taken of the AZ-102 glass, the AZ-102 surrogate glass, and a standard glass (LRM glass).[2] The AZ-102 surrogate glass was prepared from an AZ-102 simulated supernatant (See Ref. 1) using the same glass forming minerals, the same heating profile, and the same cooling profile as the AZ-102 glass. The purpose of the AZ-102 surrogate glass was to act as a control for identification of any unexpected events or observations. Results from the dissolution and analysis of the crucible scale glasses for the metal analytes are presented in this section. These results were determined from dissolving nominally 0.5 g of pulverized glass (-200 mesh) in 0.1 L of total solution by either the acid dissolution method or the peroxide fusion method. Boron values from peroxide fusion were used in the acid dissolution data since boron is used in the acid dissolution process. Nickel and sodium values from acid dissolution were used in the peroxide fusion data since sodium is used in the peroxide fusion process performed in nickel crucibles. Table II shows the oxide components of the glasses for the acid dissolution (top) and the peroxide fusion (bottom). The metal analyses were performed with Inductively Coupled Plasma-Atomic Emission Spectroscopy (ICP-ES). Potassium values were measured by Atomic Absorption Spectroscopy (AAS). For most analytes, results given in Table II were within three standard deviations of the targets. Totals given in Table II suggest analyses were successful in determining all major glass constituents present above 0.5 wt percent. Densities of the AZ-102 glasses and the LRM glass were estimated by determining the volume displacement of water from a known mass of glass. These measurements used approximately 2 grams of crushed glass in 5-mL graduated cylinders that were partially filled with deionized water. The volume of water was measured before and after addition of the glass to the cylinders. The density for the AZ-102 radioactive glass is 2.59 g/cc and for the surrogate AZ-102 glass is 2.54 g/cc. The measured density value for the LRM glass of 2.50 g/cc compares well with the published value of 2.516 +/- 0.009 g/cc.[2]

RADIONUCLIDES MEASURED IN AZ-102 GLASSES

Radionuclides were measured in the product AZ-102 glasses using radiochemical methods of gamma scan, total alpha, and total beta analyses performed on the dissolved glasses without any separation techniques. Individual radionuclides were determined by separation and counting methods. Radionuclide analyses results are shown in Table III for the primary LAW radionuclides specified in the WTP Contract[3] (Cs-137, Sr-90 and Tc-99) as well as the overall total alpha and total beta activities in the glass. Data for both the actual radioactive glass and the surrogate glass are shown. The surrogate glass was analyzed for radionuclides to investigate any significant contamination of the glasses during the vitrification process. Measured density values for the radioactive AZ-102 glass and the surrogate glass were used to calculate the final column data for Table III in units of curies per cubic meter of glass (Ci/m^3). Table III data show that the measured values for both Cs-137 and for Sr-90 were near the expected values of ~ 1.2 Ci/m^3 in the glass. This concentration of Cs-137 in the AZ-102 glass of ~ 1 Ci/m^3 is less than the WTP Contract specification of <3 Ci/m^3. However, the measured value of 1 Ci/m^3 is above the engineering limit for Cs-137 in LAW glass of <0.3 Ci/m^3. The measured value for Sr-90 in the AZ-102 glass of 1.2 Ci/m^3 is well below the WTP Contract specification of <20 Ci/m^3. Measured values for Tc-99 of <0.002 are also well below the WTP Contract specification of < 0.1 Ci/m^3. One measured value for Tc-99 in radioactive AZ-102 glass peroxide fusion of 0.16 Ci/m^3 is suspected to be in error. This sample is currently being reanalyzed. Finally, Table III data for measured total alpha activity in the glass indicates that the sum of alphas is less than 5 nCi/g in the radioactive AZ-102 glass. This value is well below the WTP Contract specified value for < 100 nCi/g transuranics in LAW glass.

RESULTS OF PCT LEACH TEST

This section shows the results of the standard ASTM C 1285–97 test on the radioactive AZ-102 glass and the surrogate AZ-102 glass. This test is commonly called the Product Consistency Test (PCT) and is performed at 90°C. The procedure for PCT-A of the ASTM C 1285-97 was strictly followed for this test. Triplicate samples of the AZ-102 glass and, as prescribed by the procedure, triplicate blanks were used. The standard glasses, Low Activity Reference Material (LRM)[2] and Analytical Reference Material

(ARM)[4] were also leached in the test with the AZ-102 glass. In the contract, SRTC was required to subject the AZ-102 glass to the PCT and report the results for B, Si, and Na for the AZ-102 glass. Section 2.2.2.17.2 of Mod. No. A029 of the contract specifies that in the PCT, the glass shall have a normalized mass loss less than 2 g/m^2 (2 grams of glass per square meter of exposed surface area of glass tested in a $90°C$ PCT) based on each of the elements B, Si, and Na. The LRM glass and the standard (ARM) glass were also tested with the AZ-102 glass to confirm that the test conditions for the PCT were properly controlled. Table IV gives average concentrations in parts per million (ppm) of B, Si, and Na, in the final leachates after the tests from quenched glass and canister-cooled glass. The averages of the final pH values of the leachates are also presented. Concentrations have been corrected for the acidification dilutions of the leachates as required by the ASTM procedure. Normalized mass losses for B, Si and Na are also shown. Normalized mass losses are the best indication of the durability of a glass in a PCT. Normalization accounts for the concentration of an element in the glass. Normalized release is a measure of the total mass of glass leached in a PCT based on a specific element in the glass. The last row of the table presents the consensus results of the PCT of a round robin on the LRM glass involving six different laboratories.[2] Results for the blanks indicate that contamination of the leachates from possible impurities in the water or on the stainless steel vessels was negligible. Results for the standard ARM-1 glass were compared to a control chart based on results for previous Product Consistency Tests on this standard glass.[5] This comparison is part of the ASTM procedure. Results were between the lower and upper control limits indicating that all the test conditions were properly controlled. Standard solutions containing B, Si, and Na were submitted for analysis with the leachates. Measured results agreed within 10% of the known values indicating that the analyses were sufficiently accurate. Thus the results of the PCT are acceptable. The final pH is an approximate indication of the durability of the glass in a PCT. The higher the final pH, the lower the durability. Measured concentrations are a much more accurate indication. Based on the results in Table IV, the AZ-102 glass appears slightly more durable that the LRM glass. The average normalized releases calculated from the PCT data and the measured composition of the glasses indicate that for all three elements of the glasses (AZ-102 Rad., AZ-102 Surrogate and LRM), the releases are less than 2 g glass/m^2.

TCLP RESULTS

Results for the TCLP (USEPA SW-846 Method 1311) tests performed by BWXT Services, Inc., Lynchburg, VA, on the radioactive AZ-102 glass are shown in Table V. These results show that all metals were well below the limits for characteristically hazardous material and the all metals were well below the Universal Treatment Standard (UTS) limits except for thallium. Thallium was not detected in any of the TCLP leachate samples. However, the glass analysis results for thallium by atomic absorption spectroscopy were low enough (<3 mg/kg) to show that the AZ-102 glass waste form did not contain enough thallium to fail a TCLP test assuming all of the thallium in the glass were to leach out.

CONCLUSIONS

This proof-of-technology demonstration for the Hanford River Protection Project (RPP) Waste Treatment Plant (WTP) has successfully produced crucible-scale glass using a reformulated recipe for the AZ-102 LAW sample. This reformulation

(LAWB88) involved decreasing Fe, Li, Mg and Ti, and increasing B, Ca, Si and Zn from the original AZ-102 formulation (LAWB53). Glass monoliths produced in this study with rapid cooling contained no visible crystalline phase on any surface of the glass. Dissolution of these reformulated, rapidly cooled AZ-102 LAW glasses and analyses of the resulting dissolved glasses provided measured glass species at close to target values for both the radioactive and the surrogate AZ-102 glass. The target waste loading in these reformulated AZ-102 glasses was nominally 5 wt% Na_2O with no sodium added as glass forming minerals. Measured densities for the AZ-102 glass were ~ 2.6 g/cc and the measured density of the LRM glass was ~ 2.5 g/cc compared to a reference value of 2.51 g/cc. Radionuclide measurements of the dissolved glasses also show that Cs-137, Sr-90 and Tc-99 were all at predictable levels in the glass and all radionuclides were below the BNI-WTP contract upper limits. The Cs-137 levels in this AZ-102 LAW glass are below the contract specification of 3 Ci/m^3, but the measured levels of 1 Ci/m^3 are above the current engineering target of 0.3 Ci/m^3. Transuranic levels in the AZ-102 glass were measured to be below 5 nCi/g, which is well below the contract specification of 100 nCi/g for LAW glasses. The product durability of these AZ-102 reformulated glasses have been tested by the PCT and the leach results indicate that all species are released at normalized levels below the BNI-WTP contract upper limit specification of 2 g/m^2. The reference LRM glass and the PCT reference ARM glass leach results have been compared to previous PCT results. These data show that the PCT was performed with proper control of all test conditions. TCLP results indicate that the AZ-102 glass waste form passes the RCRA-metal characteristically hazardous material limits and TCLP releases are also below the UTS limits.

ACKNOWLEDGMENTS

This paper was prepared in connection with work done under Contracts No. AC09-96SR18500 and DOE IWO M0SRLE60 with the U.S. Department of Energy.

REFERENCES

[1] C. L. Crawford et al., "Interim Report for Crucible Scale Active Vitrification of Waste Envelope B (AZ-102) (U), WSRC-TR-2001-00395, SRT-RPP-2001-00153, Rev. 0, March 25, 2002. See http://www.osti.gov/

[2] Ebert, W.L., and Wolfe, S.F., "Round-Robin Testing of a Reference Glass for Low-Activity Waste Forms", ANL-99/22, Argonne National Laboratory, October, 1999; (b) Ebert, W.L., and Wolfe, S.F., "Dissolution Test for Low-Activity Waste Product Acceptance", Argonne National Laboratory, Proceedings of Spectrum '98, Denver, Colorado, Sept. 13-18, 1998, pp. 724-731; (c) Peeler, D., Cozzi, A.D., Best, D.R., Coleman, C.J., and Reamer, I.A., "Characterization of the Low Level Waste Reference Glass (LRM)", WSRC-TR-99-00095, Rev. 0, March 30, 1999.

[3] BNI-WTP Contract, Contract No. DE-AC27-01RV14136, Mod. No. A029, Section C: Statement of Work, September 2003. See http://www.hanford.gov/orp/

[4] G. B. Mellinger and J. L. Daniel, "Approved Reference and Testing Materials for Use in Nuclear Waste Management Research and Development Programs," PNL-4955-1, Pacific Northwest Laboratory, December, 1983.

[5] C. M. Jantzen, N. E. Bibler, D. C. Beam, C. L. Crawford and M. A. Pickett, "Characterization of the Defense Waste Processing Facility (DWPF) Environmental Assessment (EA) Glass Standard Reference Material (U)," WSRC-TR-92-346, Rev. 1, June 1, 1993.

Table I. Comparison of Original (LAWB53) and Revised (LAWB88) AZ-102 Glass Formulation (Glass Former Additives), wt% Oxide

Component	LAWB53	LAWB88	Change
Al₂O₃	6.10	6.48	0.38
B₂O₃	10.03	12.98	2.95
CaO	6.75	7.97	1.22
Fe₂O₃	5.34	2.20	-3.14
Li₂O	5.86	4.69	-1.17
MgO	3.00	1.41	-1.59
SiO₂	48.92	50.00	1.08
TiO₂	1.41	0.00	-1.41
ZnO	3.19	4.87	1.68
ZrO₂	3.19	3.19	0.00
Total	93.79	93.79	

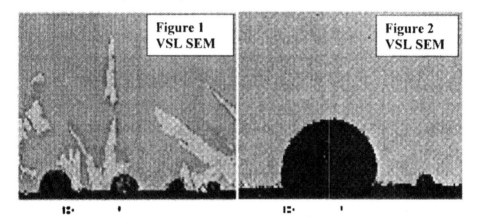

Figure 1: VSL-produced SEM backscattered electron image of LB53SRCC1 (LAWB53 after canister-cooling) crystalline bottom layer, which shows the platinum-nucleated pyroxene phase reaching a few volume%.

Figure 2: VSL-produced SEM backscattered electron image of LB88SRCC2 (LAWB88 after canister-cooling) bottom layer, which shows no pyroxene.

Table II. Results for Analyses of Acid Dissolved and Peroxide Dissolved Glasses

	AZ-102 Rad oxide wt%	AZ-102 Rad-Dup oxide wt%	AZ-102 Rad Target oxide wt%	AZ-102 Surrogate oxide wt%	AZ-102 Surrogate Duplicate oxide wt%	AZ-102 Surrogate Target oxide wt%	LRM oxide wt%	LRM Dup oxide wt%	LRM Target oxide wt%
Al2O3	5.55	5.01	6.5	6.01	5.99	6.43	9.33	9.12	9.5
B2O3	12.40	12.46	13	12.46	12.59	13	7.18	7.41	7.9
BaO	0.01	0.01	-	0.01	0.01	-	0.01	0.01	0.01
CaO	6.65	6.15	8	6.93	6.85	7.97	0.51	0.49	0.54
CdO	0.01	0.01	-	0.01	0.01	-	0.16	0.15	0.16
Cr2O3	0.08	0.08	0.06	0.07	0.07	0.06	0.19	0.19	0.19
CuO	0.06	0.07	-	0.05	0.05	-	0.03	0.03	-
Fe2O3	2.39	2.32	2.2	2.20	2.19	2.2	1.41	1.40	1.42
K2O	0.19	0.15	0.21	0.14	0.16	0.2	1.35	1.31	1.48
La2O3	0.01	0.01	-	0.01	0.01	-	0.01	0.01	0.02
Li2O	4.74	4.61	4.69	4.93	4.89	4.69	0.09	0.09	0.11
MgO	1.88	1.75	1.41	1.53	1.52	1.41	0.10	0.10	0.1
MnO2	0.01	0.01	-	0.01	0.01	-	0.09	0.09	0.08
MoO3	0.06	0.06	-	0.04	0.04	-	0.12	0.12	0.1
Na2O	4.40	4.60	5	5.11	5.04	5.08	20.74	20.19	20.03
NiO	0.02	0.02	-	0.19	0.20	-	0.17	0.17	0.19
P2O5	0.21	0.22	0.02	0.41	0.41	0.02	0.50	0.48	0.53
PbO	0.05	0.05	-	0.11	0.10	-	0.08	0.08	0.1
SiO2	57.04	55.84	50	51.34	49.20	50	58.92	57.93	54.3
SnO2	0.40	0.42	-	0.08	0.08	-	0.07	0.07	0.03
SrO	1.23	1.14	-	0.09	0.09	-	0.09	0.09	-
TiO2	0.17	0.17	-	0.16	0.16	-	0.10	0.10	0.11
UO2	0.33	0.31	-	0.17	0.17	-	0.01	0.01	-
ZnO	5.19	5.11	4.87	4.78	4.74	4.88	0.01	0.00	-
ZrO2	3.31	3.18	3.19	3.10	3.07	3.19	0.95	0.93	0.93
Cl	0.01	0.01	0.01	-	-	-	0.07	0.07	0.07
F	0.05	0.05	0.05	0.10	0.10	0.1	0.86	0.86	0.86
SO3	0.86	0.86	0.86	0.77	0.77	0.77	0.30	0.30	0.3
TOTALS	107.40	104.67	100.07	100.82	98.53	100	103.45	101.80	99.06

	AZ-102 Rad oxide wt%	AZ-102 Rad-Dup oxide wt%	AZ-102 Rad Target oxide wt%	AZ-102 Surrogate oxide wt%	AZ-102 Surrogate Duplicate oxide wt%	AZ-102 Surrogate Target oxide wt%	LRM oxide wt%	LRM Dup oxide wt%	LRM Target oxide wt%
Al2O3	6.44	6.40	6.50	6.39	6.31	6.43	9.41	9.64	9.54
B2O3	12.44	12.44	13.00	12.57	12.44	13.00	7.16	7.39	7.90
BaO	0.01	0.01	-	0.01	0.01	-	0.00	0.00	0.01
CaO	6.90	6.92	8.00	6.81	6.80	7.97	0.49	0.50	0.54
CdO	0.00	0.00	-	0.00	0.04	-	0.15	0.16	0.16
Cr2O3	0.07	0.07	0.06	0.06	0.06	0.06	0.19	0.20	0.19
CuO	0.05	0.05	-	0.05	0.05	-	0.03	0.03	-
Fe2O3	2.23	2.23	2.20	2.24	2.19	2.20	1.37	1.42	1.42
K2O	0.22	0.23	0.21	0.22	0.22	0.20	1.31	1.33	1.48
La2O3	0.01	0.01	-	0.01	0.01	-	0.01	0.01	0.02
Li2O	4.89	4.87	4.69	4.87	4.87	4.69	0.10	0.10	0.11
MgO	1.57	1.58	1.41	1.61	1.59	1.41	0.10	0.11	0.10
MnO2	0.02	0.02	-	0.02	0.02	-	0.07	0.08	0.08
MoO3	0.02	0.02	-	0.01	0.00	-	0.09	0.10	0.10
Na2O	4.40	4.60	5.00	5.11	5.04	5.08	20.74	20.19	20.03
NiO	0.02	0.02	-	0.00	0.01	-	0.17	0.17	0.19
P2O5	0.18	0.19	0.02	0.17	0.18	0.02	0.32	0.34	0.53
PbO	0.04	0.05	-	0.04	0.04	-	0.10	0.10	0.10
SiO2	49.57	51.28	50.00	51.29	49.14	50.00	58.92	57.93	54.26
SnO2	0.08	0.08	-	0.07	0.07	-	0.01	0.02	0.03
SrO	1.25	1.24	-	1.24	1.24	-	0.13	0.13	-
TiO2	0.16	0.16	-	0.16	0.16	-	0.10	0.10	0.11
UO2	0.14	0.14	-	0.14	0.14	-	0.00	0.00	-
ZnO	5.35	5.37	4.87	5.28	5.18	4.88	0.01	0.01	-
ZrO2	3.23	3.22	3.19	3.25	3.21	3.19	0.88	0.90	0.93
Cl	0.01	0.01	0.01	-	-	-	0.07	0.07	0.07
F	0.05	0.05	0.05	0.10	0.10	0.10	0.86	0.86	0.86
SO3	0.86	0.86	0.86	0.77	0.77	0.77	0.30	0.30	0.30
TOTALS	100.22	102.10	100.07	101.73	99.12	100.00	103.09	102.16	99.06

Table III. Radionuclide Analyses for Cs-137, Sr-90, Tc-99, Total Alpha and Total Beta

		Cs-137			Sr-90				Tc-99			
		uCi/mL	Ci/g	Ci/m³	dpm/mL	uCi/mL	Ci/g	Ci/m³	dpm/mL	uCi/mL	Ci/g	Ci/m³
Peroxide	Rad	1.96E-03	3.74E-07	9.65E-01	5.03E+03	2.27E-03	4.32E-07	1.12E+00	< 8.76E+00	3.95E-06	7.52E-10	1.95E-03
Fusions	Rad	1.96E-03	3.86E-07	9.99E-01	8.15E+03	3.67E-03	7.24E-07	1.87E+00	< 2.55E-01	1.15E-05	2.27E-09	5.88E-03
	Surrogate	< 8.10E-06	1.46E-09	3.77E-03	< 9.77E+02	4.40E-04	7.91E-08	2.05E-01	< 7.17E+00	3.23E-06	5.81E-10	1.50E-03
	Surrogate	< 7.30E-06	1.42E-09	3.66E-03	< 9.58E+02	4.32E-04	8.38E-08	2.17E-01	< 8.39E+00	3.78E-06	7.34E-10	1.90E-03
	ReagBlnk	< 7.72E-06	-	-	< 1.12E+03	5.04E-04	-	-	9.36E+00	4.22E-06	-	-
Acid	Rad	2.00E-03	3.86E-07	9.99E-01	4.23E+03	1.91E-03	3.68E-07	9.52E-01	< 5.68E+00	2.56E-06	4.94E-10	1.28E-03
Dissolutions	Rad	1.85E-03	3.63E-07	9.39E-01	4.10E+03	1.85E-03	3.62E-07	9.37E-01	< 7.61E+00	3.43E-06	6.73E-10	1.74E-03
	Surrogate	< 7.30E-06	1.44E-09	3.73E-03	< 1.13E+03	5.09E-04	1.01E-07	2.60E-01	4.18E-01	1.88E-05	3.72E-09	9.63E-03
	Surrogate	< 6.38E-06	1.25E-09	3.24E-03	< 9.98E+02	4.50E-04	8.82E-08	2.28E-01	6.25E+01	2.82E-05	5.52E-09	1.43E-02
	ReagBlnk	< 8.82E-06	-	-	< 1.09E+03	4.91E-04	-	-	< 5.04E+00	2.27E-06	-	-

		Alpha				Beta			
		dpm/mL	uCi/mL	Ci/g	Ci/m³	dpm/mL	uCi/mL	Ci/g	Ci/m³
Peroxide	Rad	17.2	7.75E-05	1.48E-09	3.82E-03	1.42E+04	6.40E-03	1.22E-06	3.15E+00
Fusions	Rad	16.8	7.57E-05	1.49E-09	3.86E-03	1.42E+04	6.40E-03	1.25E-06	3.26E+00
	Surrogate	< 8.16	3.68E-05	6.61E-10	1.71E-03	< 18.4	8.29E-06	1.49E-09	3.85E-03
	Surrogate	< 8.38	3.77E-06	7.33E-10	1.89E-03	< 29.2	1.32E-05	2.55E-09	6.60E-03
	ReagBlnk	< 10.2	4.59E-05	-	-	< 24.6	1.11E-05	-	-
Acid	Rad	39.8	1.79E-05	3.46E-09	8.95E-03	1.53E+04	6.89E-03	1.33E-06	3.44E+00
Dissolutions	Rad	31	1.40E-05	2.74E-09	7.09E-03	1.47E+04	6.62E-03	1.30E-06	3.36E+00
	Surrogate	< 8.58	3.86E-06	7.64E-10	1.98E-03	< 29.6	1.33E-05	2.64E-09	6.82E-03
	Surrogate	< 7.17	3.23E-06	6.34E-10	1.64E-03	< 19.2	8.65E-06	1.70E-09	4.39E-03
	ReagBlnk	< 8.19	3.68E-06	-	-	< 20.8	9.37E-06	-	-

Table IV. Results from 90°C PCT for Quenched / Canister-Cooled AZ-102 Glass

Sample ID		B	Si	Na	pH (b)
Blanks(a)	(ppm)	<0.031	<0.018	0.225	6.97
ARM(a)	(ppm)	19.8 / 17	67.1 / 62.1	40.1 / 36.1	9.8 / 9.4
AZ-102(a) Surrogate	(ppm)	17.7	45.3	15.29	9.0
Norm.Release	g glass/m^2	0.219±0.004	0.097±0.001	0.203±0.005	-
AZ-102(a) Radioactive	(ppm)	16.1 / 12.8	42.4 / 37.15	11.8 / 10.14	8.9 / 8.9
Norm.Release	g glass/m^2	0.199±0.002/ 0.159±0.006	0.091±0.001/ 0.079±0.002	0.159±0.002/ 0.137±0.006	-
LRM(a)	(ppm)	23.7 / 19.6	78.5 / 72.5	149.5 / 132.4	10.2 /10.1
Norm.Release	g glass/m^2	0.506±0.015/ 0.420±0.04	0.173±0.006/ 0.160±0.03	0.482±0.016/ 0.427±0.002	-
LRM(c)	(ppm)	26.7	82.0	159.7	11.7

a) Based on triplicate tests; b) Initial pH = 6.57; c) published consensus LRM[2]

Table V. TCLP Results in mg/L

Analytes	Sample	Duplicate	Characteristic Limit	UTS Limit
Antimony(Sb)	0.25	<0.2	-	1.15
Arsenic(As)	<0.03	<0.03	5	5
Barium(Ba)	0.25	0.21	100	21
Beryllium(Be)	<0.002	<0.002	-	1.22
Cadmium(Cd)	<0.02	<0.02	1	0.11
Chromium(Cr)	<0.08	<0.08	5	0.6
Lead(Pb)	<0.4	<0.4	5	0.75
Mercury(Hg)	<0.002	<0.002	0.2	0.025
Nickel(Ni)	<0.06	<0.06	-	11
Selenium(Se)	<0.1	<0.1	1	5.7
Silver(Ag)	<0.06	<0.06	5	0.14
Thallium(Tl)	<0.5	<0.5	-	0.2
Vanadium(V)	0.028	0.043	-	1.6
Zinc(Zn)	1.5	2.1	-	4.3

Glass Formulation and Property Models

PRELIMINARY GLASS DEVELOPMENT AND TESTING FOR IN-CONTAINER VITRIFICATION OF HANFORD LOW-ACTIVITY WASTE

J.D. Vienna, D.-S. Kim, M.J. Schweiger,
P. Hrma, J. Matyáš, J.V. Crum, and D.E. Smith
Pacific Northwest National Laboratory
Richland, Washington

INTRODUCTION

Roughly 200,000 m^3 of high-level waste (HLW) are stored at the Hanford site. This waste will be separated into HLW and low-activity waste (LAW) fractions and each fraction will be immobilized for final storage/disposal. The US Department of Energy (DOE) Office of River Protection (ORP) is constructing a Waste Treatment and Immobilization Plant (WTP) which will be capable of separating the waste, vitrifyingthe entire HLW fraction of the waste and vitrifying roughly 50% the LAW fraction. The remaining fraction of LAW will be immobilized by one of a number of possible technologies. ORP, through its contractor CH2M Hill Hanford Group, is currently evaluating options for LAW immobilization to supplement the WTP. One possible option is In-Container Vitrification (ICV) of the LAW.

ICV is a technology developed by AMEC, GeoMelt Division, for treatment of hazardous, radioactive, and mixed wastes. The ICV process, as applied to Hanford LAW, includes the blending of liquid waste with additives (primarily composed of local soil) and drying to a granular state. The dried material is loaded into a refractory lined steel box and melted by passing a current through the material between two graphite electrodes. The box containing the molten waste/additive mixture is allowed to cool, backfilled, and disposed.

The purpose of this study was to develop a glass composition suitable for the demonstration of ICV on Hanford LAW at full scale. Testing included crucible-scale tests with simulants and actual Hanford LAW. Following the crucible-scale tests, engineering-scale and large-scale melts were performed with LAW simulants. This paper discusses the formulation and testing of glass compositions for ICV of Hanford LAW at crucible scale. The results from process scale-up tests are reported elsewhere.

FORMULATION AND TESTING OF PRELIMINARY GLASSES

In the ICV process, the major additive to the waste is soil. Sixteen preliminary glass compositions were formulated in four groups to investigate the impacts of waste loading, soil composition variation (only Al_2O_3 and Fe_2O_3 content), and additional additive composition and concentration on key glass properties. The representative LAW composition used for testing (Table 1) was supplied by CH2M Hill. Glass formulations were focused on achieving target Na_2O loadings in the vitrified product ranging from 17 to 26 mass% with Hanford soil (of the composition listed in Table 1) as the major additive. Glasses were formulated in four general sets:

1. Glasses -01 through -04 were soil-LAW mixtures that ranged from 17 to 26 mass% Na_2O.

2. Glasses -05 through -08 contained 20 and 23 mass% Na_2O with additives of 5 mass% ZrO_2 or 2.5 mass% ZrO_2 + 2.5 mass% B_2O_3 in addition to soil.

3. Glasses -09 through -12 were formulated with varied concentrations of Al_2O_3 and Fe_2O_3 keeping the same proportions of soil, LAW, and ZrO_2 as in glass 7.

4. Glasses -13 through -16 contained 5 mass% B_2O_3 and varied ZrO_2 concentration. Glass -15 also contained P_2O_5, La_2O_3, and TiO_2 as additional additives, and glass 16 contained 3 mass% SiO_2 as an additive. Glasses -13 through -15 contained 20 mass% Na_2O and glass -16 contained 17 mass% Na_2O.

The resulting 16 preliminary glass compositions are listed in Table 2. These glasses were fabricated with a target Fe(II):Fe ratio of roughly 50% because the ICV process is expected to operate in a relatively reducing mode.

Table 1. Soil and Waste-Simulant Compositions (mass fractions)

	Soil	LAW
Al_2O_3	0.1396	0.0188
CaO	0.0550	
Cl		0.0090
Cr_2O_3		0.0046
F		0.0035
Fe_2O_3	0.0928	
K_2O	0.0248	0.0034
MgO	0.0143	
Na_2O	0.0321	0.8983
P_2O_5	0.0029	0.0202
$ReO_2^{(a)}$		0.0001
SO_3		0.0418
SiO_2	0.6242	
TiO_2	0.0143	

$^{(a)}$ Re was added to the simulant as Re_2O_7 as a Tc surrogate.

Batch chemicals were weighed to within ±1 relative% precision. Batches were mixed in an agate milling chamber in the Angstrom milling machine for approximately 4-min and melted in Pt-10%Rh crucibles at temperatures ranging from 1250 to 1510°C. The crucible was covered with a lid with a hole through which argon was introduced to prevent oxidation. After 1 h of melting, the glass was quenched on a stainless steel plate. The glass was homogenized by grinding into a fine powder in a tungsten-carbide milling chamber in the Angstrom milling machine for 4 min. The ground glass was remelted in a Pt-10%Rh crucible with a tight lid under argon at the same temperature as the first melt. Each melt was divided into three portions which were quenched on a steel plate (Q), annealed, or slow cooled (SC) according to the slowest cooling schedule expected in a full size ICV glass package. The details of glass compositions, melting temperatures, redox control, and SC schedule are reported elsewhere.[1]

Table 2. Compositions of 16 Preliminary ICV Glasses

Glass	-01	-02	-03	-04	-05	-06	-07	-08	-09	-10	-11	-12	-13	-14	-15	-16
Al_2O_3	12.04	11.62	11.20	10.78	10.90	10.90	10.48	10.48	8.48	12.48	10.80	10.16	9.89	9.46	9.46	9.88
B_2O_3	0.00	0.00	0.00	0.00	0.00	2.50	0.00	2.50	0.00	0.00	0.00	0.00	5.00	5.00	5.00	5.00
CaO	4.62	4.43	4.24	4.05	4.15	4.15	3.96	3.96	4.09	3.83	4.08	3.83	3.75	3.58	3.58	3.77
Cl	0.14	0.17	0.21	0.24	0.18	0.18	0.21	0.21	0.21	0.21	0.21	0.21	0.18	0.18	0.18	0.15
Cr_2O_3	0.07	0.09	0.11	0.12	0.09	0.09	0.11	0.11	0.11	0.11	0.11	0.11	0.09	0.09	0.09	0.08
F	0.06	0.07	0.08	0.09	0.07	0.07	0.08	0.08	0.08	0.08	0.08	0.08	0.07	0.07	0.07	0.06
Fe_2O_3	7.80	7.48	7.16	6.84	7.00	7.00	6.68	6.68	6.90	6.45	4.68	8.68	6.33	6.04	6.04	6.36
K_2O	2.14	2.07	1.99	1.92	1.94	1.94	1.86	1.86	1.92	1.80	1.92	1.81	1.76	1.68	1.68	1.76
MgO	1.20	1.15	1.10	1.05	1.08	1.08	1.03	1.03	1.06	0.99	1.06	1.00	0.97	0.93	0.93	0.98
Na_2O	17.00	20.00	23.00	26.00	20.00	20.00	23.00	23.00	23.00	23.00	23.00	23.00	20.00	20.00	20.00	17.00
P_2O_5	0.57	0.63	0.69	0.75	0.61	0.61	0.67	0.67	0.68	0.67	0.68	0.67	0.60	0.59	1.59	0.53
ReO_2	0.01	0.01	0.01	0.01	0.01	0.01	0.01	0.01	0.01	0.01	0.01	0.01	0.01	0.01	0.01	0.01
SiO_2	52.48	50.32	48.16	46.00	47.07	47.07	44.92	44.92	46.44	43.41	46.35	43.47	42.55	40.61	40.61	45.75
SO_3	0.67	0.81	0.95	1.10	0.82	0.82	0.96	0.96	0.96	0.97	0.96	0.97	0.83	0.83	0.83	0.69
TiO_2	1.20	1.15	1.10	1.05	1.08	1.08	1.03	1.03	1.06	0.99	1.06	1.00	0.97	0.93	1.93	0.98
La_2O_3	0.00	0.00	0.00	0.00	0.00	0.00	0.00	0.00	0.00	0.00	0.00	0.00	0.00	0.00	2.00	0.00
ZrO_2	0.00	0.00	0.00	0.00	5.00	2.50	5.00	2.50	5.00	5.00	5.00	5.00	7.00	10.00	6.00	7.00

Glass samples with these two temperature histories (i.e., Q/annealed and SC) were tested for key properties including: vapor hydration test (VHT) response, product consistency test (PCT) response, toxicity characteristic leaching procedure (TCLP) response, density (ρ), and crystal-phase identification.

All glasses except for -04 passed the 2 g/m^2 constraint for Hanford LAW glasses for PCT for both Q and SC samples; glasses 13 through 16 passed by nearly an order of magnitude. Comparing normalized releases showed that the normalized release of Na (r_{Na}) is the most conservative indication of glass dissolution in PCT conditions. Figure 1 shows the effect of SC treatment on the r_{Na} values for ICV glasses. SC treatment results in slightly lower r_{Na} values as compared to Q glass indicating that the SC treatment had no adverse effect on the PCT response.

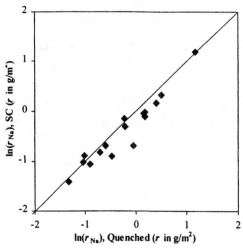

Figure 1. Effect of Slow Cooling on r_{Na}

The 14 day VHT responses of all preliminary glasses were measured on both Q and SC samples at 200°C. Glasses numbered -01 to -04, -07, -08, and -10 to -12, and -06 (Q), completely corroded during the 14-day VHTs. Glasses -13, -14, and -16 were tested for times of 7, 14, and 28 days and had average corrosion rates lower than 50 g/(m²·d) for Q and SC samples and for all durations tested. The VHT responses of these three glasses are compared to data from literature in Figure 2 showing that the VHT response of typical ICV glass samples was well below the constraint, better than the glass that formed the basis of the 2001 Hanford Site performance assessment (LAW-ABP1)[2], and at least as well as typical WTP glasses.[3] Figure 3 compares average (e.g., assuming constant rate from zero time to 14 days) corrosion rates (r_a) of Q and SC-treated samples. The general trends suggest that the higher the concentrations of Al_2O_3 and Na_2O, the higher the VHT alteration while the higher the ZrO_2 and B_2O_3, the lower the VHT alteration. These results are consistent with those reported earlier, which found Al_2O_3 to have a non-linear effect on VHT response.[4]

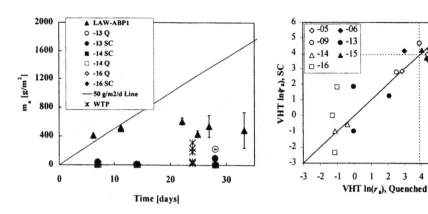

Figure 2. Comparison of VHT Responses for ICV Glasses -13, -14, and -16 with those from LAW-ABP1 and Typical WTP Glasses

Figure 3. Comparison of VHT Corrosion Rates in Q and SC Samples

The TCLP was performed on selected preliminary ICV glasses. The releases of Cr and B were measured in these glasses. The Cr is the only regulated component in these preliminary ICV glasses. The normalized B release (r_B) can be used as a representative measure of glass dissolution in the TCLP condition.[5] As expected from low Cr_2O_3 in glass (0.09 to 0.12 mass%), all the glasses, even the glass with 26 mass% Na_2O (-04), exhibited Cr release well below the Uniform Treatment Standards limit of 0.6 mg/L. The ICV glasses had equal or lower r_B than typical WTP LAW glasses.[3,6] This result implies that the ICV glasses have a low risk of failing the TCLP requirements to obtain a variance to land disposal restrictions.

The melt viscosity (η) and conductivity (ε) of glasses -13 (20% Na_2O) and -16 (17% Na_2O) were measured as functions of temperature. The responses can be summarized by the Arrhenius relationships: $\ln[\varepsilon\ (S/m)] = 8.4 - 7528/T(K)$ and $\ln[\varepsilon\ (S/m)] = 8.3 - 7766/T(K)$ for conductivity and $\ln[\eta\ (Pa\cdot s)] = -13.424 + 23771/T(K)$ and $\ln[\eta\ (Pa\cdot s)] = -13.112 + 24056/T(K)$ for viscosity (for glasses -13 and -16, respectively). The ICV process typically operates at viscosities near 10 Pa·s which corresponds to 1238°C and 1288°C (for glasses -13 and -16, respectively). In the ICV process, the temperature is adjusted by power input through graphite electrodes and is not limited by electrode lifetime issues.

Only glasses -04, -10, -12, and -14 showed signs of phase changes upon SC. In glasses -04, -10, and -12, large numbers of crystals appeared at the SC sample surfaces within ~1.5 mm from the Pt-glass interface. A few crystals, below the detection limit of XRD, were seen in the bulk glass. Table 3 lists mass fractions of crystalline phases identified in this layer: nepheline [$NaAlSiO_4$], combeite [$Na_4Ca_4(Si_6O_{18})$], and baddeleyite [ZrO_2]. In glass -14, ZrO_2 crystals partially settled to a ~5-mm layer at the crucible bottom. Mass fractions of a ZrO_2 phase in the bulk sample and the crucible bottom area are also included in Table 3.

Table 3. Crystalline Phases, in Mass%, Determined by XRD in SC Samples within ~1.5-mm of Pt-Glass Interface

Glass	Nepheline ($NaAlSiO_4$)	Combeite [$Na_4Ca_4(Si_6O_{18})$]	Baddeleyite (ZrO_2)
-04-SC, Pt-glass interface area	1.15	0	0
-10-SC, Pt-glass interface area	6.90	1.95	0.35
-12-SC, Pt-glass interface area	0.35	0	0.27
-14-SC, bulk glass	0	0	0.49
-14-SC, crucible bottom area	0	0	5.36

BASELINE GLASS SELECTION AND TESTING RESULTS

The glass with the best mix of properties is -13. This glass contains 20 mass% Na_2O, 7 ZrO_2, and 5 B_2O_3; has outstanding PCT and VHT responses; and does not contain crystals after SC heat treatment. This made it suitable for scale-up and radioactive testing of the ICV process. Table 4 lists a series of glass compositions based on glass -13 with waste loadings ranging between 16.4 and 24.4 mass% (e.g., between 17 and 24 mass% Na_2O). These compositions were fabricated using local soil and liquid LAW simulant.

The PCT, VHT and TCLP responses of the glasses as functions of Na_2O content are shown in Figures 4 through 6. The r_{Na} values were found to be a conservative measure of glass release rates. The normalized releases by PCT were far below the imposed limit of 2 g/m^2 for all glasses. Releases increase with increased waste loading and are well predicted using the models reported previously.[7] The VHT responses also increase with increasing waste loading and passed the imposed limit of 50 $g/m^2/d$ for all but the highest Na_2O glass (e.g., 24%). The TCLP responses of the glasses were relatively insensitive to waste loading and averaged near 40 mg/L for r_B. The slow cooling of glasses tended to improve the chemical durability of these glasses. No secondary phases were identified in any of the samples with Na_2O content higher than 17%. A small (~2μm) well dispersed phase was found in low concentrations in slow cooled samples of both 17% Na_2O glasses.

A radioactive glass was fabricated at crucible scale using actual Hanford LAW with a target composition the same as BL-20 glass (Table 4). The PCT and VHT responses and densities of the two glasses are within one to ten relative percent of each other, with the exception of the PCT normalized boron release of the SC samples which were 33% different (with higher release from the simulant glass). These results show excellent agreement between the properties of simulated and actual LAW glasses.

To test the impact of redox on VHT response, the 22 mass% Na_2O glass was prepared in three different oxidation-reduction states. To control the atmosphere, glasses were melted in a sealed alumina tube with openings for inlet and outlet gas tubes and for the thermocouple. The VHT alteration mass decreased linearly with the increasing fraction of Fe(II) for all times tested (7, 14, and 28 days). This result implies that Fe(III) tends to increase the VHT alteration rate which is not consistent with earlier work by Vienna et al. that showed the final corrosion rate (r_∞) decreased with increasing Fe(II)/Fe fraction.[4] However, this effect of Fe(II)/Fe on r_∞ was only measured on one glass sample and further testing may be require to verify this result and determine its applicability to other compositions.

Table 4. Composition of Baseline Glasses (mass% constituents)

	BL-20	BL-17a	BL-17b	BL-22	BL-24
Soil	68.2	68.5	71.6	65.9	63.6
Waste	19.8	16.5	16.4	22.1	24.4
B_2O_3	5.0	5.0	5.0	5.0	5.0
ZrO_2	7.0	7.0	7.0	7.0	7.0
SiO_2	0.0	3.0	0.0	0.0	0.0

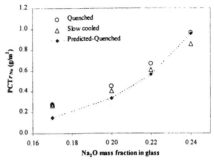

Figure 4. r_{Na} (g/m²) as a Function of Na_2O Concentration

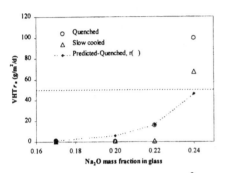

Figure 5. VHT Alteration Rate (g/m²/d) as a Function of Na_2O Concentration

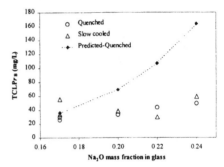

Figure 6. TCLP r_B (mg/L) as a Function of Na_2O Concentration

SUMMARY AND DISCUSSION

A glass was developed for demonstration of the ICV process with Hanford LAW. This glass (BL-20) easily met all the durability constraints placed on a Hanford LAW glass. The main chemical additive is local soil, with ZrO_2 (7%) and B_2O_3 (5%) as minor additives. The glass was tested with variations in redox and waste loading and found to be tolerant to relatively broad ranges of both and still met all the product requirements. The use of LAW simulants for measurement of glass properties was verified through fabrication and testing of an actual LAW glass of the same target composition as the baseline glass.

It's been demonstrated on crucible scale that the ICV process can successfully treat Hanford LAW with 20% waste loading. Tests at engineering- and full-scale with simulated Hanford LAW were successfully performed and resulting properties were consistent with those of the crucible-scale work.

ACKNOWLEDGMENT

This study was funded by AMEC, GeoMelt Division, through a U.S. Department of Energy (DOE) contract with CH2M Hill Hanford Group (CHG). The authors thank Leo Thomson (AMEC), Pat Lowery (AMEC), Dennis Hamilton (CHG), and Rick Raymond (CHG) for support and guidance during this study. David Peeler and David Best from the Savannah River Technology Center performed most of the chemical analyses for this study. Pacific Northwest National Laboratory is operated for the DOE by Battelle under Contract DE-AC06-76RL01830.

REFERENCES

[1] DS Kim, JD Vienna, P Hrma, MJ Schweiger, J Matyáš, JV Crum, DE Smith, GJ Sevigny, WC Buchmiller, JS Tixier, Jr., JD Yeager, KB Belew, *Development and Testing of ICV Glasses for Hanford LAW*, PNNL-14351, Pacific Northwest National Laboratory, Richland, Washington (2003).

[2] FM Mann, KC Burgard, WR Root, RJ Puigh, SH Finfrock, R Khaleel, DH Bacon, EJ Freeman, BP McGrail, SK Wurstner, and PE Lamont, *Hanford Immobilized Low-Activity Waste Performance Assessment: 2001 Version*, DOE/ORP-2000-24, Rev. 0, U. S. Department of Energy, Office of River Protection, Richland, Washington (2001)

[3] IS Muller, AC Buechele, and IL Pegg, *Glass Formulation and Testing with RTP-WTP LAW Simulants-Final Report*, VSL-01R3560-2, The Catholic University of America, Washington, D.C. (2001).

[4] JD Vienna, P Hrma, A Jiricka, DE Smith, TH Lorier, IA Reamer, and RL Schulz, *Hanford Immobilized LAW Product Acceptance Testing: Tanks Focus Area Results*, PNNL-13744, Pacific Northwest National Laboratory, Richland, Washington (2001).

[5] DS Kim and JD Vienna Kim, D-S. and J. D. Vienna, *Model for TCLP Releases from Waste Glasses*, PNNL-14061, Rev. 1, Pacific Northwest National Laboratory, Richland, Washington (2003).

[6] IS Muller and IL Pegg, *Glass Formulation and Testing with TWRS LAW Simulants*, Final Report for GTS Durateck Inc. and BNFL Inc., Catholic University of America, Washington D.C. (1998).

[7] JD Vienna, DS Kim, and P Hrma, *Database and Interim Glass Property Models for Hanford HLW and LAW Glasses*, PNNL-14060, Pacific Northwest National Laboratory, Richland, Washington (2002).

EVALUATION OF MELT RATE THROUGH HIGHER WASTE LOADING

Troy H. Lorier and Psaras L. McGrier
Westinghouse Savannah River Company
999-W, Rm. 401
Aiken, SC 29808

ABSTRACT

This report describes scoping investigations into higher waste loading (WL) effects on melt rate for sludge batch 2 (SB2) at the Westinghouse Savannah River Company (WSRC). Previous baseline tests in support of the melt rate program were performed at 25% WL. Waste loadings for these higher waste-loading tests ranged from 27% to 41% (in 2% increments), and all utilized Frit 320 and simulated SB2 Sludge Receipt and Adjustment Tank (SRAT) product. The 41% WL bounds projected maximum waste loadings based on implementation of the new liquidus temperature (T_L) model. The results and trends of these scoping studies indicate that within the WL range tested, melt rate decreases as WL increases (see Table I and Figure I). However, the net effect is that the amount of waste processed per hour (throughput) increases relative to the 25% WL baseline up to 35% WL, based on the linear melt rate data (see Table II and Figure II). Operation beyond 37% WL may not be advantageous unless the cost savings associated with the reduced number of canisters outweighs the throughput reduction.

INTRODUCTION

Fundamental work is needed to better understand the effects of feed chemistry, rheology, and process chemistry on melting behavior for Department of Energy (DOE) high-level waste (HLW) glasses. Melt rate furnace testing is being conducted with the goal of improving melt rate, but also to establish an improved understanding of parameters controlling melt rate. Presently, dry-fed melt rate furnace (MRF) tests are being used to obtain relative melt rate behavior, and a slurry-fed melt rate furnace (SMRF) is being developed to increase confidence in the present tests and to reduce the risk that no major differences are observed from dry-fed to slurry-fed systems. Melting behavior, foaming, and changes to the acid addition chemistry will be evaluated using both melt rate furnaces, while the effects of slurry feeding and the associated cold-cap chemistry will be assessed with the SMRF.

Glass melting is a complex process that involves a number of reactions and transformations, and its rate or behavior can only be described considering all related processing properties. Given that, one needs to establish or identify the relevant processes that impede melt rate for a given system and focus research and development efforts on potential mitigation techniques. Melt rate is defined as how quickly feed materials are converted to a liquid melt (i.e., the rate of the batch-to-glass conversion process) which is ultimately linked to glass throughput.

The current focus of this research is to enhance the basic understanding of the role of glass chemistry and/or acid addition strategy changes on the overall melting process for the Defense Waste Processing Facility (DWPF). More specifically, by controlling the chemistry of the incoming feed materials (e.g., in particular the frit composition for a given sludge) or by adjusting chemical processing strategies (e.g., formic or nitric acid relative to the current flowsheet), the conversion rate of slurry-fed raw materials into a molten state can be increased. Intermediate reaction products and/or the development of an insulating foamy layer which can impede the melting rate can be avoided by altering the reaction pathway within the cold cap. If successful, the result is an enhanced melting process which increases throughput without compromising the quality of the final waste form.

Increased melting efficiencies decrease overall operational costs by reducing the immobilization campaign time for a particular waste stream and the entire DWPF process. For melt rate limited systems, a small increase in melting efficiency translates into significant savings by reducing operational costs. For example, a 10% increase in melter throughput (melt rate) would reduce the DWPF's overall processing time by almost two years, translating to an $860M savings in life cycle costs ([1]Savannah River Site High Level Waste System Plan, Rev. 12).

A study performed in 2001 concluded that a change from Frit 200 to Frit 320 would increase the melt rate in the DWPF for SB2, without decreasing the WL ([2]Lambert et. al, 2001). Since those initial crucible-scale and 4" dry-feed melt rate furnace (MRF) tests, Frit 320 has been tested in the Minimelter ([3]Miller, 2002) and further in the 4" MRF ([4]Stone and Josephs, 2001), all of which showed similar trends in melt rate. For this particular study however, higher WL of SB2 with Frit 320 was the focus. Frit 320 was tested in the 4" MRF for waste loadings ranging from 27 to 41%.

Increasing the WL in the glass is desirable to reduce the amount of glass generated and ultimately the number of canisters. Dramatic cost savings can also be realized through increased waste loadings. It has been estimated that a 1% increase in WL would translate into more than $330M total cost savings to the DWPF (assuming the same fixed annual operating cost described above and a 1% increase from 29% to 30% WL). Given that WL has typically been limited by predictions of T_L, the Savannah River Technology Center (SRTC) developed a new T_L model. This new model allows for higher waste loadings to be achieved not only for SB2, but for future sludge batches as well ([5]Brown et. al, 2001). Implementation of this new T_L model in the DWPF process control system is currently in progress. The effects of implementing the new T_L model and Frit 320 offer a significant advantage in terms of WL projections for SB2 – increasing from a maximum WL of ~31% with the old T_L model and Frit 200 to ~39% when using the new T_L model and Frit 320. It should be noted that the WL projections are based on nominal sludge compositions and do not take into account any compositional variation in the sludge.

Current plans are to implement Frit 320 in 4th quarter FY02 without a change in WL to assess the impact of this frit change on melt rate relative to Frit 200. Once that has been established, plans are to increase waste loadings in incremental steps up to the point allowed by the new T_L model. Frit 320 was developed for lower waste loadings (25%) and prior to development of the new liquidus model, and may not be the preferred frit for higher waste loadings. For a specific frit, lower melt rates are generally expected at higher waste loadings. Therefore, the work described in this report was undertaken to provide an initial evaluation of the effect of increased WL on melt rate and overall waste throughput.

DISCUSSION

All tests for the higher waste-loading series were completed with melter feed prepared from SB2 simulant. The feed preparation process converted the Tank 8 and Tank 40 blend into feed for the 4" MRF. The feed preparation process was composed of three steps: 1) adding trim chemicals to the sludge simulant, 2) processing the sludge simulant through a SRAT process, and 3) combining the SRAT product with Frit 320 and removing water by drying in an oven. The amounts of the SRAT product and Frit 320 in Step 3 were based on the % WL being evaluated. All feed preparation methods are discussed in [6]Stone and Lambert (2001).

The melt rate furnace test methods (equipment description, equipment operation, etc.) are discussed in Stone and Josephs (2001). For each higher waste-loading test performed, four thermocouples were placed in the beaker at ½", 1", 2", and 5" from the bottom. For consistency and comparison purposes, the run time was 42 minutes for each run, this time being determined by testing performed in 2001 (Stone and Josephs, 2001).

Temperature profiles generated during each melter run were very useful in determining the melting and cold cap behaviors for each batch. For instance, the plots for waste loadings up to 35% show there is significant rise and fall in temperature throughout the entire run, especially at heights of ½" and 1" from the bottom of the beaker. This would indicate that heat is being transferred through the batch well, and the off-gases being produced are escaping past the cold cap. Beginning at 37% WL, and especially at 39% and 41% WL, there is a significant rise in temperature which does not dissipate quickly, indicating significant volume expansion and/or foam formation. This creates an insulating effect, which greatly slows down the transfer of heat to the unreacted material above. The cold cap and melting behavior of the batch are adversely affected, and more foam and bridging ultimately result. Bridging is when an air gap develops between the glass pool and the melting feed and does not dissipate rapidly (Stone and Josephs, 2001).

With a short residence time in the furnace and increased waste loadings, it was not expected that each batch would melt to completion. The first indication of unreacted feed after 42 minutes arose in the 31% WL run. As the WL increased, more unreacted feed remained on top of the batch after 42 minutes. Also, larger bubbles remained trapped during higher WL runs, creating an insulating layer that ultimately slowed down melt rate, resulting in less glass being produced (more unreacted batch remaining).

The method used to determine the melt rate was to measure the height of the glass pool formed during each test and divide by the run time (consistent with that used by Stone and Josephs, 2001). The beakers were sectioned in half and the height of the glass pool was measured from the bottom of the beaker to the point where the glass was no longer free of bubbles.

Two melt rates were calculated for each run – a linear melt rate and a volumetric melt rate. The linear method involves measuring the glass height at ¼" intervals across the beaker, and then averaging the values to obtain an average glass pool height. The average glass pool height was divided by the run time to obtain the melt rate result in inches per hour. The volumetric method involves calculation of the volume of each concentric ring represented by the glass pool height at ¼" intervals, and then summing the volume of the rings to obtain the volume of the glass produced during the run. The glass volume was then divided by the run time to obtain a melt rate in cubic inches per hour. This melt rate data is presented in Table I and Figure I.

Table I. Semi-quantitative Melt Rates for Each Waste Loading.

Waste loading	Linear melt rate (in./hr.)	Volumetric melt rate (in³/hr.)
25%	1.04	12.8
27%	1.04	13.1
29%	0.92	12.1
31%	0.93	11.7
33%	0.81	12.4
35%	0.89	11.6
37%	0.82	10.6
39%	0.56	8.8
41%	0.58	8.7

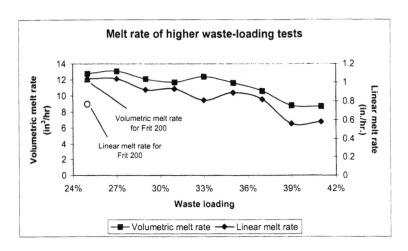

Figure I. Plot of the Melt Rate Data for Frit 320 with SB2 as a Function of Waste Loading (Data from Table I).

As shown by the data, the general trend is that melt rate decreases as WL increases (for both measurement techniques). By increasing the ratio of waste to frit, less flux (less alkali) is added to the system (Stone and Josephs, 2001). An increase in alkali correlates strongly to an increase in melt rate, whether it is present in the frit or in the waste (Lambert et. al, 2001). Also, the linear and volumetric melt rates for Frit 200 and SB2 at 25% WL have been added to Figure 3 to emphasize the effect of switching from Frit 200 to Frit 320.

Even though melt rate may decrease as WL increases as indicated above, the waste throughput per unit time may not. The data in Table II and Figure II show the dependence of waste throughput on WL. Waste throughput was calculated via Equation 1, and assuming the amount of waste glass processed per hour by the DWPF is approximately 170 pounds.

$$Throughput_{WL\#2} = 170\left(\frac{LMR_{WL\#2}}{LMR_{25\%WL}}\right)(WL\#2) \qquad (1)$$

The waste loading of interest is WL#2, Throughput$_{WL\#2}$ is the waste throughput at that specified waste loading, and LMR is the linear melt rate (values listed in Table I). For example, the waste throughput at 33% WL is 43.7 pounds per hour, which was calculated as follows:

$$Throughput_{33\%WL} = 170\left(\frac{0.81}{1.04}\right)(0.33)$$

Table II. Waste Throughput as a Function of Waste Loading.

Waste loading	Waste throughput (lb./hr)
25%	42.5
27%	45.9
29%	43.6
31%	47.1
33%	43.7
35%	50.9
37%	49.6
39%	35.7
41%	38.9

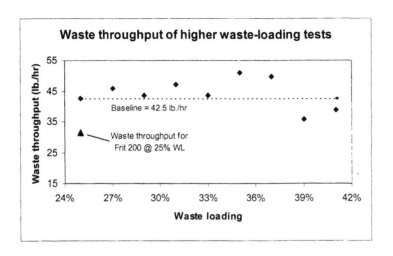

Figure II. Plot of Waste Throughput for Frit 320 with SB2 as a Function of Waste Loading (Data from Table II).

The waste throughput for Frit 200 with SB2 at 25% WL was added to Figure II to show the advantage of just switching from Frit 200 to Frit 320. Also shown in Figure II is a dashed line indicating the baseline waste throughput for the Frit 320/SB2 system. Up to a WL of 37%, the waste throughput is greater than the baseline value of 42.5 lb./hr. However, waste throughput values for the 39% and 41% waste loadings dip below the baseline. This trend indicates that WL can be increased for SB2 with Frit 320, but caution must be used in doing so since melt rate and overall throughput decrease at higher waste loadings. However, it should be noted that the waste throughput values for all Frit 320/SB2 waste loadings are greater than the current Frit 200 based system. To determine if operation above 37% WL would be warranted, a cost analysis would need to be done to evaluate the competing effects of decreased throughput versus an overall reduction in the number of canisters produced.

CONCLUSIONS

As WL was increased (from 25 to 41%) during this testing of SB2 with Frit 320, melt rate was found to decrease in the overall WL range tested and melting behavior deteriorated. Waste throughput, however, exceeded the 25% WL baseline case until a critical waste loading was reached, at which point melting behavior deteriorated sufficiently to significantly decrease melting rate. For the SB2/Frit 320 system, this occurred in the 37 to 39% WL range. It should be noted that the waste throughput values for all Frit 320/SB2 waste loadings are greater than the current Frit 200 based system (at 25% WL). Selected tests with the slurry-fed melt rate furnace will be performed to verify the results of the dry-fed testing. To determine if operation above 37% WL would be warranted, a cost analysis would need to be performed to evaluate the competing effects of decreased throughput versus an overall reduction in the number of canisters produced.

RECOMMENDATIONS

1. In transitioning from Frit 200 to Frit 320 in 2nd quarter FY03, it is recommended that processing with Frit 320 begin at the same waste loading as when processing with Frit 200. Five melter volumes (~75K pounds of glass produced) should be processed at this waste loading. At this point the DWPF should then process five melter volumes each at a WL below and above that original WL (a large enough increment that the DWPF can detect and to show a noticeable difference in melt rate) in order to determine the melter's optimum level of operation. Observations of the electrode power and dome heater power, along with visual observations of the cold cap and melt behavior (use of a boroscope), should be made since higher waste loadings may have detrimental effects on waste throughput and melting behavior.

2. Although the ultimate objective is to increase both waste loading and melt rate, one should not lose sight of total waste throughput, which may require a compromise between the two controlling factors. The concept that reduced melt rates at higher waste loading is unacceptable should be tempered with an evaluation of the total waste throughput. More specifically, during an assessment of the impacts of waste loading on melt rate, decisions on frit selection or targeted waste loading should not be made solely on the relative melt rate. Conversely, a decision on waste loading should not be made solely on the liquidus model. The decision should consider the total sludge throughput per unit time.

3. Given the key criteria of waste loading and melt rate can be competing, the basis for not only developing but ultimately selecting a frit for any sludge batch is complex. It is recommended that the selection process should not be made based on a single criterion but on a collection of criteria that provide insight into the economics of pretreatment requirements, processing issues, and ultimately total number of canisters produced. A balanced approach is suggested for both the development and selection of a frit for specific sludge batches.

REFERENCES

[1]Westinghouse Savannah River Company (WSRC). 2001. Savannah River Site High Level Waste System Plan (HLW), HLW-2001-00040, Revision 12. Aiken, SC.

[2]Lambert, D.P., T.H. Lorier, D.K. Peeler, and M.E. Stone. 2001. *Melt Rate Improvement for DWPF MB3: Summary and Recommendations (U)*, WSRC-TR-2001-00148, Westinghouse Savannah River Company, Aiken, SC.

[3]Miller, D.H. 2002. *Summary of Results from Minimelter Run with Macrobatch 3 Baseline Feed Using Frit 320 (U)*, WSRC-TR-2002-00188, Westinghouse Savannah River Company, Aiken, SC.

[4]Stone, M.E., and J.E. Josephs. 2001. *Melt Rate Improvement for DWPF MB3: Melt Rate Furnace Testing (U)*, WSRC-TR-2001-00146, Westinghouse Savannah River Company, Aiken, SC.

[5]Brown, K.G., C.M. Jantzen, and G. Ritzhaupt. 2001. *Relating Liquidus Temperature to Composition for the Defense Waste Processing Facility (DWPF) Process Control*, WSRC-TR-2001-00520, Westinghouse Savannah River Company, Aiken, SC.

[6]Stone, M.E., and D.P. Lambert. 2001. *Melt Rate Improvement for MB3: Feed Preparation*, WSRC-TR-2001-00126, Westinghouse Savannah River Company, Aiken, SC.

SPINEL CRYSTALLIZATION IN HLW GLASS MELTS: CATION EXCHANGE SYSTEMATICS AND THE ROLE OF Rh_2O_3 IN SPINEL FORMATION

Sezhian Annamalai, Hao Gan, Malabika Chaudhuri, Wing K. Kot, and
Ian L. Pegg
Vitreous State Laboratory
The Catholic University of America
Washington, DC, 20064

ABSTRACT

Iron-chrome spinel is one of the most prevalent liquidus phases in typical high-level waste (HLW) glasses, often limiting waste loadings. The spinels that form exhibit extensive solid solution ranges accommodating various divalent and trivalent cations. The present study examined the dominant cation exchange in spinel and the role of Rh^{3+} in spinel formation in HLW glasses. The spinel phase crystallized from the HLW glasses studied can be roughly represented by a pseudo-binary between two complex end member species, Ni-Fe^{3+}-rich spinel and (Fe^{2+}, Zn, Mn^{2+})-Cr-rich spinel. The composition of the spinel in the HLW glass system varies with system composition and temperature. The Cr-rich species is favored at higher temperature. For the first time, Rh^{3+} has been identified as an important spinel constituent near the liquidus temperatures in HLW glass melts. The Rh^{3+} concentration in the spinel phase coexisting with the HLW glass melts is observed to be enriched by factors of several hundred over its concentration in the melt. Formation of such Rh-rich spinels leads to significant upward shifts in the liquidus temperature compared to that of the corresponding Rh-free version. Liquidus temperature measurements collected on Rh-free HLW glasses and models that are built on them will, therefore, tend to underestimate the true liquidus temperature of actual Rh-containing waste glasses.

INTRODUCTION

High-level nuclear waste (HLW) streams often contain significant amounts of Fe, Ni, Mn, and Cr. The waste loading that can be achieved when such wastes are vitrified is often limited by crystallization of spinel phases containing these elements. While the presence of such phases rarely affects the quality of the glass product, it can present a significant processing concern since such phases can sediment and collect in the melter, ultimately limiting melter life. For this reason, HLW glasses have traditionally been designed to have a liquidus temperature that is lower than the nominal processing temperature by some safety margin. However, since in fact, insoluble phases such as noble metals are generally present in high-level waste glass melts, the imposition of such a "zero-tolerance" constraint for spinels (or other liquidus phases) is not only logically inconsistent but unnecessarily restrictive with respect to waste loadings[1]. This is especially true for melters in which the melt pool is actively rather than convectively mixed. Since the basic practical issue is sedimentation and collection, depending on their size and density, non-zero amounts of such crystals can often be maintained in suspension and, therefore, tolerated. These considerations have led to the use of an "operational liquidus" constraint in which a certain non-zero fraction of crystals is tolerated at a given temperature; specifically, glasses designed for DuraMelter vitrification systems have employed a limit of <1 vol% crystals at a reference temperature (typically 950°C) below the operating temperature[1,2]. Such a constraint also has the advantage of robustness since, as the results of the present work show, while minor and even trace components can shift the liquidus temperature significantly, their impact on the volume fraction of crystals produced is naturally limited by simple mass balance.

Spinel exhibits extensive solid solution ranges at typical vitrification temperatures, accommodating various divalent and trivalent cations in tetrahedral and octahedral coordinations. Such cations are generally readily available in typical HLW glass melts. The flexibility in the composition of the spinel phase and the typically rather low solubility of the Fe-Cr-rich spinel phase often leads to spinel-limited waste loadings. The melting temperatures of the various spinel end members differ considerably and the melting point changes through the solid solution series[3]. Furthermore, since the melt exchanges cations with the coexisting spinel crystal as the system and environment changes, the composition of the crystalline and melt phases need to be considered simultaneously. The objective of the present study is to investigate the exchange relationship between various spinel components. As a specific case, in view of its presence as a minor or trace component in HLW streams, the role of Rh^{3+} in the formation of spinel phases in HLW glasses was investigated for the first time and its potential impact on the liquidus temperature of HLW glasses was assessed.

EXPERIMENTAL PROCEDURE

The basic experimental approach involved glasses of various compositions that were heat treated to form spinel phases that were then characterized. Glass powders (~200 mesh) and any necessary chemical additives to modify the composition were mixed in a mortar and pestle using ethyl alcohol. After drying in an oven at 110°C, the mixture was transferred to a 5 ml Pt/Au crucible, covered with a Pt/Au lid, and subjected to isothermal heat treatment for 24-72 hours; longer-time tests confirmed that the composition of spinel crystals did not change significantly. All heat-treatments were performed in ambient atmosphere. At the end of the heat treatment, the sample crucible was quenched in cold water, and the sample was subjected to microstructural, compositional, and phase characterization using optical microscopy, scanning electron microscopy coupled with energy dispersive X-ray spectroscopy (SEM/EDS), and X-ray diffraction (XRD). The composition of the homogeneous glass samples was analyzed by X-ray fluorescence spectroscopy (XRF) or acid dissolution followed by direct current plasma atomic emission spectroscopy (DCP).

Rh-bearing spinel was synthesized using the following method. Appropriate amounts of powders of $Al(OH)_3$, $Cr(NO_3)_3 \cdot 9H_2O$, Fe_2O_3, MnO, NiO, ZnO and $Rh(NO_3)_3 \cdot 2H_2O$ were weighed and mixed in ethyl alcohol. The mixture was then dried for one hour at $110^\circ C$ and one hour at $180^\circ C$ before being ground to pass through a 200-mesh sieve and pressed into pellets of 8 mm diameter. The pellets were then heat treated at $1145^\circ C$ in Pt/Au crucibles for different lengths of time. At the end of each heat treatment, the sample crucibles were quenched in water.

XRD powder patterns were collected using a Thermo-ARL X'TRA X-Ray Diffractometer from 10 to 80° 2-theta angle with a collection time of 1 second for each step (0.05°). MDI/JADE 5 was used for phase identification analysis. A JEOL JSM-5910LV/Oxford INCA300 was used for SEM/EDS analysis of the synthesized Rh-bearing spinel samples. A JEOL JSM-35C/Thermo-Noran Vantage was used for SEM/EDS analysis of spinel-saturated glass samples.

RESULTS AND DISCUSSION

Spinel is an isotropic crystal consisting of a cubic close-packed array of oxygen ions in which, for normal spinel, divalent cations are tetrahedrally coordinated and trivalent cations are octahedrally coordinated[3] (in the case of inverse spinel, the opposite is true for the divalent cations and half of the trivalent cations). The ratio in a typical unit cell is one divalent cation (A site) to two trivalent cations (B site) to four oxygen anions (O), as expressed generally as AB_2O_4. Typical divalent cations found in the spinels formed in HLW glasses are Fe, Mg, Mn, Ni, Zn, and typical trivalent cations are Cr, Fe. In the present work, Rh^{3+} has been identified for the first time as an important spinel component in HLW glasses; Rh^{3+} resides in the B site. Cation exchange occurs for both spinel sites between the melt and spinel. Cation substitution in two spinel sites has been investigated for a data set of 54 spinel

Table I. Composition range of HLW glass systems studied.

Oxide	Min (Wt %)	Max (Wt %)
Al_2O_3	0	11
B_2O_3	5	20
Cr_2O_3	0	0.45
Fe_2O_3	3	13
Li_2O	2	7
MgO	0	5
MnO	0	5
Na_2O	4	20
NiO	0	2
Rh_2O_3	0	0.08
SiO_2	34	53
ZnO	0	2
ZrO_2	0	6

saturation results for waste glasses spanning a wide range of the compositions (Table I). The dominant trends of the cation exchanges are described below.

Cation Exchange in the Trivalent Site of Cr-Fe Spinel

Table II summarizes the ranges of the cation occupation for two spinel sites. It is clear that Cr^{3+} and Fe^{3+} (calculated by spinel stoichiometry) are the most important trivalent cations in HLW glass samples. A negative correlation between the two trivalent cations in spinels with changes in the composition of the glass is expected and observed (Figure 1). It is, nevertheless interesting to note that, Al^{3+}, another important spinel component, is not incorporated into the spinel to any significant extent and Mn, although present in spinel, shows a strong preference for the A site (Table II). The substitution of Cr for Fe was observed with increasing temperature for a fixed glass composition for numerous HLW glasses; this behavior is also apparent in data reported by Vienna et al[4].

Table II. Cation occupancy in the divalent (A) and trivalent sites (B) of a spinel phase. Total of 54 spinel samples from 38 distinct glasses have been analyzed. The Mn^{2+}, Mn^{3+}, Fe^{2+} and Fe^{3+} fractions are calculated based on the stoichiometry of an ideal spinel.

Cation	Divalent Site (A)		Trivalent Site (B)	
	Max (Molar Fraction)	Min (Molar Fraction)	Max (Molar Fraction)	Min (Molar Fraction)
Cr^{3+}	-	-	0.96	0.01
Fe^{2+}	0.91	0	-	-
Fe^{3+}	-	-	0.99	0
Mn^{2+}	0.76	0		
Mn^{3+}	-	-	0.16	0
Ni^{2+}	0.97	0	-	-
Zn^{2+}	0.61	0	-	-
Rh^{3+}	-	-	0.22	0

Cation Exchange in the Divalent Site of Cr-Fe Spinel

Table II indicates that Fe^{2+}, Ni^{2+}, Mn^{2+} and Zn^{2+} are the major divalent cations incorporated into the spinels forming in these HLW glasses. The substitution of (Fe^{2+}, Mn^{2+}, Zn^{2+}) as a group for Ni^{2+} (Figure 2) correlates with the substitution of Cr^{3+} for Fe^{3+} over the wide range of glass compositions investigated (Figures 3 and 4). Apparently, the cation exchanges in the A and B sites are not independent. The coupled cation substitutions in two spinel sites suggests two major end-member spinel species in the HLW glass system with Ni^{2+}-Fe^{3+}-spinel on one end and (Fe^{2+}, Zn^{2+}, Mn^{2+})-Cr^{3+} spinel on the other end.

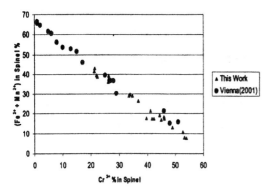

Figure 1. Cr^{3+} vs. $(Fe^{3+} + Mn^{3+})$ in spinel.

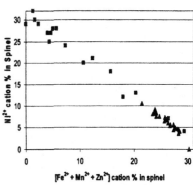

Figure 2. Ni $^{2+}$ vs. $(Fe^{2+} + Mn^{2+} + Zn^{2+})$ in spinel.

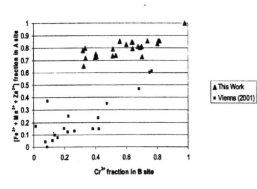

Figure 3. Cr $^{3+}$ vs. $(Fe^{2+} + Mn^{2+} + Zn^{2+})$ in spinel.

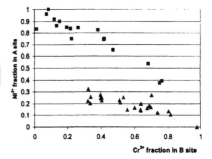

Figure 4. Cr $^{3+}$ vs. Ni $^{2+}$ in spinel.

The evolution of the spinel phase has not only been observed with changes in glass composition but also with changes in temperature at a fixed system composition (Figure 5). The "pseudo activity" of the cations in the spinel versus inverse temperature indicates a roughly linear relationship between the two parameters and also that the $(Fe^{2+}, Mn^{2+}, Zn^{2+})$-$Cr^{3+}$ end member spinel is favored at higher temperatures.

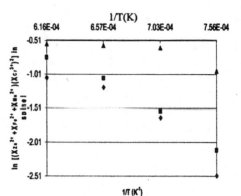

Figure 5. $Ln[(X_{Zn}^{2+} + X_{Fe}^{2+} + X_{Mn}^{2+})*(X_{Cr}^{3+})^2]$ in spinel vs. inverse temperature for three HLW glasses.

Role of Rh₂O₃ in Spinel Formation in HLW Glasses

Since rhodium is a fission product, it is present to some extent in high-level wastes. However, its concentration is typically very low, in HLW glasses, usually below 0.1 wt% Rh_2O_3. In the present work, Rh^{3+} has, for the first time, been identified as an important spinel component in the HLW glass system. As shown in Table III, Rh_2O_3 makes up a large portion of the spinel crystals formed in the HLW glasses investigated. Because of the large and well-defined crystalline phase present in the samples examined by SEM/EDS (Figure 6), analytical interference is not likely to be significant for the compositional analysis by EDS. However, in order to confirm that this Cr-rich spinel composition can incorporate as

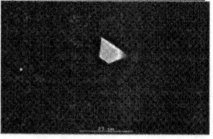

Figure 6. Micrographs of regular spinel (top) and Rh-bearing spinel (bottom).

much Rh_2O_3 as indicated by SEM/EDS analysis, a series of polycrystalline samples were synthesized from reagent grade chemicals on the basis of the chemical composition obtained from EDS analysis of the Rh-rich complex oxide phase formed in the saturation experiments (Table III). SEM/EDS analysis revealed that the synthesized materials, with the exception of a very small amount of a metallic phase (Pt-Rh alloy), is made entirely of a single type of crystalline phase (Figure 7). This phase is identical in both composition and morphology to that present in heat treated HLW glass samples (Figure 6 and Table III). The XRD powder patterns of the polycrystalline samples match very well with the Fe-rich spinel, magnetite (Figure 8). It is therefore concluded that Rh-rich spinel can indeed form from HLW glasses containing as little as 0.05 wt% Rh_2O_3.

Table III. Composition analysis of Rh-bearing spinel crystals in HLW glass sample heat-treated at 1145°C for 24 hours and the target composition used for synthesis of the Rh-rich spinel.

Oxide, wt%	Crystal 1	Crystal 2	Crystal 3	Average	Target composition for spinel synthesis
Al_2O_3	0.6	1.0	0.8	0.8	0.8
Cr_2O_3	14.2	13.8	14.0	14.0	14.2
Fe_2O_3	38.9	37.4	37.1	37.8	38.2
MnO	3.9	4.0	4.0	4.0	4.0
NiO	12.2	12.2	11.9	12.1	12.2
ZnO	8.2	8.2	7.5	8.0	8.1
Rh_2O_3	21.7	23.1	22.4	22.4	22.6
Others	0.3	0.4	2.3	0.9	0

While the extent of the incorporation of Rh into these spinels is surprising and previously unrecognized in HLW glass systems, it is not surprising that Rh can be incorporated into spinels since Rh_2O_3 has a corundum-type structure and the ionic radius of Rh^{3+} is similar to that of Fe^{3+}. Rh-rich spinel phases have been reported previously in other systems[5-7]. It is, however, quite remarkable to observe such a high level of enrichment of Rh_2O_3 in a spinel phase coexisting with melts containing only 0.05 wt% Rh_2O_3. It is evident that Rh^{3+} can be enriched in the spinel phase by a factor of several hundred times its

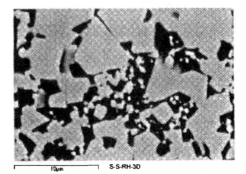

Figure 7. SEM micrograph of Rh-rich spinel synthesized from heating the mixed chemical powders at 1145°C for 3 days.

concentration in the coexisting HLW glass melt. The preference of an element between the two coexisting phases is readily compared using a simple partitioning factor, ("enrichment factor"), defined as the ratio of the wt% of the oxide of interest in the spinel and to that in the coexisting glass melt. Table IV lists the enrichment factors of the major spinel components observed in the HLW glasses investigated here; these glasses contain 0.01 – 0.08 wt% Rh_2O_3 and were tested over the temperature range of 850 - 1145°C. Based on these enrichment factors, the order of increasing affinity for the spinel phase is: (Fe_2O_3, MnO, and ZnO) < NiO < Cr_2O_3 < Rh_2O_3; the variation in enrichment factors spans several orders of magnitude.

In view of the high concentration of Rh in the coexisting spinel phase, it would be surprising if the liquidus temperature were not affected and that is indeed the case: measurements on HLW glasses with and without Rh, showed shifts in the liquidus temperature of up to 100°C.

Finally, the Rh content in spinel present in HLW melts is a strong function of temperature. In a similar manner to that observed for Cr^{3+}, but at a faster rate, the Rh content in the spinel decreases with decreasing temperature. For example, for one particular glass, the Rh_2O_3 wt% in spinel decreased from 23 wt% at 1175°C to merely 1 wt % at 945°C (Figure 9). Mass balance based on the measured phase compositions indicates that this is predominantly an effect of temperature rather than simply depletion of Rh from the melt.

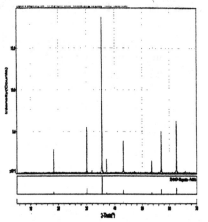

Figure 8. XRD spectrum of synthesized Rh-bearing spinel (top) compared to that of magnetite (bottom).

Table IV. Partitioning (enrichment factors) of selected elements between the spinel phase and coexisting HLW glass melts at various temperatures. Data from heat treatment experiments of 31 glasses.

	NiO	MnO	ZnO	Fe_2O_3	Cr_2O_3	Rh_2O_3
1050°C	16-52	3	5 - 8	3 - 5	167-221	46-184
1150°C	10-37	2 - 5	5 - 9	3 - 4	130-210	427-447
1250°C	8 - 11	2-3	6 - 8	2 - 3	150-210	--
1350°C	7 - 10	2 - 3	6 - 7	2 - 3	165-171	--

CONCLUSIONS

The spinel phase crystallized from the HLW glasses studied can be roughly represented by a pseudo-binary between two complex end member species, $Ni-Fe^{3+}$-rich spinel and (Fe^{2+}, Zn, Mn^{2+})-Cr-rich spinel. The composition of the spinel in the HLW glass system varies with system composition and temperature. The Cr-rich species is favored at higher temperature. For the first time, Rh^{3+} has been identified as an important spinel constituent near the liquidus temperatures in HLW glass melts. The Rh^{3+} concentration in the spinel phase coexisting with the HLW glass

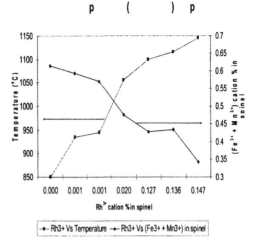

Figure 9. Rh^{3+} vs. temperature (square) and Rh^{3+} vs. $(Fe^{3+} + Mn^{3+})$ (diamond).

melts is observed to be enriched by factors of several hundred over its concentration in the melt. Formation of such Rh-rich spinel leads to significant upward shifts in the liquidus temperature compared to that of the corresponding Rh-free version. Liquidus temperature measurements collected on Rh-free HLW glasses and models that are built on them will, therefore, tend to underestimate the true liquidus temperature of actual Rh-containing waste glasses. In contrast, HLW glasses designed using an "operational liquidus" constraint (e.g., 1 vol% crystals at 950°C) are essentially unaffected since (i) although the spinel phase is highly enriched in Rh, its volume fraction, which is limited by mass balance, is very small as a result of the very low concentration of Rh in the original glass; and (ii) Rh enrichment decreases with temperature, essentially disappearing by 950°C.

ACKNOWLEDGEMENTS

The authors would like to thank Dr. Biprodas Dutta, Steven Grant, and Isidro Carranza for help with sample preparation, Dr. Andrew Buechele for helpful discussions, and Catherine Paul for help with manuscript preparation.

REFERENCES

[1]W. K. Kot and I. L. Pegg, "Glass Formulation and Testing with RPP-WTP HLW Simulants," Final Report, VSL-01R2540-2, Rev. 0, Vitreous State Laboratory, The Catholic University of America, Washington, D.C., February 16, 2001.

[2]S.S. Fu and I.L. Pegg, "Glass Formulation and Testing with TWRS HLW Simulants," Final Report, Vitreous State Laboratory, The Catholic University of America, Washington, DC, January 1998.

[3]W. A. Deer, R. A. Howie, and J. Zussman, "An Introduction to the Rock-Forming Minerals," Longman (1992).

[4]J.D. Vienna, P. Hrma, J.V. Crum, M. Mika, "Liquidus Temperature-Composition Model for Multi-Component Glasses in the Fe, Cr, Ni, and Mn Spinel Primary Phase Field," J. Non-Cryst. Solids **292** 1-24 (2001).

[5]R.J. Hill, J.R. Craig, and G.V. Gibbs, "Systematics of the Spinel Structure Type," Phys. Chem. Minerals **4**, 317-339 (1979).

[6]C. J. Capobianco and M.J. Drake, "Partitioning of Ru, Rh, and Pd Between Spinel and Silicate Melt and Implications for Platinum Group Element Fractionation Trends," Geochim. Cosmochim. Acta, **54**, 869-874 (1990).

[7]C.J. Capobianco R.L. Hervig and M.J. Drake, "Experiments on Crystal/Liquid Partitioning of Ru, Rh, and Pd for Magnetite and Hematite Solid Solutions Crystallized from Silicate Melt," Chemical Geology, **113**, 23-43, (1994).

COMPOSITION EFFECTS ON THE VAPOR HYDRATION OF WASTE GLASSES

Andrew C. Buechele, Frank Lofaj, Isabelle S. Muller, Cavin T.F. Mooers, and Ian L. Pegg
Vitreous State Laboratory, The Catholic University of America
Washington, DC 20064

ABSTRACT

The Vapor Hydration Test (VHT) is a highly accelerated test that measures the rate at which glass alters when it is exposed to a high-temperature humid environment. The VHT performance is strongly affected by the glass composition, often in a complex fashion. In this work, the effects of composition on VHT response were investigated for typical high-sodium borosilicate waste glasses as well as for a simplified alkali-alumina-borosilicate system. In a typical waste glass composition, VHT durability was improved by substituting potassium for lithium. Increasing the sulfur content glass generally improved VHT durability. Tin appeared to increase VHT durability substantially when present at 1 wt% and higher, and to a lesser degree below the 1 wt% level. The effect of the alkaline earth elements on VHT durability varied progressively with their atomic number, with the higher the atomic number, the poorer the durability. In the simplified glass, VHT durability improved progressively as additions of iron, magnesium, and calcium were made, respectively. Progressive improvement was also noted as partial substitutions were made of potassium for sodium.

INTRODUCTION

The Vapor Hydration Test (VHT) is one of the tests employed in assessing the durability of waste glasses and is part of the product quality requirements for vitrified low-activity nuclear waste (LAW) at the Hanford site[1,2]. Details of the test procedure have been reported previously[3]; in summary, the test involves exposing a glass coupon to saturated water vapor at elevated temperature, usually 200°C, for a predetermined time. Glass durability on the VHT is determined by measuring the thickness of the reacted layer that forms on the glass. This can be done either by direct measurement, or by measuring the thickness of the remaining glass, subtracting it from the original coupon thickness, and dividing by two. The conditions of the test result in a highly accelerated rate of reaction of water with the glass surface and bulk, often resulting in extensive formation of alteration phases.

Small changes in glass composition often produce substantial effects on the VHT durability. In this work, the effects of composition on VHT durability were investigated for typical high-sodium borosilicate waste glasses as well as for a simplified alkali-alumina-borosilicate system. The simplified glass was formulated, melted, and tested to investigate VHT durability variations produced by alkali substitution and by additions of iron, magnesium, and calcium.

Table I. Compositions of simplified glasses (target values in wt%).

	Series 1				Series 2				Series 3			
	Lo1A	Lo1B	Lo1C	Lo1D	Lo2A	Lo2B	Lo2C	Lo2D	Lo3A	Lo3B	Lo3C	Lo3D
Al_2O_3	6.2	5.8	5.7	5.6	6.2	5.8	5.7	5.6	6.2	5.8	5.7	5.6
B_2O_3	8.9	8.3	8.2	8.1	8.9	8.3	8.2	8.1	8.9	8.3	8.2	8.1
CaO	0	0	0	1.8	0	0	0	1.8	0	0	0	1.8
K_2O	0.5	0.5	0.5	0.5	2.5	2.3	2.3	2.3	8.1	7.6	7.5	7.3
Fe_2O_3	0	6.5	6.5	6.3	0	6.5	6.5	6.3	0	6.5	6.5	6.3
MgO	0	0	0.5	0.5	0	0	0.5	0.5	0	0	0.5	0.5
Na_2O	20	18.7	18.4	18.1	18	16.8	16.6	16.3	12.4	11.6	11.4	11.2
SiO_2	64.4	60.2	59.4	58.3	64.4	60.2	59.4	58.3	64.4	60.2	59.4	58.3

PROCEDURE AND RESULTS

All glasses for this study were formulated and melted from raw chemicals. The waste glass compositions were typical of those currently being considered for vitrification of high-sodium low-activity wastes at the Hanford site[2]. The target compositions of the simplified glasses are shown in Table I. Independent chemical analysis verified that the actual compositions were close to the target values. Preparation of the coupons for the VHT and the test itself, as well as the analysis of the coupons after the test, were all done as described in previous work[3].

Figure 1 shows SEM micrographs of sectioned glass coupons after 24 days at 200°C on the VHT. The three coupons shown are from glasses of identical composition except that each contained a different alkaline earth element, Ca, Sr, or Ba, at the 7.8 wt% level as oxide. The extent of reaction, as measured by the thickness of the altered layer, increases drastically as the atomic number of the alkaline earth element increases.

Several elements have been found to have beneficial effects on VHT durability, including iron, sulfur, and tin[2]. Tin exhibited a particularly pronounced effect on VHT durability, and therefore a series of glass compositions were selected to study the effect more systematically. A typical LAW glass formulation LLG-10, containing about 20 wt% Na_2O and 8 wt% CaO, was modified by substituting 5 wt% SnO_2 for the equivalent amount of CaO to produce a new glass designated LLG-11. However, a considerable amount of undissolved SnO_2 was observed in LLG-11. The results of a standard 24-day 200°C VHT are shown in Figure 2a and 2b. The average layer thickness observed on LLG-10 was 68 µm while the layer thickness on LLG-11 was ≤ 3µm. A systematic matrix of glasses

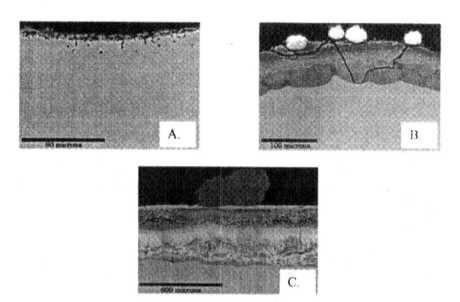

Figure 1. SEM micrographs of VHT layers formed on glasses containing 7.8 wt%: A. CaO, B. SrO, C. BaO. (Balance: 20 wt% Na_2O, 42 wt% SiO_2, and others.)

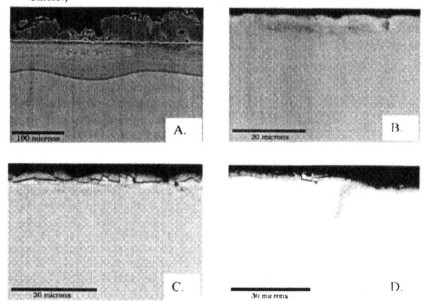

Figure 2. VHT layers on glasses: A. LLG-10 (0 wt% SnO_2), B. LLG-11 (5 wt% SnO_2), C. LG-12 (3 wt% SnO_2), and D. LLG-13 (1 wt% SnO_2).

based on LLG-10 was then made to try to determine the lowest level of SnO_2 that would confer significant protection, and also to test the possibility that the improved durability had been accomplished by the reduction of CaO rather than by the addition of SnO_2. Table II

Table II. Compositions, first series Sn glasses (balance: 20 wt% Na_2O, 42 wt% SiO_2, and others.)

	Glass Composition (wt%)						
LLG-	10	11	12	13	14	15	16
CaO	7.8	2.8	5	6.8	7.8	7.8	7.8
Fe_2O_3	7.4	7.4	7.4	7.4	6.4	4.4	2.4
SnO_2	0	5	3	1	1	3	5

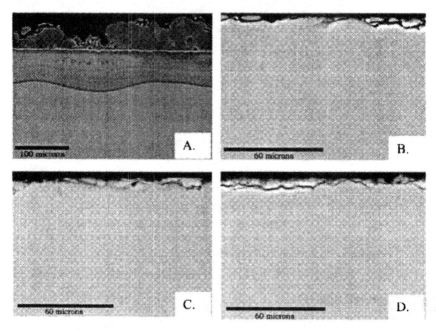

Figure 3. VHT layers on glasses: A. LLG-10 (0 wt% SnO_2), B. LLG-16 (5 wt% SnO_2), C. LG-15 (3 wt% SnO_2), and D. LLG-14 (1 wt% SnO_2).

summarizes how this was done. First, substitutions of 3 wt% and 1 wt% were made for an equivalent amount of CaO in the original formulation of LLG-10 to produce LLG-12 and LLG-13, respectively. Then, the CaO was restored to its original value of 7.8 wt%, and SnO_2 at levels of 1, 3, and 5 wt% was substituted for equivalent amounts of Fe_2O_3 in LLG-10 to produce LLG-14, LLG-15, and LLG-16 respectively. Since it has been observed that Fe_2O_3 usually improves VHT durability, it was reasoned that any improvement of durability seen in the glasses in which iron had been replaced by tin could safely be attributed to the tin.

All of these glasses exhibited layer thicknesses of $\leq 3\mu m$ on a standard 24 day 200°C VHT. No measurable difference in layer thickness could be found on any of the tin-containing glasses regardless of SnO_2 concentration, as shown in Figures 2 and 3. Based on these findings, it can be concluded that, in this formulation, as little as 1 wt% SnO_2 is sufficiently to almost completely suppress alteration after 24 days at 200°C in the VHT. Lower concentrations of SnO_2 in this formulation have not been tested.

Two other typical LAW glasses, LLG-20 with about 12 wt% Na_2O (balance: 47 wt% SiO_2 and others) and LLG-30 with about 24 wt% Na_2O (balance: 41 wt% SiO_2 and others) and different overall compositions from LLG-10, were modified by simple addition of 0.5 wt% SnO_2 to produce glasses LLG-21 and LLG-31 respectively. Table III

Table III. Layer thicknesses on second series Sn glasses.

Coupon	Layer Thickness (μm)	
	LLG-30 (No SnO_2)	LLG-31 (0.5 wt% SnO_2)
1	89	9.3
2	61	5.7
3	80	17.8
Mean	77±19	11±5

shows the results of a triplicate VHT of LLG-30 and the tin-containing version, LLG-31, after 5 days on the VHT, which clearly show the protective effect of tin.

The glass formulation for LLG-20 was modified by the addition of 0.5 wt% SnO_2 and the overall composition renormalized to produce LLG-21. Figure 4 plots the results of the VHT at 6, 12, 24, and 48 days. Once again, the protective effect of tin is evident but the extent appears to vary with the test duration.

The final set of glasses to be discussed here is a set of simplified glasses whose compositions are shown in Table I. This test matrix was designed to start with a five-component, high-sodium, low-potassium glass containing no iron, magnesium, or calcium, and then make successive additions of iron, magnesium, and calcium. To these four glasses, modifications were made in two stages by replacing some of the sodium with additional potassium. The results of a 24 day, 200°C VHT are summarized in Table IV. It can be seen in the first row of Table IV that the addition of iron and then magnesium makes successive improvements in the VHT durability. The effect of the calcium addition at the highest sodium content seems to be a decrease in durability, but this changes as some of the sodium is replaced by potassium and an increase in durability is then seen with the calcium addition. It is immediately clear from Table IV that replacement of some of the sodium by potassium significantly improves VHT durability, and that the degree of improvement is directly related to the amount of sodium replaced by potassium. The first two glasses in the first column actually liquefied during the VHT and dripped from the support wire in the VHT vessel. However, with the increase of the potassium level to 8 wt%, the Glass Lo3A was stabilized sufficiently to retain its monolithic integrity throughout the test, although it did develop a 1000-μm reacted layer, meaning that only a very thin layer of unreacted glass remained at the center of the coupon. The micrographs in Figure 5 display layer thicknesses on glasses in the intermediate-potassium Lo2 series showing how progressive additions of iron, magnesium, and calcium improve durability.

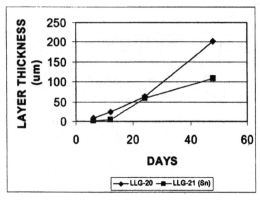

Figure 4. VHT Layer growth on LLG-20 and -21.

DISCUSSION

Improvements in VHT durability were observed with additions of Fe, Mg, Ca, and Sn, although these effects are themselves likely to be dependent on glass composition. In view of the complex series of processes that ensue in the VHT, including water adsorption, alkali diffusion, water diffusion[4], pH rise, matrix hydrolysis, and secondary phase formation, glass composition changes can affect

Table IV. Layer thicknesses on simplified glasses in μm.

	-A	-B	-C	-D
Lo1	Melted	765	399	508
Lo2	Melted	444	162	100
Lo3	1000	114	7	2

the measured VHT alteration rate in a variety of ways. Replacement of Na by other cations may enhance durability as a result of lower mobility in the glass network that they modify. Sulfur additions may reduce VHT rates by moderating the pH in the adsorbed water layer. The improvement in durability conferred by the addition of tin is not fully understood, although Ellison, et al.[5,6] indicate that there is a relation between the stabilization of high-valence, low solubility cations in four-fold coordination in glass through the peralkaline effect, and an increase in resistance to aqueous corrosion. A glass is said to be peralkaline when the molar concentration of alkalis exceeds the combined molar concentration of aluminum and boron. Ellison et al.[6] note that the durability of Savannah River waste glasses increased as a function of the concentration of iron-rich simulated waste[7]. These glasses contain alkalis well in excess of aluminum and boron and the increase in durability is attributed by Ellison et al. to the stabilization of the low solubility iron in four-fold coordination in the glass by the excess alkali. Several studies have shown that the solubilities of +3 and +4 cations in silicate glasses have a strong dependence on the alkali/aluminum ratio.[8-10] The solubility of such cations tends to be relatively constant when the alkali/aluminum ratio is less than one, but increases rapidly when the ratio exceeds one, and in an approximately linear fashion with the excess of alkali. This increase in solubility is attributed to the stabilization in four-fold coordination by the excess alkali mentioned earlier. A

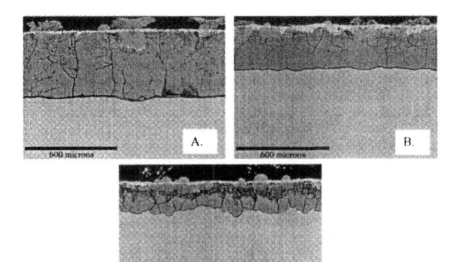

Figure 5. VHT layers on glasses in the Lo2 series (~2.5 wt% K_2O): A. Lo2B (6.5 wt% Fe_2O_3), B. Lo2C (=0.5 wt% MgO), C. Lo2D (+1.8 wt% CaO).

peralkaline effect was observed by Ellison, et al.[6] for Sn^{4+} in K_2O-Al_2O_3-SiO_2 glasses; the glasses to which tin was added in the present study can be said to be peralkaline. In the simplified glass system studied, the increase in durability imparted by the introduction of iron may also be attributable to the peralkaline effect. The solubility products for hydroxides of Fe^{3+} and Sn^{4+} are quite small and are listed in Table V.

Table V also lists the solubility products for three alkaline earth hydroxides. The solubilities increase as the atomic numbers, which may indicate a connection between this increasing solubility and the decreasing durability observed in the alkaline-earth series of glasses as the atomic number increased. The size of the ions and their mobility in the network probably also play a role in determining VHT durability.

Table V. Solubility products.

$Ba(OH)_2$	5.0×10^{-3}
$Sr(OH)_2$	3.2×10^{-4}
$Ca(OH)_2$	1.3×10^{-6}
$Fe(OH)_3$	4×10^{-38}
$Sn(OH)_4$	1.0×10^{-57}

REFERENCES

[1]Bechtel National, Inc. "Design, Construction, and Commissioning of the Hanford Tank Waste Treatment and Immobilization Plant," Contract Number:

DE-AC27-01RV14136, U.S. Department of Energy, Office of River Protection 12/11/00, (2000).

[2]I.S. Muller, A.C. Buechele, and I.L. Pegg, "Glass Formulation and Testing with RPP-WTP LAW Simulants," Final Report, VSL-00R3560-2, February 23, 2001, (2001).

[3]A.C. Buechele, F. Lofaj, C. Mooers, and I.L. Pegg,"Analysis of Layer Structures Formed During Vapor Hydration Testing of High-sodium Waste Glasses," *Ceramic Transactions* **132**, 301-9, (2002).

[4] T. Schatz, A.C. Buechele, C. Mooers, R. Wysoczanski, and I.L. Pegg,"Vapor Phase Hydration of Glasses in H_2O and D_2O," *Ceramic Transactions*, in press (2003).

[5]A.J.G. Ellison, J.J. Mazer, and W.L. Ebert, *Effect of Glass Composition on Waste Form Durability: a Critical Review*, Report No. ANL-94/28, Argonne National Laboratory, Argonne, IL, (1994).

[6]A.J.G. Ellison, P.C. Hess, G.C. and Naski, "Cassiterite Solubillity in High-Silica K_2O-Al_2O_3-SiO_2 Liquids," *J. Am. Ceram. Soc.*, **81** [12], 3215-20, (1998).

[7]W.D. Rankin, and G.G. Wicks, "Chemical Durability of Savannah River Plant Waste Glass as a Function of Waste Loading," *J. Am. Ceram. Soc.*, **66** [6], 417-20, 1983.

[8]J.E. Dickenson, Jr., and P.C. Hess, "Rutile Solubility and Titanium Coordination in Silicate Melts," *Geochim. Cosmochim. Acta,* **49**, 2289-96, (1985).

[9]E.B. Watson, "Zircon Saturation in Felsic Liquids: Experimental Results and Applications to Trace Element Geochemistry," *Contrib. mineral. Petrol.*, **70**, 407-19, (1986).

[10]A.J.G. Ellison, and P.C. Hess, "Solution Behavior of +4 Cations in High Silica Melts: Petrologic and Geochemical Implications," *Contrib. mineral. Petrol.*, **94**, 343-51, (1986).

GLASS COMPOSITION-TCLP RESPONSE MODEL FOR WASTE GLASSES

Dong-Sang Kim and John D. Vienna
Pacific Northwest National Laboratory, Richland, WA 99352

ABSTRACT
A first-order property model for normalized Toxicity Characteristic Leaching Procedure (TCLP) release as a function of glass composition was developed with data collected from various studies. The normalized boron release is used to estimate the release of toxic elements based on the observation that the boron release represents the conservative release for those constituents of interest. The current TCLP model has two targeted application areas: (1) delisting of waste-glass product as radioactive (not mixed) waste and (2) designating the glass wastes generated from waste-glass research activities as hazardous or non-hazardous. This paper describes the data collection and model development for TCLP releases and discusses the issues related to the application of the model.

INTRODUCTION

The United States Department of Energy's (DOE's) Hanford Nuclear Reservation is the current home of approximately 200,000 m^3 of mixed waste stored in 177 waste tanks. The waste material contains hazardous wastes and constituents regulated under *Resource Conservation and Recovery Act of 1976* (RCRA) and are subject to the land disposal restrictions (LDR) contained in 40 CFR 268[1] and WAC 173-303.[2] The immobilized high-level waste (IHLW) is intended for disposal at the proposed geologic repository where hazardous wastes will not be accepted.

One of the criteria typically used to determine whether a material is toxic or not is the release of toxic elements under the conditions of the Environmental Protection Agency (EPA) Toxicity Characteristic Leaching Procedure (TCLP)[3] (e.g., 40 CFR 261[4]). The strategy for the successful compliance with regulations is based in part on the development and application of TCLP response models developed from the evaluation of non-radioactive glasses made from simulated Hanford tank waste.

Currently, glass waste is produced in the glass laboratories at the Pacific Northwest National Laboratory (PNNL) and other facilities for the research of nuclear waste vitrification technologies. If no associated TCLP data exist, the glasses produced in these research facilities must be designated based on their composition and are mostly disposed of as hazardous waste even if the concentration of the toxic constituents is very low. For a small quantity of glasses containing toxic metals, it is typically more economical to dispose the glass as hazardous waste because of the relative high cost of TCLP test. We propose that a TCLP release model be applied to designate the glass wastes while accounting for the appropriate uncertainties in model prediction.

Literature data on TCLP releases were collected for this study. These TCLP release data were initially evaluated to determine the element that can be used as a representative release for glass dissolution under TCLP condition. This paper reports the results of data evaluation and summarizes the development of the first-order model from collected data. Application of the model for the designation of glass wastes is also discussed.

DATA EVALUATION AND SCREENING

A database containing a total of 251 glasses with TCLP release data from 9 different studies[5-13] was compiled for this study. Although, the same TCLP method, SW-846 method 1311,[3] was used for the extractions performed in all these studies, subtle differences in results from different studies can be expected, which can make combining data less ideal for developing property models.[*]

The normalized elemental releases, r_i (in mg/L), are defined as:

$$r_i = \frac{c_i}{f_i} \qquad (1)$$

where c_i (mg/L) is the TCLP release of i-th element and f_i is the mass fraction of i-th element in glass. Figure 1 is the plot $\ln(r_i)$ versus $\ln(r_B)$ for eight RCRA elements of interest (Ag, As, Ba, Cd, Cr, Ni, Pb, and Se) plus Tl and Zn for the glasses collected from Kot and Pegg.[10] One of the RCRA elements, mercury, is not included because there are no glasses containing HgO as a glass component. Figure 1 shows that the normalized releases of all 10 elements are similar to or lower than that of boron except for three glasses that are obvious outliers. The glasses with lower TCLP releases show larger scatter of data, which is presumably caused by the larger uncertainty (sample collection, handling, and analytical error) involved in the leachate analysis and glass preparation when the elements are present in low concentration and glass dissolution is slow.

The elements in Figure 1 were divided into two groups based on their normalized releases compared to normalized boron release: (1) congruent (Ba, Cd, Ni, and Zn) and (2) incongruent (Ag, As, Cr, Pb, Se, and Tl). A similar groupings can also be made to all other elements reported in the study by Kot and

---

[*]The differences in TCLP response between studies come primarily from differences in particle sizing prior to the extraction which can cause variation in particle size distribution of the samples.

Pegg[10] but not included in Figure 1, i.e., some are released at the similar level as boron and some at a lower level than boron. The exception was alkali elements, K and Li (excluding Na because it is used in TCLP leachate)—the most mobile elements, that show sometimes higher normalized releases than boron, which indicates that alkali elements may be released selectively by ion exchange. It is also possible that K and Li may be present as impurities in extraction solution.

The similar plot for Los Alamos National Laboratory (LANL) glasses[11] is in Figure 2. Figure 2 also includes the releases of two alkali elements (K and Li) to illustrate the higher release of alkali elements than of boron. Figure 2 agrees well with the tendency observed in Figure 1 in that Ni is congruent and Cr is incongruent (no comparison is possible for Pb because it has only one data point). Other studies did not have both B

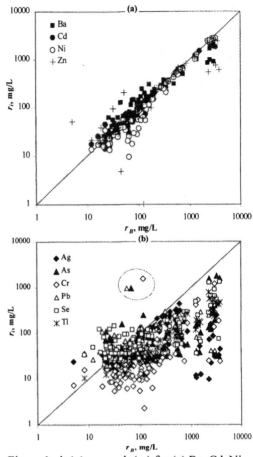

Figure 1. $\ln(r_i)$ versus $\ln(r_B)$ for (a) Ba, Cd, Ni, and Zn and (b) Ag, As, Cr, Pb, Se, and Tl in the glasses studied by Kot and Pegg[10]

release and RCRA element releases together for similar analysis (some studies had B release only and some had RCRA element releases without B release).

The evaluation based on Figures 1 and 2 indicates that normalized boron release can be used to estimate the extent of glass dissolution under TCLP conditions in a similar way as used in the PCT.[14]

Of the 251 glasses collected, 221 glasses had boron-release data. These 221 glasses were evaluated for concentration distribution, which identified six glasses that had an extreme concentration of SiO_2, B_2O_3, or CaO, i.e., far away from the concentration range of majority of glasses. These glasses were removed from the data set. During the model fitting, four glasses that constantly had large residuals in $\ln(r_B)$ regardless of model components chosen were removed as outliers. The remaining 211 glasses were used to develop the model.

MODEL DEVELOPMENT

For model calculation, the glass compositions were first converted into mole fractions of oxides (and halogens) by standard methods. Several components were listed in the database with multiple oxidation states. These components were combined into groups of like metal oxides, e.g., MnO_x for MnO and MnO_2, Sb_2O_x for Sb_2O_3 and Sb_2O_5, and so on. The resultant compositions were then normalized to one. In addition, the lanthanide oxides and Y_2O_3 were combined to form a single component – LN_2O_3:

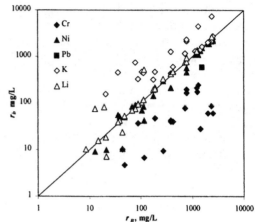

Figure 2. $\ln(r_i)$ versus $\ln(r_B)$ in the glasses studied by Vienna et al.[11]

$$LN_2O_3 = Ce_2O_3 + Eu_2O_3 + Gd_2O_3 + La_2O_3 + Nd_2O_3 + Pr_2O_3 + Sm_2O_3 + Y_2O_3$$

because these components were in insufficient concentrations in most glasses and did not, by themselves, have sufficient variation to justify fitting separate coefficients for them. However, combined they were in sufficient concentration and showed sufficient variation for a combined coefficient. Generally, their effects on glass properties are similar and vary only with ionic radius; however, this variation with radius was not accounted for in this study.

After forming the combined components, the 211 glass compositions were expressed as normalized mole fractions of 44 components. Those components with concentrations greater than 1.5 mole% in at least one glass and with a reasonable distribution were considered as possible model components—leaving 19 components. Only a limited number of components can be included in modeling. In treating the components that are not included in the model components, two different approaches can be applied: (1) to use the Others component as a sum of all the components not included in the model or (2) to use compositions normalized to sum to one after deleting the components not included in the model. The approach (1) using Others component was adopted for this study as was used for models in the recent interim model report for Hanford waste glasses.[15]

The natural logarithm of normalized boron release (r_B) was modeled as a linear function of glass composition:

$$\ln[r_B] = \sum_{i=1}^{N} r_{B,i} x_i \qquad (2)$$

where $r_{B,i}$ is the model coefficients for i-th glass component, x_i is the mole fraction of i-th component in glass, and N is the number of components. Then, the TCLP release of each RCRA element is calculated by the relation:

$$c_i = r_B f_i \qquad (3)$$

The calculated c_i is based on the observation that $r_i \leq r_B$ for all RCRA elements as discussed earlier.

The model coefficients from 211 glasses were initially calculated using 20 components (19 components plus Others), which were identified as the maximum number of components that have reasonable ranges of mole fractions and distributions of mole-fraction values within the range to consider including them in the model. Then the next models were calculated based on the reduced number of components removing the component(s) of higher uncertainty or of less importance. These models were evaluated in terms of their model statistics relative to the total number of model components, which resulted in the selection of 14-cmponent model as the final model. Detailed discussion on the model selection process is described in Kim and Vienna.[16]

The initial model calculations and application to the glasses archived at PNNL identified the need for improvements in the model validity range to cover the compositions of the majority of glasses prepared in the PNNL glass development lab. Therefore, it was decided to expand the database by including TCLP data for more glasses. Twenty glasses with a higher concentration of Al_2O_3 and ZrO_2 were collected from three studies[17-19] that didn't include TCLP analyses. The TCLP releases of these glasses were measured for inclusion in the final model. Aluminum oxide and ZrO_2 were selected because these components had caused the largest number of glasses in the PNNL archive to be outside of the initial model validity range.

The 231 glasses discussed above were used to develop the final model. Summary statistics from the regression analysis for the final 14-component model is summarized in Table I. The plot of calculated $\ln(r_B)$ versus measured $\ln(r_B)$ is in Figure 3. The composition ranges of these 231 glasses expressed in mole fraction of 14 model components are also included in Table I.

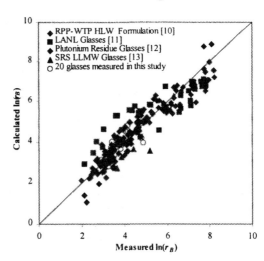

Figure 3. Calculated versus measured $\ln(r_B)$ for glasses used in the final model

Table I. Model coefficient, standard error, model statistics, and mole fraction range for components in 231 glasses for the final TCLP r_B model

Component	Coefficient	Std Error	Composition Range Min	Max
Al_2O_3	-11.830	1.637	0	0.1296
B_2O_3	14.155	0.767	0.0165	0.2217
CaO	14.266	0.982	0	0.2269
Fe_2O_3	-9.869	2.027	0	0.0845
K_2O	29.025	2.465	0	0.1097
Li_2O	10.456	0.937	0	0.2051
MgO	12.980	1.032	0	0.1960
Na_2O	18.440	0.874	0	0.2581
SiO_2	-1.270	0.358	0.3331	0.6458
ZrO_2	-10.114	2.685	0	0.0908
LN_2O_3	-98.649	7.577	0	0.0231
MnO_x	15.308	6.942	0	0.0795
SrO	8.975	5.238	0	0.1023
Others	9.696	1.116	0.0004	0.1685
# of Glasses	231			
R^2	0.870			
R^2 (Adjusted)	0.862			
s (RMSE)	0.583			

MODEL APPLICATIONS

Table II summarizes the regulatory limits used by EPA for designation of regulatory wastes[4] and the land-disposal restrictions (LDR) universal treatment standard (UTS).[1] UTS limits are lower than the toxicity limits for all elements except for As and Se.

Table II. RCRA toxicity limits by TCLP and "maximum" glass oxide concentrations (see text below)

Element	Ag	As	Ba	Cd	Cr	Hg	Ni	Pb	Se
RCRA Toxicity Limit (mg/L)	5	5	100	1	5	0.2	--	5	1
RCRA UTS Limit (mg/L)	0.14	5	21	0.11	0.6	0.025	11	0.75	5.7

The normalized release, r_i, represents the amount of glass dissolved in the TCLP condition if it is assumed that the glass dissolves congruently, i.e., all the elements in glass release at the same rate. If the glass sample were entirely dissolved in TCLP and the dissolution were congruent, then $r_i = 50,000$ for all elements. Similar to PCT, boron is believed to be the most appropriate element to represent the glass network dissolution—neither participation in precipitation of mineral phases nor preferential release of elements like alkalis by ion exchange (see Hrma et al.[14] for example).

The calculated c_i using Equation (3) is based on the observation that generally $r_i \leq r_B$ for all RCRA elements, which provides a reasonable estimation for congruent elements (Ba, Cd, Ni, and Zn) and a conservative estimation for those "incongruent" elements (Ag, As, Cr, Pb, and Se). It will be necessary to develop separate models for these incongruent elements instead of using Equation (3) if they become a limiting factor that causes the many glasses to fail the TCLP requirements. Use of separate models for specific components will remove the conservatism involved in Equation (3) and decrease the prediction error. The development of separate models would require larger number of data than currently available.

Figure 4 shows a plot of calculated versus measured releases of cadmium for 231 glasses used in the model development as an example. Also included in Figure 4 are the lines representing the value of RCRA toxicity limit, $\ln(c_{Cd})$ = 0. In terms of pass/fail classification, the data points in the top-left rectangle separated by a line of toxicity limit represent the false classification of non-hazardous glasses as hazardous ("false positive"), and the bottom right represents the false classification of hazardous

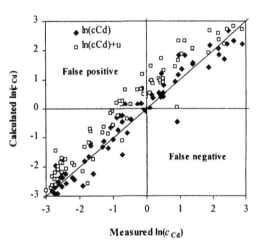

Figure 4. Plot of calculated vs. measured TCLP releases of Cd

glasses as non-hazardous ("false negative"). In order to decrease the probability of "false negative" classification, it is necessary to apply the prediction uncertainty, u. The method for uncertainty calculation will be dependent on the nature of application and governing regulations. Figure 4 shows that there are one glass in the "false negative" region and four glasses in the "false positive" if the classification is based on as-predicted values. However, by applying the upper bound ($\ln(c_{Cd})+u$) of the confidence band ($\ln(c_{Cd})\pm u$) as a limit (here, u was assumed as a constant value arbitrary chosen for the purpose of this illustration), the one glass in the "false negative" region is eliminated but the four glasses in false positive" region increases to 12.

SUMMARY

Evaluation of data collected from various studies and development of a first-order property model for normalized TCLP boron release as a function of glass composition were described. It was shown that the normalized boron release can be used to estimate the release of toxic elements based on the observation that the boron release represents the conservative release for those constituents of interest. The current TCLP model has two targeted application areas: (1) delisting of waste-glass product as radioactive (not mixed) waste and (2) designating the glass wastes generated from waste-glass research activities as hazardous or non-hazardous. Currently, the first order model prediction with 90% one-sided confidence band is being applied for the designation of glass wastes generated at PNNL.

ACKNOWLEDGMENT

The authors are grateful to Steve Lambert (Numatec Hanford Company), Kami Baisch (Pacific Northwest National Laboratory, PNNL), Gregg Bartel-Bailey (PNNL), Kim Fowler (PNNL), Scott Cooley (PNNL), and Bob Daubt (PNNL) for helpful discussions and suggestions; Connie Herman (Westinghouse Savannah River Company, WSRC) and Ian Pegg (Catholic University of America) for supplying TCLP data; and SK Sundaram (PNNL), David Blumenkranz (Waste Treatment Plant Project, WTP), Bill Holtzscheiter (WSRC), Noel Smith-Jackson (Washington State Department of Ecology), and Jacob Reynolds (WTP) for helpful comments. This work was funded by Battelle through the PNNL P2 Pays Program and the DOE Office of Environmental Management through the Tanks Focus Area Immobilization Program. Pacific Northwest National Laboratory is operated for the U.S. Department of Energy (DOE) by Battelle under Contract DE-AC06-76RL01830.

REFERENCES

[1]Environmental Protection Agency. 40CFR Protection of Environment, Chapter I - Environmental Protection Agency, Subchapter I - Solid Wastes, Part 268 - Land Disposal Restrictions, as amended.

[2]Washington Department of Ecology. Washington Administrative Code (WAC). Chapter 173-303 - Dangerous Waste Regulations, as-amended.

[3]Environmental Protection Agency, SW-846 Test Methods for Evaluation of Solid Waste Physical/Chemical Methods, Method 1311 Toxicity Characteristic Leaching Procedure, Rev.0, US EPA, Washington, D.C., 1992.

[4]Environmental Protection Agency (EPA). 40CFR Protection of Environment, Chapter I - Environmental Protection Agency, Subchapter I - Solid Wastes, Part 261 - Identification and Listing of Hazardous Waste, as amended.

[5]Ferrara DM, CL Crawford, BC Ha, and NE Bibler. 1998. "Vitrification of Three Low-Activity Radioactive Waste Streams from Hanford." In: *Proceedings of the International Conference on Decommissioning and Decontamination and on Nuclear and Hazardous Waste Management*, Vol. 1, 706–713.

[6]Fu SS, H Gan, IS Muller, IL Pegg, and PB Macedo. 1997. "Optimization of Savannah River M-Area Mixed Waste for Vitrification." In: *MRS Symposium Proceedings*, Vol. 465, 139-146.

[7]Muller IS, and IL Pegg. 1998. "Glass Formulation and Testing with TWRS LAW Simulants," Final Report for GTS Durateck Inc. and BNFL Inc., Catholic University of America, Washington D.C.

[8]Fu SS, and IL Pegg. 1998. "Glass Formulation and Testing with TWRS HLW Simulants, VSL Final Report." Vitreous State Laboratory, The Catholic University of America, Washington D.C.

[9]Muller IS, AC Buechele, and IL Pegg. 2001. "Glass Formulation and Testing with RPP-WTP LAW Simulants," VSL-01R3560-2, Vitreous State Laboratory, The Catholic University of America, Washington D.C.

[10]Kot WK, and IL Pegg, 2001. "Glass Formulation and Testing with RPP-WTP HLW Simulants," VSL-01R2540-2, Vitreous State Laboratory, The Catholic University of America, Washington D.C.

[11]Vienna JD, RP Thimpke, GF Piepel, DE Smith, ML Elliott, RK Nakaoka, and GW Veazey. 1998. *Glass Development for Treatment of LANL Evaporator Bottoms waste,* PNNL-11865, Pacific Northwest National Laboratory, Richland, Washington.

[12]Bulkley SA, and JD Vienna. 1997. "Composition Effects on Viscosity and Chemical Durability of Simulated Plutonium Residue Glasses," *Mat. Res. Soc. Symp. Proc.* 465:1243–50.

[13]Herman CA.2002. TCLP Data on SRS LLMW Glasses, Unpublished work.

[14]Hrma, PR, GF Piepel, MJ Schweiger, DE Smith, DS Kim, PE Redgate, JD Vienna, CA LoPresti, DB Simpson, DK Peeler, and MH Langowski. 1994. *Property/Composition Relationships for Hanford High-Level Waste Glasses Melting at 1150°C.* PNL-10359, Vol. 1 and 2, Pacific Northwest Laboratory, Richland, Washington.

[15]Vienna JV, D-S Kim, and P Hrma. 2002. "Database and Interim Glass Property Models for Hanford HLW and LAW Glasses," PNNL-14060, Pacific Northwest National Laboratory, Richland, Washington.

[16]Kim, D-S and JD Vienna. 2003. "Models for TCLP Releases from Waste Glasses," PNNL-14061, Rev.1, Pacific Northwest National Laboratory, Richland, Washington.

[17]Crum JV, MJ Schweiger, P Hrma, and JD Vienna. 1997. "Liquidus Temperature Model for Hanford High-Level Waste Glasses with High Concentrations of Zirconia," *Mat. Res. Soc. Symp. Proc.* 465:79–85.

[18]Li H, JD Vienna, P Hrma, DE Smith, and MJ Schweiger. 1997. "Nepheline Precipitation in High-Level Waste Glasses: Compositional Effects and Impacts on the Waste Form Acceptability," *Mat. Res. Soc. Symp. Proc.* 465:261-268.

[19]Vienna JD, P Hrma, JV Crum, and M Mika. 2001. "Liquidus Temperature-Composition Model for Multi-Component Glasses in the Fe, Cr, Ni, and Mn Spinel Primary Phase Field," *J. Non-Cryst. Solids* 292:1-24.

Alternate Waste Forms and Processes

IRON PHOSPHATE GLASS FOR IMMOBILIZATION OF HANFORD LAW

C.W. Kim, D. Zhu, and D.E. Day
Graduate Center for Materials Research, University of Missouri-Rolla, MO 65409

D.-S. Kim and J.D. Vienna
Pacific Northwest National Laboratory, Richland, WA 99352

D.K. Peeler
Westinghouse Savannah River Company, Aiken, SC 29808

T.E. Day and T. Neidt
MO-SCI Corporation, Rolla, MO 65402

ABSTRACT

Three iron phosphate glasses containing up to 35 wt% of a high sulfur Hanford LAW simulant were successfully melted in electric furnaces at 1150-1250°C for 2-3 hours. No sulfate salt segregation was found in the glass when examined by SEM. The glass retained up to 73% of the sulfur originally present in the waste, which was equivalent to 2.4 wt% SO_3 on a target glass oxide basis. This suggests that the waste loading in the iron phosphate glasses will not be limited by the SO_3 content of the LAW as it typically is by the baseline technology for Hanford LAW treatment. The chemical durability of the iron phosphate glasses was determined by the product consistency test (PCT) and the vapor hydration test (VHT). The mass release of sodium from both annealed and canister centerline cooled (CCC), partially crystallized iron phosphate wasteforms after the PCT at 90°C for 7 days was below the current DOE specification for LAW. The VHT alteration rates (200°C for 7 days) of the iron phosphate glasses for both annealed and CCC treated samples were also significantly lower than the current DOE limit and those of standard borosilicate glasses. All of the iron phosphate wasteforms met all of the existing requirements for aqueous chemical durability and the crystallization that occurred during CCC treatment did not reduce the chemical durability. Iron phosphate glasses were successfully melted in a hot-wall induction furnace and in a microwave oven. These melting processes avoid any corrosion problems associated with metal electrodes needed for joule heated melting.

INTRODUCTION

The U.S. Department of Energy (DOE)'s Hanford Site in Washington State produced more than 55 million gallons of radioactive waste, which is stored in 177 underground storage tanks [1]. The waste will be retrieved from these tanks and separated into low-activity waste (LAW) and high-level waste (HLW) fractions which will be separately vitrified in the Hanford Tank Waste Treatment Plant (WTP).

The Hanford LAW composition is high in sodium and sulfur (Table 1). The loading of this waste in the borosilicate (BS) waste glass is largely determined by the allowable fraction of sulfur in the waste. If sulfur loading were not to limit the loading of LAW in glass then the amount of glass produced at Hanford could be reduced by as much as 50%. The purpose of the present study is to investigate the feasibility of vitrifying the Hanford LAW in iron phosphate glasses with an aim to producing a wasteform having a higher waste loading combined with acceptable chemical durability.

Table 1. Summary of Hanford LAW composition (wt% non-volatile oxides and halogens) and simplified composition was used in present study.

Oxide (wt%)	Min	Ave	Max	Simplified*
Al_2O_3	0.68	13.19	35.62	4.4
Cl	0.00	0.89	2.61	0.6
Cr_2O_3	0.00	0.33	1.64	0.4
F	0.00	0.94	5.38	1.6
K_2O	0.00	1.02	16.06	0.0
MoO_3	0.00	0.01	5.06	0.0
Na_2O	32.24	75.25	97.93	75.3
P_2O_5	0.29	4.08	48.72	7.7
SiO_2	0.01	0.86	7.42	0.5
SO_3	0.02	3.19	14.11	9.5
Other	0.01	0.23	1.61	0.0
Total		99.99		100.0

* LAW wastes with SO_3 concentrations above 7 wt% were averaged & simplified and used in iron phosphate glasses.

GLASS PREPARATION

The appropriate amounts of the raw materials (Table 2) were mixed in a sealed plastic container for sufficient time to produce a homogeneous mixture. The IP27LAW, IP32LAW, and IP35LAW glasses contained 27, 32, and 35 wt% of the simplified Hanford LAW (given in the 5th column of Table 1) and 73, 68, and 65 wt% Glass Forming Additives (GFA), respectively. Table 2 lists the glass forming additives and possible raw materials required to produce these iron phosphate glasses. The most convenient method of combining the majority of the GFA constituents with the Hanford LAW is to melt a glass frit. The GFA constituents, marked with an "*" in Table 2, have been successfully melted as a glass frit. The amount of P_2O_5 used in the GFA included an additional 5 wt% of the required amount in order to compensate for potential volatilization losses of P_2O_5 during melting of the frit (which will reduce P_2O_5 loss during LAW vitrification). This glass frit combined with the appropriate silica and alumina constitute the GFA.

Table 2. Composition of the iron phosphate glass (IP27LAW) containing 27 wt% LAW and the raw materials used.

Oxide (wt%)	Hanford LAW	27 wt% LAW	73 wt% GFA	Raw material for GFA	IP27LAW Glass
Al_2O_3	4.4	1.2	15.0	Al_2O_3, $Al(OH)_3$, kaolinite: $Al_2SiO_5(OH)_4$	16.2
Cl	0.6	0.2	0.0		0.2
Cr_2O_3	0.4	0.1	3.3*	Cr_2O_3, chromite: $(Mg,Fe)O-(Cr,Al)_2O_3$	3.4
F	1.6	0.4	0.7*	CaF_2	1.1
Na_2O	75.3	20.3	0.0		20.3
P_2O_5	7.7	2.1	27.1*	P_2O_5, H_3PO_4, $NH_4H_2PO_4$	29.2
SiO_2	0.5	0.1	12.3	SiO_2, kaolinite: $Al_2SiO_5(OH)_4$	12.4
SO_3	9.5	2.6	0.0		2.6
Bi_2O_3			2.7*	Bi_2O_3	2.7
Fe_2O_3			7.2*	Fe_2O_3, iron ore, iron phosphate waste**	7.2
La_2O_3			1.2*	La_2O_3	1.2
ZrO_2			2.7*	ZrO_2, $ZrSiO_4$	2.7
CaO			1.0*	CaF_2	1.0
Total	100.0	27.0	73.2		100.2

* denotes GFA constituents that can be melted as a glass frit.
** Phosphate chemical conversion coating processes are used by the metal products fabrication industry to condition metal surfaces for subsequent processes such as painting. This processing generates iron phosphate waste in the form of a sludge that can be used in iron phosphate glasses.

All these iron phosphate glasses (IP27LAW, IP32LAW, and IP35LAW) were melted in an electric furnace in air at 1150-1250°C for 2-3 hours in a dense high silica (DFC 83% silica 17% alumina) crucible. Each melt was stirred 3 to 4 times with a fused silica rod to insure chemical homogeneity. After melting, the glass was cast into bars that were annealed at 480-520°C for 5 hours and cooled to room temperature overnight in the annealing furnace. The annealing temperature was determined from differential thermal analysis (DTA) measurements. A sample of the IP27LAW melt was also cooled according to the simulated immobilized LAW canister centerline cooling (CCC) profile [2]. For the IP32LAW and IP35LAW, samples were slowly cooled (~2°C/min), designated as SC, in the furnace instead of performing the fully programmed CCC treatment. These slowly cooled IP32LAW-SC and IP35LAW-SC samples were not analyzed by SEM or XRD, but their chemical durability was measured.

RESULTS AND DISCUSSION
Glass Formation

The chemical composition of each glass was calculated from the batch composition and was also measured by inductively coupled plasma-emission spectroscopy (ICP-ES) as well as by the Leco method for sulfur analysis (Table 3). The batch and measured compositions were in reasonable agreement. A noticeable difference between batch and analyzed compositions is the higher SiO_2 concentration in each glass. A possible explanation for the higher silica concentration is the slight dissolution of the dense high silica crucible in which the glasses were melted.

Table 3. Batch and analyzed (ICP-ES) compositions (wt%) of iron phosphate glasses containing 27, 32, and 35 wt% LAW (IP27LAW, IP32LAW, and IP35LAW, respectively).

Oxide (wt%)	IP27LAW		IP32LAW		IP35LAW	
	Batch	ICP-ES	Batch	ICP-ES	Batch	ICP-ES
Al_2O_3	16.2	18.5	15.4	14.3	14.9	16.4
Cl	0.2	NM	0.2	NM	0.2	NM
Cr_2O_3	3.4	3.0	3.2	2.3	3.0	2.5
F	1.1	NM	1.1	NM	1.1	NM
Na_2O	20.3	19.2	24.1	24.5	26.4	25.1
P_2O_5	29.2	27.8	27.7	26.1	26.8	26.2
SiO_2	12.4	15.4	11.7	14.6	11.2	15.7
SO_3	2.6	1.1*	3.0	2.1*	3.3	2.4*
Bi_2O_3	2.7	NM	2.5	NM	2.4	NM
Fe_2O_3	7.2	7.1	6.7	6.9	6.4	6.7
La_2O_3	1.2	NM	1.1	NM	1.1	NM
ZrO_2	2.7	NM	2.5	NM	2.4	NM
CaO	1.0	1.3	1.0	1.6	1.0	1.2
Total	100.2	93.4	100.2	92.4	100.2	96.2

* Analyzed by Leco (Leco and ICP-ES analyses were conducted by Acme Analytical Laboratories Ltd. in Vancouver, Canada.).
NM = not measured.

No evidence of salt segregation was found either from the visual appearance of the crucible or SEM analysis or powder XRD of the quenched melt (Figure 1). There was no observable sulfate "gall" layer on the surface of the molten iron phosphate glass, as has been reported [3] for borosilicate melts that contained > 1 wt% SO_3. Leco analysis indicated that 42 to 73% of the SO_3 originally present in the waste was retained in the glasses (Table 3). The sulfur retention in iron phosphate glasses was previously found to be compositionally dependent [4]. The quenched samples of IP27LAW and IP35LAW had trace amounts of crystalline $(Cr,Fe)_2O_3$ (solid solution of Cr_2O_3 and Fe_2O_3) and the CCC treated IP27LAW sample contained crystalline sodium iron phosphates ($NaFeP_2O_7$ and $Na_3Fe_2(PO_4)_3$) as well as crystalline $(Cr,Fe)_2O_3$. The IP32LAW sample was not analyzed by SEM or XRD.

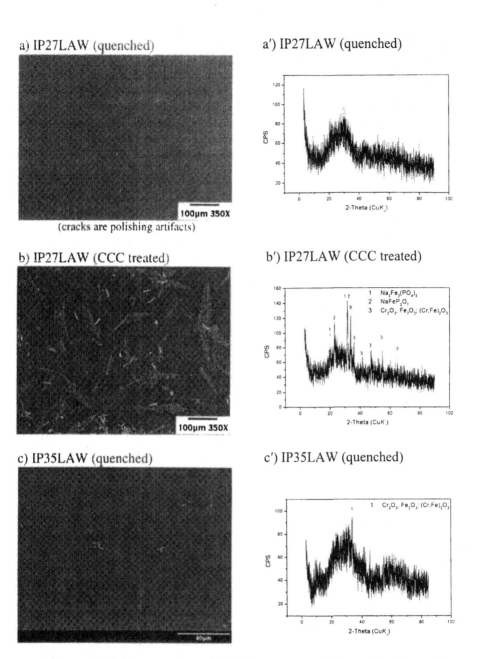

a) IP27LAW (quenched)

100µm 350X

(cracks are polishing artifacts)

a') IP27LAW (quenched)

b) IP27LAW (CCC treated)

100µm 350X

b') IP27LAW (CCC treated)

1 $Na_3Fe_2(PO_4)_3$
2 $NaFeP_2O_7$
3 Cr_2O_3, Fe_2O_3; $(Cr,Fe)_2O_3$

c) IP35LAW (quenched)

40µm

c') IP35LAW (quenched)

1 Cr_2O_3, Fe_2O_3; $(Cr,Fe)_2O_3$

Figure 1. SEM micrographs and XRD patterns of iron phosphate glasses containing 27 and 35 wt% LAW (IP27LAW and IP35LAW, respectively). The slowly cooled IP35LAW sample and the IP32LAW samples (quenched, slowly cooled) were not analyzed by SEM or XRD.

Physical Properties

The density of each sample was measured at room temperature by the Archimedes' method using deionized water as the suspending medium. The density was 2.74 to 2.76 g/cm^3 for the glass samples and less than 2% higher for the CCC treated or slowly cooled samples (Table 4). The thermal expansion coefficient (TEC) was determined by dilatometry. The liquidus temperature (T_L) was measured per ASTM C 829-81 [5] procedures in a temperature gradient furnace using a platinum tray in which glass particles were fused to form a thin (~3 mm) layer of melt. The TEC and T_L for the iron phosphate glasses are summarized in Table 4.

Table 4. Density, thermal expansion coefficient (TEC), and liquidus temperature (T_L) of iron phosphate wasteforms.

Wasteform	Density (g/cm^3)		TEC for 35-300°C	T_L
	Glass	CCC/SC*	(x 10^{-7}/°C)	(°C)
IP27LAW	2.76±0.01	2.82±0.01	138±3	762±5
IP32LAW	2.77±0.01	2.82±0.01	NM	NM
IP35LAW	2.74±0.01	2.77±0.01	NM	736±5

* CCC treated for IP27LAW and Slowly Cooled (~2°C/min) for IP32LAW and IP35LAW samples.
NM = not measured.

Chemical Durability

The chemical durability of the iron phosphate glasses was determined by the product consistency test (PCT) following the procedures in ASTM C 1285-97 [6]. After completion of the PCT, the concentration of ions in the leachate was measured by ICP-ES.

The normalized elemental mass release, r_i, was calculated from the concentration of each element in the leachate and from the mass fraction of element in the glass. The mass release of Na, Si, P, Al, Cr, Fe, and Ca from the iron phosphate wasteforms is summarized in Table 5. The current PCT specification for Hanford LAW borosilicate glass is that the r_i of Na, Si, and B should be less than 2 g/m^2 from a 7-day test at 90°C [7] (Table 5). Given that the iron phosphate glasses in this study do not contain boron, no boron value can be given. The r_i's for i = sodium and silicon for the annealed and CCC/SC treated samples of IP27LAW and IP32LAW are well below the current DOE specification for LAW and the Environmental Assessment (EA) glass values. The chemical durability of the IP35LAW is not quite as good as that of the IP27LAW and IP32LAW, but is still comparable to that of the EA glass [8]. The mass release of the IP27LAW-CCC and IP32LAW-SC samples suggests that any crystallization that occurred during CCC treatment or slowly cooling at ~2°C/min did not significantly change the chemical durability of these iron phosphate glasses. In fact, the r_{Na} and r_P values for the IP32LAW-SC sample are smaller than those for the normally annealed sample. The amount of crystals present in the SC and CCC iron phosphate samples was not determined.

Table 5. Normalized elemental mass release (g/m^2) from iron phosphate wasteforms, DOE limit for Hanford LAW borosilicate glass [7], and Environmental Assessment (EA) glass value [8] after PCT in deionized water at 90°C for 7 days.

Wasteform	r_{Na}	r_{Si}	r_P	r_{Al}	r_{Cr}	r_{Fe}	r_{Ca}
IP27LAW	0.51	0.19	0.24	0.21	< 0.01	0.03	< 0.01
IP27LAW-CCC*	0.48	0.22	0.21	0.23	< 0.01	0.02	< 0.01
IP32LAW	1.81	0.16	1.43	0.61	0.09	0.03	< 0.01
IP32LAW-SC**	1.02	0.21	0.83	0.53	0.09	0.02	< 0.01
IP35LAW	4.85	0.08	3.37	1.06	0.85	0.15	0.01
IP35LAW-SC**	5.70	0.11	4.09	1.39	0.83	0.80	0.03
DOE limit for LAW	2	2	NA	NA	NA	NA	NA
EA glass value	6.67	1.96	NA	NA	NA	NA	NA

* CCC treated sample.
** Slowly Cooled sample.

The chemical durability of the iron phosphate glasses was also evaluated by the vapor hydration test (VHT). A complete description of the VHT can be found in reference [9], whose procedures were used in the present work. Optical micrographs of a cross section of the VHT samples after 7-day tests at 200°C are shown in Figure 2. The VHT corrosion rate (mass of specimen dissolved per unit surface area) was calculated from the initial specimen thickness and the thickness of the glass remaining at the end of the test [9]. The VHT corrosion rates (estimated by assuming a constant rate from time = 0) for the three annealed glasses and for the CCC treated or slowly cooled samples are considerably smaller than the current DOE limit of 50 $g/m^2/day$ [7] (Table 6). These results show that the iron phosphate wasteforms have an outstanding chemical resistance to the humid conditions existing at 200°C for 7 days.

Table 6. VHT corrosion rates $(g/m^2/day)$ for iron phosphate wasteforms containing up to 35 wt% LAW tested at 200°C for 7 days. DOE limit for corrosion rates as given in reference [7].

Test	VHT Corrosion Rate $(g/m^2/day)$						
	IP27LAW		IP32LAW		IP35LAW		DOE
	Annealed	CCC*	Annealed	SC**	Annealed	SC**	Limit
7-day	5.9	10.1	28.9	31.8	35.4	38.2	50 [7]

* CCC treated sample.
* Slowly Cooled sample.

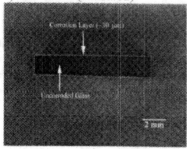

a) IP27LAW (annealed)

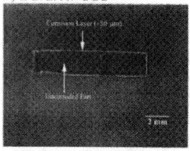

a') IP27LAW-CCC

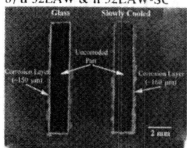

b) IP32LAW & IP32LAW-SC

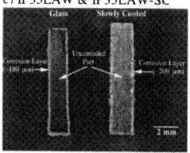

c) IP35LAW & IP35LAW-SC

Figure 2. Optical micrographs of the cross section of iron phosphate wasteforms containing 27, 32 and 35 wt% LAW after VHT at 200°C for 7 days.

The corrosion layers (products) on the surface of the annealed and CCC treated IP27LAW samples after the VHT (200°C, 7 days) were identified by XRD to be mainly the zeolite, $Na_4(Al_4Si_{12}O_{32})(H_2O)_{14}$. The corrosion products on the IP32LAW and IP35LAW wasteforms were not determined by XRD, but are expected to be similar to that for the IP27LAW samples due to the similarities in composition and alteration extent.

Alternative Melting Methods

Previous studies [4,10] have shown that the corrosion of typical joule heated melter glass contact materials may not be a significant issue with high soda iron phosphate glasses. However, it is of use to consider other melting methods so that the most effective melter types can be used to produce this glass.

Small amounts (300 g) of the iron phosphate glass, IP35LAW, were successfully melted in a hot-wall induction furnace (Figure 3). The batch was contained in a dense high silica crucible inside a graphite susceptor and melted at 1150°C for 2 hours at a power level of 12 to 18 kW, depending on the stage of melting. The chemical durability and other properties of the glasses melted in the hot-wall induction furnace were the same as those of the same glass conventionally melted in an electric furnace. The successful melting of the iron phosphate glass in the hot-wall induction furnace is encouraging since this method eliminates the need for metal electrodes in the melt, as in joule heated melting and the stirring of the fluid melt by the magnetic field rapidly homogenizes the melt, thereby reducing the melting time.

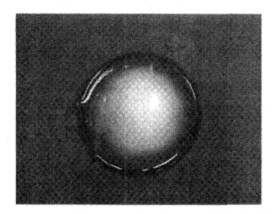

Figure 3. Iron phosphate glass wasteform (300 g) containing 35 wt% Hanford LAW that was melted at 1150°C for 2 hours in a hot-wall induction furnace.

Microwave melting is another alternative technology for vitrifying iron phosphate glass wasteforms. In collaboration with the Energy and Nuclear Research Institute, Brazil, small amounts (50 to 100 g) of the IP35LAW glass have been successfully melted, starting from a cold batch, in an ordinary microwave oven (1100 watts). Compositions containing significant amounts of alkalis, such as the soda (~75 wt%) in the Hanford LAW, are an advantage since this improves the coupling of the microwave energy to the melt and promotes rapid melting. Microwave melting also eliminates the need for metal electrodes in the melt, but a refractory crucible is required (alumina and silica work well). The chemical durability and other properties of the iron phosphate glasses prepared with microwave heating are the same as those of the same glass melted conventionally.

SUMMARY AND CONCLUSIONS

The present investigation has demonstrated several advantages for vitrifying the Hanford LAW in iron phosphate glass wasteforms (IP27LAW, IP32LAW, and IP35LAW). A major advantage is a high waste loading of the simulated LAW, from 27 to 35 wt% depending upon the desired chemical durability. The glassy and partially crystallized iron phosphate wasteforms satisfy DOE's PCT and VHT requirements for aqueous chemical durability. In addition, there was no indication of sulfate salt segregation, as seen in most borosilicate glasses, so it is unlikely that the LAW waste loading in iron phosphate glasses will be limited by the SO_3

content of the LAW. Furthermore, the iron phosphate glass retained a large percentage (up to 73%, depending on composition) of the sulfur originally present in the waste, which was equivalent to 2.4 wt% SO_3 on a target glass oxide basis.

The successful melting of the iron phosphate glass in the hot-wall induction furnace and microwave oven is encouraging since both procedures avoid any potential corrosion problems associated with the metal electrodes needed for joule heated melting.

ACKNOWLEDGEMENTS

This work was supported by the U.S. Department of Energy (DOE) Office of Science and Technology through the Tanks Focus Area under contract DE-AC06-76RL01830, and by the DOE Environmental Management Science Program (EMSP) under contract DE-FG07-96ER45618.

REFERENCES

[1] U.S. DOE, "Summary Data on the Radioactive Waste, Spent Nuclear Fuel, and Contaminated Media Managed by the U.S. Department of Energy", U.S. Department of Energy, Washington D.C., 2001.
[2] G.L. Smith, L.R. Greenwood, G.F. Piepel, M.J. Schweiger, H.D. Smith, M.W. Urie, and J.J. Wagner, "Vitrification and Product Testing of AW-101 and AN-107 Pretreated Waste", PNNL-13372, Pacific Northwest National Laboratory, Richland, WA, 2000.
[3] G.K. Sullivan, M.H. Langowski, and P. Hrma, "Sulfate Segregation in Vitrification of Simulated Hanford Nuclear Waste", Ceramic Transactions 61 (1995) 187-193.
[4] D.S. Kim, W.C. Buchmiller, M.J. Schweiger, J.D. Vienna, D.E. Day, C.W. Kim, D. Zhu, T.E. Day, T. Neidt, D.K. Peeler, T.B. Edwards, I.A. Reamer, and R.J. Workman, "Iron Phosphate Glass as an Alternative Waste-Form for Hanford LAW", PNNL-14251, Pacific Northwest National Laboratory, Richland, WA, 2003
[5] American Society for Testing and Materials (ASTM), "Standard Practices for Measurement of Liquidus Temperature of Glass by the Gradient Furnace Method", C 829-81, 1995.
[6] American Society for Testing and Materials (ASTM), "Standard Test Methods for Determining Chemical Durability of Nuclear, Hazardous, and Mixed Waste Glasses: The Product Consistency Test (PCT)", C 1285-97, 1998.
[7] U.S. Department of Energy (DOE), "Design, Construction, and Commissioning of the Hanford Tank Waste Treatment and Immobilization Plant", DOE Office of River Protection, Richland, WA; Contract with Bechtel National, Inc., San Francisco, CA, Contract No.: DE-AC27-01RV14136, 2001.
[8] C.M. Jantzen, N.E. Bibler, D.C. Beam, C.L. Crawford, and M.A. Pickett, "Characterization of the Defense Waste Processing Facility (DWPF) Environmental Assessment (EA) Glass Standard Reference Material (U)", WSRC-TR-92-346, Rev.1, Westinghouse Savannah River Company, Aiken, South Carolina,1993.
[9] J.D. Vienna, P. Hrma, A. Jiricka, D.E. Smith, T.H. Lorier, I.A. Reamer, and R.L. Schulz, "Hanford Immobilized LAW Product Acceptance Testing: Tanks Focus Area Results", PNNL-13744, Pacific Northwest National Laboratory, Richland, WA, 2001.
[10] F. Chen and D.E. Day, "Corrosion of Selected Refractories by Iron Phosphate Melts", Ceramic Transactions 93 (1999) 213-220.

CHARACTERIZATION AND PERFORMANCE OF FLUIDIZED BED STEAM REFORMING (FBSR) PRODUCT AS A FINAL WASTE FORM

C.M. Jantzen
Savannah River Technology Center
Westinghouse Savannah River Co.
Aiken, SC 29808

ABSTRACT

A demonstration of Fluidized Bed Steam Reforming (FBSR) was recently completed on a Hanford Low Activity Waste (LAW) simulant. This technology produced stable mineral phases (feldspathoids) when co-fired with clay. The mineral phases are cage structured and were determined to retain anions such as $SO_4^=$ as well as cations such as Re (simulant for Tc) in the mineral cages. The mineral phases are produced at moderate steam reformer operating temperatures between 650-750°C. The FBSR mineral waste form exhibited incongruent leaching characteristics during Product Consistency Testing (PCT or ASTM C1285). The radionuclides ([133]Cs and Re as simulants for ^{137}Cs and ^{99}Tc) are released in significantly lower concentrations than Na. In addition, the Na release is less than the 2 g/m^2 Hanford contract requirement for vitrified LAW. FBSR mineral waste forms are EPA regulatory compliant at the Universal Treatment Standard (UTS) making delisting an attractive option for this waste form.

INTRODUCTION

The Hanford LAW is a basic high Na^+ molarity (~8.1M) sodium nitrate – sodium hydroxide solution that also contains significant amounts of sulfate, chloride, fluoride and organic compounds as well as certain heavy metals and radionuclides. LAW is the low activity salt supernate fraction of Hanford High Level Liquid Waste (HLLW). In December 2001, Fluiduized Bed Steam Reforming (FBSR), was investigated for stabilization of LAW waste by THOR Treatment Technologies (TTT) using the patented THermal Organic Reduction (THOR[sm]) process. This process, developed by Studsvik, utilizes

---

pyrolysis*/steam reforming to destroy both organics and nitrates/nitrites in the waste. The FBSR demonstrations were performed on ~150 gallons of a Hanford AN-107 LAW simulant [1, 2] simultaneously being used for a vitrification demonstration. The radionuclide ^{99}Tc was simulated with Re while ^{137}Cs was simulated with stable cesium ([133]Cs).

The non-radioactive AN-107 simulant of 8.1M Na^+ was successfully tested in a 6-inch TTT pilot scale facility.[†] Other demonstrations performed by TTT showed that LAW waste could be transformed into Na_2CO_3, $NaAlO_2$, or Na_2SiO_3 feed material for the LAW Hanford melter (Table I). Addition of no solid co-reactant yielded a sodium carbonate product. Sodium will combine with carbon dioxide in the reformer gases to provide a sodium carbonate product. The generation of sodium carbonate in this type of application has been known since the 1950s in fluid bed denitration systems [1]. Addition of an $Al(OH)_3$ co-reactant will provide an $NaAlO_2$ product, addition of SiO_2 will provide an Na_2SiO_3 product, while addition of clay will provide a final mineral waste form product (Table I). The latter has been shown to perform well as a final waste form [2,3]. Testing on Hanford LAW surrogates has shown that over 95% of the sulfur compounds, fluorides and chlorides in the waste feed react in the steam reformer with the clay co-reactant and become an integral part of the final mineral waste product structure [2].

Table I. THOR[sm] Pilot Scale Demonstrations with Simulated LAW Wastes

# Pilot Runs	Solid Additive	Mineral Product	Purpose
5	Clay	Feldspathoid minerals (nepheline and sodalite) that stabilize problematic anions such as Cl, F, and SO_4	Stabilization of LAW or salt supernates as a final mineral waste form
3	Sand	Sodium silicate	Pretreatment of LAW for vitrification and/or recycle of melter off-gas blowdown.
2	$Al(OH)_3$	Sodium aluminate	Pretreatment of LAW for vitrification and/or recycle of melter off-gas blowdown.
3	None	Sodium carbonate	Pretreatment of LAW for vitrification and/or recycle of melter off-gas blowdown.

---

*   Pyrolysis chemically decomposes organic materials by heat in the absence of oxygen
†   The solution was diluted to 5.2M Na^+ to homogenize the feed before processing due to the observation of precipitated solids in the feed tank

EXPERIMENTAL

Scoping FBSR tests were performed by TTT between December 6, 2001 and December 20, 2001 using the Studsvik THOR^{sm} process. The FBSR waste forms were made from Hanford AN-107 LAW simulant. The results of the final waste form testing performed by TTT under reducing FBSR conditions will be summarized in this report (Scoping Tests 1 and 2 and Production Run 2). Additional testing performed under oxidizing FBSR conditions will be discussed comparatively since there was only one set of tests run under these conditions. The Na-Al-Si (NAS) mineral waste form was produced during the scoping tests SCT-01 and SCT-02 and during the two production runs PR-01 and PR-02 (Table II).

Table II Fluidized Bed Steam Reformer (FBSR) Demonstration Program

Product	Operating Mode	Scoping Test No.	Production Run No.
$Na_2O\text{-}Al_2O_3\text{-}2SiO_2$	Reducing	SCT-01 (18.6 hrs) SCT-02 (4.8 hrs)	PR-02 (21.7 hrs)
$Na_2O\text{-}Al_2O_3\text{-}2SiO_2$	Oxidizing	None	PR-01 (23.3 hrs)

The first four days of SCT-01 testing were short test series utilized to optimize the direct conversion of nitrates and nitrites to nitrogen gas in the FBSR. The type of carbon reductant and catalyst used was varied to provide high NO_x conversions. The final SCT-01 test run demonstrated the Reformer's ability to reliably and continuously maintain NO_x levels in the FBSR off-gas at less than 500ppm.

The SCT-02 test run was short as was terminated when a carbon reductant injection/transfer line plug occurred. The unplugging efforts allowed excess oxygen to enter the bed and cause poor fluidization and an excess of oxygen in one area of the bed. The non-uniform fluidization caused an agglomeration of the waste form to occur and the SCT-02 test was terminated after only 4.8 hours. The carbon transfer line pluggage is considered anomalous because full scale FBSR units use a transfer system with a higher purge gas flow and an in-bed downcomer that eliminates additive transfer line plugging events that occasionally occurred throughout the demonstration tests.

During the SCT-02 test run, coal was used to assist in denitrtion of the waste at temperatures between 715-735°C. Clay, small amounts of excess SiO2, and iron oxide were the only waste form additives. The waste loading achieved with SCT-02 was 27 wt% since 73 wt% additives were used. This corresponds to an Na_2O loading of ~20 wt% in the mineral waste form.

The PR-01 and PR-02 demonstration runs successfully produced an NAS mineral waste product without the presence of agglomeration or significant operational problems. These runs was terminated when the planned simulant waste feed volume was processed as planned.

Sample SCT02-098 from the SCT-02 demonstration was dissolved using a lithium borate fusion and a second dissolution performed via an Na_2O_2 fusion (ASTM C1463). Each dissolution was performed twice on two different days. Each dissolution was analyzed twice, once with no dilution and once with a 10X dilution. All four replicate analyses were averaged. A glass standard, Batch 1 glass, was analyzed simultaneously for quality assurance. The sample was analyzed for anions (phosphorous and sulfur) and cations by both Inductively Coupled Plasma Emission Spectroscopy (ICP-ES) and Inductively Coupled Plasma Mass Spectroscopy (ICP-MS) (ASTM C1463). In addition, the sample was dissolved in a Na_2O_2 dissolution with a water uptake and analyzed for additional anions by Ion Chromotography (IC) (ASTM D4327). The PCT leachates were analyzed for cations by both Inductively Coupled Plasma Emission Spectroscopy (ICP-ES) (ASTM C1109) and Inductively Coupled Plasma Mass Spectroscopy (ICP-MS).

Dissolution of sample SCT02-098-FM was also performed by H_2SO_4/HF in the presence of NH_4VO_3 followed by colorimetric determination of Fe^{2+} and total iron (ΣFe) in order to determine the REDuction/OXidation (REDOX) state of the sample in terms of the $Fe^{2+}/\Sigma Fe$ ratio [4]. A standard glass, the EA glass [5], with a known and reproducible REDOX state, was used during the analysis for quality assurance.

X-Ray Diffraction (XRD) analyses were performed by TTT and confirmed at SRTC for sample SCT02-098-FM. XRD was performed at SRTC both before and after durability testing.

Durability testing of the FBSR sample SCT02-098-FM from Scoping Test 02 was performed at SRTC. Durability testing was performed using ASTM C1285-97 (PCT-A test protocol). The PCT-A test was run for 7 days at 90°C in stainless steel vessels. Triplicate samples were tested along with two standard glasses; the ARM-1 standard and the Environmental Assessment (EA) glass [5] standard used to assess the durability of HLW vitrified waste forms. Testing of the FBSR final waste form using the EPA TCLP protocol was performed by Evergreen Analytical, an EPA certified laboratory, under subcontract to TTT. The results of the testing of samples from the FBSR process fabricated under both reducing and oxidizing conditions are summarized in this report.

RESULTS AND DISCUSSION

Chemical Analysis of the FBSR Mineral Product

A complete chemical analysis of FBSR Scoping Test 02 Sample SCT02-098-FM was performed. The average of the replicate analyses is given in Table III. The replicate REDOX analyses are given in Table IV. The average REDOX of the two replicate analyses in Table IV were used to calculate the relative proportions of FeO and Fe_2O_3 in the FBSR mineral waste form as given in Table III.

The REDOX measurements in Table IV indicate that the mineral phases are not overly reduced at a $Fe^{+2}/\Sigma Fe$ of 0.15 even in the presence of coal added for denitration of the feed. REDOX ratios of 0.15 are too oxidizing for any metallic iron to be present in the FBSR product, assuming uniform oxygen potential.

The data in Table III indicate that Cs, Re, SO_4, Cr, and Pb are retained in the FBSR mineral phases and do not volatilize during processing. This was confirmed by the TTT analyses of the off-gas during processing

Based on the analysis provided in Table III, a waste loading of 27 wt% was calculated by assuming that all of the SiO_2, all of the Al_2O_3 and all of the Fe_2O_3 and FeO are waste form additives.

Table III. Chemical Analysis of FBSR Sample SCT02-098-FM

Analytic Method	Oxide	Wt% 5/31/02	Wt% 8/15/02	Analytic Method	Oxide	Wt% 5/31/02	Wt% 8/15/02
ICP-ES	Al_2O_3	31.7436	38.924	ICP-ES	P_2O_5	0.2176	0.2474
ICP-ES	B_2O_3	0.2576	BDL	ICP-ES	PbO	0.0248	0.0199
ICP-ES	CaO	0.7332	0.5681	ICP-MS	PbO	0.0175	NA
ICP-ES	Cr_2O_3	0.0716	0.0278	ICP-MS	ReO_2	0.0005	0.0006
ICP-MS	Cs_2O	0.0029	0.005	ICP-ES	SiO_2	34.8706	30.0572
ICP-ES	Fe_2O_3	5.4471	6.23	ICP-ES	SO_4	1.1175	NA
ICP-ES	FeO	0.8749	1.001	IC	SO_4	NA	2.6335
ICP-ES	K_2O	0.6975	0.6794	IC	Cl	NA	0.318
ICP-ES	La_2O_3	0.0117	NA	IC	NO_2	<0.0005	NA
ICP-ES	Na_2O	19.8156	16.7826	IC	NO_3	<0.0005	NA
ICP-ES	NiO	0.0814	0.0350		SUM	95.9681	97.525

BDL = Below Detection Limit, NA = Not Analyzed

Table IV. Replicate Redox Analyses of FBSR Sample SCT02-098-FM

Analysis	EA Standard	AN107-A	AN107-B
Fe^{+2}	0.088	0.058	0.056
$\Sigma Fe_{(total)}$	0.458	0.377	0.376
$Fe^{+2}/\Sigma Fe$	0.192	0.154	0.149

X-Ray Diffraction Analysis of the FBSR Mineral Product
The phases identified by TTT in the FBSR Sample from Scoping Test 02 (Sample SCT02-098-FM) are given in Table V. The phases identified for the same sample at SRTC are given for comparison in Table V. Analysis at SRTC indicated the presence of a minor second iron oxide phase, magnetite (Fe_3O_4) in addition to the hematite (Fe_2O_3). Nepheline (the hexagonal type) is the major component with subordinate amounts of nosean and corundum. A cubic structured nepheline was not observed in this sample but was observed in the production run of a similar material. The relative amounts of the two types of nepheline and sodalite

(nosean) will vary with optimization of waste additives, e.g., types of clay or other aluminosilicates, and processing parameters.

Table V Phases Identified in FBSR Scample SCT02-098-FM

Mineral Phases Identified by TTT	Mineral Phases Identified by SRTC before PCT-A Testing	Mineral Phases Identified by SRTC after PCT-A Testing
$Na_8Al_6Si_6O_{24}(SO_4)$ (Nosean)	$Na_8Al_6Si_6O_{24}(SO_4)$ (Nosean)	$Na_8Al_6Si_6O_{24}(SO_4)$ (Nosean)
$NaAlSiO_4$ (Nepheline)	$NaAlSiO_4$ (Nepheline)	$NaAlSiO_4$ (Nepheline)
Al_2O_3 (Corundum)	Al_2O_3 (Corundum)	Al_2O_3 (Corundum)
Fe_2O_3 (Hematite)	Fe_2O_3 (Hematite)	Fe_2O_3 (Hematite)
	Fe_3O_4 (Magnetite)	Fe_3O_4 (Magnetite)

The sodium aluminosilicate (NAS) mineral phase assemblage(s) given in Table V are anhydrous feldspathoid phases such as sodalite. The sodalite family of minerals (including nosean) are unique because they have cage-like structures formed of aluminosilicate tetrahedra. The remaining feldspathoid minerals, such as nepheline, have a silica "stuffed derivative" ring type structure. The cage structures are typical of sodalite and/or nosean phases where the cavities in the cage structure retain anions and/or radionuclides which are ionically bonded to the aluminosilicate tetrahdra and to sodium. The cage structured feldspathoid system of minerals has the basic structural framework formula $Na_6[Al_6Si_6O_{24}]$. The square brackets in the formula are used to delineate the alumina:silica ratio of the aluminosilicate cage structure which is 1:1.

The feldspathoid mineral, sodalite has the formula $Na_8[Al_6Si_6O_{24}](Cl_2)$. The cage is occupied by two sodium and two chlorine ions in natural sodalites [6]. The formula can also be written as $Na_6[Al_6Si_6O_{24}]\bullet(2NaCl)$ to indicate that two NaCl are ionically bonded in the cavities of the cage structure while the remaining Na:Si:Al have a 1:1:1 stoichiometry [6]. When the 2NaCl are replaced by Na_2SO_4, the mineral phase is known as nosean , $(Na_6[Al_6Si_6O_{24}](Na_2SO_4))$ which is one of the feldspathoid cage structured minerals found in the FBSR waste form. Since the Cl^-, $SO_4^=$, and/or S_2 are chemically bonded inside the sodalite cage structure, these species do not readily leach out of the respective FBSR waste form mineral phases.

Other minerals in the sodalite family, namely hauyne and lazurite which are also cage structured minerals, can accommodate either SO_4 or S_2 depending on the REDOX of the sulfur during the steam reforming process. Regardless of the FBSR REDOX the feldspathoid minerals can accommodate sulfur as either sulfate or sulfide. Sodalite minerals are known to accommodate Be in place of Al and S_2 in the cage structure along with Fe, Mn, and Zn, e.g. helvite $(Mn_4[Be_3Si_3O_{12}]S)$, danalite $(Fe_4[Be_3Si_3O_{12}]S)$, and genthelvite $(Zn_4[Be_3Si_3O_{12}]S)$ [6]. These cage-structured sodalites were minor phases in High Level Waste

(HLW) supercalcine waste forms[*] [7] and were found to retain Cs, Sr, and Mo into the cage-like structure, e.g., Mo as $(Na_6Al_6Si_6O_{24})$ $(NaMoO_4)_2$ [7]. In addition, sodalite structures are known to retain B [8], Ge[9], I[9,6], and Br[9,6] in the cage like structures. Indeed, waste stabilization at Argonne National Laboratory-West (ANL-W) currently uses a glass-bonded sodalite ceramic waste form (CWF) for disposal of electrorefiner wastes for sodium-bonded metallic spent nuclear fuel from the EBR II fast breeder reactor [10,11].

A second feldspathoid mineral found in the FBSR waste form is nepheline $(NaAlSiO_4)$ [12]. Nepheline is a hexagonal structured feldspathoid mineral. The ring structured aluminosilicate framework of nepheline forms cavities within the framework. There are eight large (nine-fold oxygen) coordination sites and six smaller (8-fold oxygen) coordination sites [6] The larger sites nine-fold sites can hold large cations such as Cs, K, and Ca while the smaller sites accommodate the Na. The K analogue is known as leucite $(KAlSi_2O_6)$. In nature, the nepheline structure is known to accommodate Fe, Ti and Mg as well.

The remaining aluminosilicate mineral found in the FBSR waste form is a sodium rich cubic structured nepheline derivative $(Na_2O)_{0.33}Na[AlSiO_4]$ (PDF#39-0101). This nepheline derivative structure has large (twelve-fold oxygen) cage like voids in the structure [13]. This cage structured nepheline is not known to occur in nature but the large cage like voids should be capable of retaining large radionuclides, especially monovalent radionuclides such as Cs.

Durability Testing of the FBSR Mineral Product

The PCT-A was performed in triplicate on sample SCT02-098-FM in conjunction with glass durability standards, e.g. the ARM-1 and EA glasses. Stainless steel vessels (304L) were used as specified in the PCT-A leaching protocol for the first set of tests. These tests were repeated in Teflon® vessels to demonstrate that Re release was independent of vessel type. The logarithm of the NL_i was taken for each replicate and then averaged per ASTM C1285-02.

The PCT responses measured in this study for Sample SCT02-098-FM and the standard glasses tested are summarized in Table VI. It is evident that the leach testing was in control as the response for the EA glass is within the allowable standard deviations of the reference response shown in Table VI.

The leaching of Sample SCTO2-098-FM demonstrates that the normalized Na release is 1.74 g/m^2 within the 2 g/m^2 Hanford specification. In addition, it is obvious that the AN-107 FBSR waste form leaches incongruently instead of congruently[*] as most vitrified waste forms. Incongruent dissolution of a waste

---

[*]   Supercalcines were the high temperature silicate based "natural mineral" assemblages proposed for HLW waste stabilization in the United States (1973-1985).

[*]   Congruent dissolution of a waste form is the dissolving of species in their stoichiometric amounts. For congruent dissolution, the rate of release of a radionculide from the waste form is proportional to both the dissolution rate of the waste form and the relative abundance of the radionculide in the waste form. Thus for borosilicate glass ^{99}Tc is released at the same rate, congruently, as Na, Li and B.

form means that some of the dissolving species are released preferentially to others. Incongruent dissolution is often diffusion-controlled and can be either surface reaction-limited under conditions of near saturation or mass transport-controlled. Preferential phase dissolution, ion-exchange reactions, grain-boundary dissolution, and dissolution-reaction product formation (surface crystallization and recrystallization) are among the more likely mechanism of incongruent dissolution, which will prevail, in a complex polyphase ceramic waste form [14].

Incongruent dissolution is only detrimental to a waste form if a radionuculide species is released preferentially to a matrix element. In the FBSR final waste form the radionuclide release (Cs and Re) is retarded when compared to the matrix element, Na, release (Table VI) or conversely, Na is released from one of the phases preferentially compared to the nosean phase which retains the Re. This finding is noteworthy because the Hanford specification for Na release for vitrified waste forms is an indicator for the congruent release of ^{99}Tc since Na and B and ^{99}Tc are all released at similar stoichiometric rates (congruently) from vitrified waste forms [10, 11,15,16,17,18,19,20,21,22,23,24].

The incongruent release of Cs and Re is not attributed to the use of stainless steel vessels as numerous studies have shown that neither Cs nor ^{99}Tc have an affinity for stainless steel vessels [19,15]. In addition, the PCT triplicate analyses were rerun in Teflon® vessels and the Re release indicated in Table VI is comparable to the Re release measured in this study in the stainless steel vessels. The solids remaining after PCT testing were analyzed by X-ray Diffraction and all of the same phases were present as before PCT testing (see Table V).

Table VI PCT Performance of Sample SCT02-098-FM and the Glass Standards Tested Compared to the Durability Response of Known Standards

Sample	pH	$NL_{(B)}$ g/m^2	$NL_{(Na)}$ g/m^2	$NL_{(Cs)}$ g/m^2	$NL_{(Re)}$ g/m^2	$NL_{(Si)}$ g/m^2
EA	11.64	7.76	6.05			2.21
ARM-1	10.47	0.29	0.27			0.15
AN-107 FBSR	11.95	1.27	1.74	0.16	0.29 (0.22)*	0.35
AN-107 FBSR *	11.98	BDL	f	f	0.22	0.48
EA REF	11.85	8.37	6.67			1.96
LAW REF	10.90	0.55	0.54			0.16
AN-102 RAD GLASS	10.60	0.29	0.35			0.12

* = rerun in Teflon vessels, f = analysis indicated that vessel blanks were contaminated with Na and Cs

Regulatory Testing of the FBSR Mineral Waste Form

The mineral phases formed during the FBSR process were subject to the EPA TCLP leaching protocol. All of the mineral phases, regardless of particle size, appear to have met the LDR Universal Treatment Standards (UTS) as shown in Table VII.

Delisting the final LAW waste form may also be accomplished by delisting the final waste form at the point of generation so that the UTS are not applicable. The EPA calculates delisting levels and risk levels for a given waste form using their DRAS code (EPACMTP model) for calculation of major pathways for human exposure to a given waste. If the allowable concentrations in the TCLP leachate of the waste, as calculated by DRAS, are higher than the Toxcitiy Characteristic (TC) level for the TC constituents, then the delisting level for the TC constituents can be capped at the TC regulatory limits. The UTS levels may or may not apply to a delisted waste. This is still highly debated even within the EPA. However, a waste form that meets the EPA UTS treatment standard limits could be easily delisted.

The results of the TCLP testing is provided in Table VII for FBSR products produced under reducing (Scoping Test 01 and Production Test 02) and oxidizing FBSR conditions (Production Test 01) since no TCLP testing had been performed on sample SCT02-098-FM from Scoping Test 02.

Table VII. TCLP Testing of FBSR Final Mineral Waste Form Products

Element of Concern[†]	TCLP Releases for FBSR Under Reducing Conditions (ppm)	TCLP Releases for FBSR Under Reducing Conditions (ppm)	UTS Limits Federal Register, V.63, No. 100 p.28748-9 May 28, 1998 (ppm)
Cr	0.015-0.060	0.001-0.018	0.60
Pb	0.005-0.023	0.002-0.067	0.75
Ni	0.001-3.11	0.66-2.80	11

* 1.0 or still hazardous

† As, Ba, Cd, Hg, Se, Ag, Zn, Sb, Be, Tl and V not in simulant

CONCLUSIONS

The following can be concluded about the use of Fluidized Bed Steam Reforming (FBSR) as a final waste form for Hanford's LAW waste:

- FBSR is a robust technology capable of accommodating wide ranges of feeds and additives including high concentrations of sulfate
- FBSR's ability to retain sulfate can lead to increased waste loadings and accelerated stabilization of Hanford's LAW vs. LAW vitrification

- FBSR's corrosion is incongruent and the radionuclides (Cs and Re as simulants for ^{137}Cs and ^{99}Tc) are released at a rate lower than that of Na (Na release is <2 g/m^2)
- FBSR is a medium temperature process: temperatures are low enough not to vaporize radionuclides but high enough to destroy volatile organic compounds (VOC's) in the presence of a reductant and a catalyst
- FBSR waste form mineral phases are cage like structures that trap radionuclides and anions
- FBSR waste form mineral phases alter to zeolites that have the same cage-like structures and will likely still retain the radionuclides and anions
- FBSR waste forms are regulatory compliant at the Universal Treatment Standard (UTS) making delisting of the final waste form a nearly assured option

ACKNOWLEDGEMENTS

This paper was prepared in connection with work done under Contract No. DE-AC09-96SR18500 with the U.S. Department of Energy.

REFERENCES

1 J.B. Mason, J. McKibben, J. Ryan, J. Schmoker, " **Steam Reforming Technology for Denitration and Immobilization of DOE Tank Wastes,**" Waste Mgt. 03 (February 2003).

2 C.M. Jantzen, **"Engineering Study of the Hanford Low Activity Waste (LAW) Steam Reforming Process,"** U.S. DOE Report WSRC-TR-2002-00317, Westinghouse Savannah River Co., Aiken, SC (July 12, 2002).

3 B.P. McGrail, H.T. Schaef, P.F. Martin, D.H. Bacon, E.A. Rodriguez, D.E. McCready, A.N. Primak, and R.D. Orr, **"Initial Evaluation of Steam-Reformed Low Activity Waste for Direct Land Disposal,"** U.S. DOE Report PNWD-3288, Battelle Pacific Northwest Division, Richland, WA (January 2003).

4 E.W. Baumann, **"Colorimetric Determination of Iron (II) and Iron (III) in Glass,"** Analyst, v. 117, 913-916 (1992).

5 C.M. Jantzen, N.E. Bibler, D.C. Beam, and M.A. Pickett, **"Characterization of the Defense Waste Processing Facility (DWPF) Environmental Assessment (EA) Glass Standard Reference Material,"** U.S. DOE Report WSRC-TR-92-346, Rev.1 (February, 1993).

6 W. A. Deer, R. A. Howie, and J. Zussman, **"Rock-Forming Minerals, Vol IV,"** John Wiley & Sons, Inc., New York, 435pp. (1963).

7 D.G. Brookins, **"Geochemical Aspects of Radioactive Waste Disposal,"** Springer-Verlag, New York, 347pp. (1984).

8 J.Ch. Buhl, G. Englehardt, and J. Felsche, **"Synthesis, X-ray Diffraction, and MAS n.m.r. Characteristics of Tetrahydroxoborate Sodalite,"** Zeolites, 9, 40-44 (1989).

9 M.E. Fleet, **"Structures of Sodium Alumino-Germanate Sodalites,"** Acta Cryst., C45, 843-847 (1989).

10 W. Sinkler, T.P. O'Holleran, S.M. Frank, M.K. Richmann, and S.G. Johnson, **"Characterization of a Glass-Bonded Ceramic Waste Form Loaded with U and Pu,"** Sci.Basis Nucl. Waste Mgt., XXIII, R.W. Smith and D.W. Shoesmith (Eds) Mat. Res. Soc., Pittsburgh, PA, 423-429 (2000).

11 T.L. Moschetti, W. Sinkler, T. DiSanto, M.H. Novy, A.R. Warren, D. Cummings, S.G. Johnson, K.M. Goff, K.J. Bateman, and S.M. Frank, **"Characterization of a Ceramic Waste**

Form Encapsulating Radioactive Electrorefiner Salt," Sci.Basis Nucl. Waste Mgt., XXIII, R.W. Smith and D.W. Shoesmith (Eds.), Mat. Res. Soc., Pittsburgh, PA, 577-582 (2000).

12 L.G. Berry and B. Mason, **"Mineralogy Concepts, Descriptions, Determinations,"** W.H. Freeman & Co., San Francisco, CA, 630pp (1959).

13 R. Klingenberg and J. Felsche, **"Interstitial Cristobalite-type Compounds $(Na_2O)_{0.33}Na[AlSiO_4])$,"** J. Solid State Chemistry, 61, 40-46 (1986).

14 C.M. Jantzen, D.R. Clarke, P.E.D. Morgan and A.B. Harker, **"Leaching of Polyphase Nuclear Waste Ceramics: Microstructural and Phase Characterization,"** J. Am. Ceram. Soc. 65[6], 292-300 (1982).

15 N.E. Bibler, and J.K. Bates, **"Product Consistency Leach Tests of Savannah River Site Radioactive Waste Glasses,"** Sci.Basis Nucl. Waste Mgt, XIII, Oversby, V. M. and Brown, P. W., eds., Mat. Res. Soc., Pittsburgh, PA, 1990, pp. 327–338.

16 J.K. Bates, D.J. Lam, M.J. Steindler, **"Extended Leach Studies of Actinide-Doped SRL 131 Glass,"** Sci. Basis Nucl. Waste Mgt, VI, D.G. Brookins (Ed.), North-Holland, New York, 183-190 (1983).

17 L.R. Morss, M.L. Stanley, C.D. Tatko, W.L. Ebert, **"Corrosion of Glass Bonded Sodalite as a Function of pH and Temperature,"** Sci.Basis Nucl. Waste Mgt., XXIII, R.W. Smith and D.W. Shoesmith (Eds.), Mat. Res. Soc., Pittsburgh, PA, 733-738 (2000).

18 B.P. McGrail, **"Waste Package Component Interactions with Savannah River Defense Waste Glass in a Low-Magnesium Salt Brine,"** Nuclear Technology, 168-186 (1986).

19 W.L. Ebert, S.F. Wolf, and J.K. Bates, **"The Release of Technetium from Defense Waste Processing Facility Glasses,"** Sci.Basis Nucl. Waste Mgt., XIX, W.M. Murphy and D.A. Knecht (Ed.), Mat. Res. Soc., Pittsburgh, PA, 221-227 (1996).

20 E.Y. Vernaz and N. Godon, **"Leaching of Actinides from Nuclear Waste Glass: French Experience,"** Sci.Basis Nucl. Waste Mgt., XV, C.G. Sombret (Ed) Mat. Res. Soc., Pittsburgh, PA, 37-48 (1992).

21 N.E. Bibler and A.R. Jurgensen, **"Leaching Tc-99 from SRP Glass in Simulated Tuff and Salt Groundwaters,"** Sci.Basis Nucl. Waste Mgt, XI, M.J. Apted and R.E. Westerman (Eds.), Mat. Res. Soc., Pittsburgh, PA, 585-593 (1988).

22 D.J. Bradley, C.O. Harvey, and R.P. Turcotte, **"Leaching of Actinides and Technetium from Simulated High-Level Waste Glass,"** Pacific Northwest Laboratory Report, PNL-3152, Richland, WA (1979).

23 S. Fillet, J. Nogues, E. Vernaz, and N. Jacquet-Francillon, **"Leaching of Actinides from the French LWR Reference Glass,"** Sci.Basis Nucl. Waste Mgt, IX, L.O. Werme, Mat. Res. Soc., Pittsburgh, PA, 211-218 (1985).

24 F. Bazan, J. Rego, and R.D. Aines, **"Leaching of Actinide-doped Nuclear Waste Glass in a Tuff-Dominated System,"** Sci.Basis Nucl. Waste Mgt, X, J.K. Bates and W.B. Seefeldt (Eds.), Mat. Res. Soc., Pittsburgh, PA, 447-458 (1987).

MICROSTRUCTURE OF EMULSION-BASED POLYMERIC WASTE FORMS FOR ENCAPSULATING LOW-LEVEL, RADIOACTIVE AND TOXIC METAL WASTES

Rachel Evans,[+] Anh Quach,[*] Guanguang (Gordon) Xia,[†] Brian J.J. Zelinski,[+] Wendell P. Ela,[*] Dunbar P. Birnie III,[+] A. Eduardo Sáez,[*] Harry D. Smith,[†] and Gary L. Smith[†]

[+]Department of Materials Science
and Engineering
University of Arizona
Tucson, AZ 85721

[*]Department of Chemical and
Environmental Engineering
University of Arizona
Tucson, AZ 85721

[†]Pacific Northwest National
Laboratory
902 Battelle Blvd.
Richland, WA 99352

ABSTRACT

Developed technologies in vitrification, cement, and polymeric materials manufactured using flammable organic solvents have been used to encapsulate solid wastes, including low-level radioactive materials, but are impractical for high salt-content waste streams.[1] In this work, we investigate an emulsification process for producing an aqueous-based polymeric waste form consisting of epoxy resin and polystyrene-butadiene (PSB) latex that is non-flammable, light weight and of relatively low cost. Sodium nitrate was used as a model for the salt waste. The microstructure and composition of the samples were probed using SEM/EDS techniques and salt extraction. The results show that some portion of the salt migrates towards the exterior surfaces of the waste forms during the curing process. The portion of the salt in the interior of the sample is contained in polymer corpuscles or sacs. These sacs are embedded in a polymer matrix phase that contains fine, well-dispersed salt crystals. A companion paper describes work in which small-scale samples were manufactured and analyzed using leach tests designed to measure the diffusion coefficient and leachability index for the fastest diffusing species in the waste form, the salt ions.[2]

INTRODUCTION

Many industrial processes create wastes containing high concentrations of salts mixed with toxic metals such as cadmium, lead, and arsenic. These processes include semiconductor manufacturing mining/mineral processing, agricultural desalination, glass manufacturing, and pulp and paper mills, among others. Water treatment utilities use ion removal processes such as packed-bed adsorption, membrane filtration and ion exchange to trap toxic metals and other contaminants in iron or aluminum hydroxide containing sludges or residuals that often contain large quantities of salt. Another source of salt waste includes the 200 million kilograms of chloride, sulfate, and nitrate salts, contaminated with toxic metals and low-level radioactive elements, which are distributed among the various Department of Energy sites around the nation.[1] Attempts to dispose of this low-level radioactive waste generate an additional 5 million kilograms of salt wastes per year.

Disposal of high salt containing waste is problematic because the salt component is vulnerable to leaching and high loadings tend to destabilize most matrices currently used for waste encapsulation. Encapsulation is necessary because the contaminating toxic metals must be prevented from entering the environment in an uncontrolled manner in the unlikely event that the storage facility be compromised. Currently, various types of technologies exist to encapsulate waste. Vitrification requires firing the waste at high temperatures where some toxic components may volatilize. Encapsulation using polymer matrices or composite polymers usually relies on the use of volatile and flammable organic solvents[3]. Salt loading in grout is limited because large amounts decrease durability and lead to poor leaching control.[1,4,5]

Collaborative efforts between the University of Arizona and Pacific Northwest National Laboratory have focused on developing polycerams, or organic-inorganic hybrid materials for use as waste forms. Polycerams incorporate a ceramic or oxide component into the polymeric matrix, where the scale, structure, and connectivity of the inorganic phase may vary over a wide range, resulting in significant improvements in mechanical and chemical properties, depending upon the level of connectivity between the inorganic and organic components.[6,7] Early studies confirmed that polycerams of rubber (butadiene) as the polymer component and silica as the ceramic component were capable of stabilizing wastes containing more than 10 wt% salt contaminated with toxic metals.[8] Unfortunately, these waste forms required the use of highly volatile and flammable solvents in their fabrication. To circumvent this problem, work was begun to replace the hazardous organic precursors and solvents with water-borne equivalents to develop environmentally benign aqueous-based processing routes for fabricating durable waste forms. Matrices derived from equal parts of epoxy and polystyrene/butadiene (PSB) showed the greatest promise.[9] Later work indicated that the successful encapsulation of a waste component may well depend upon the occurrence of a phase inversion to convert the initial oil-in-water (O/W) epoxy/PSB emulsion to a water-in-oil (W/O) emulsion.[10]

This paper describes the microstructure of these water-based, epoxy/polystyrene-butadiene polymer waste forms with the intent of identifying how much salt is incorporated into their structure and how this salt is distributed in the bulk of the waste form. A companion paper in this volume describes the leaching behavior of the salt component of these waste forms.[2] Most regulatory restrictions address the leaching of toxic metals and other contaminants. However, the salt is likely to be the most mobile species in these waste forms. Thus, characterization of its behavior will represent an extreme in behavior as compared to that of other slower, less mobile waste form contaminants. Future work will reintroduce an oxide component into the processing to take advantage of the benefits to be gained by using a polyceram matrix for waste encapsulation.

MATERIALS AND METHODS

To fabricate the samples, polystyrene-butadiene (PSB) latex (BASF, Styronal ND 656), epoxy resin (Buehler, Epo-Kwick Resin), and the surfactant sorbitan monooleate (Span-80) were mixed for 30 minutes to create an emulsion. To this, the cross-linking agent diethylenetriamine (DETA, 99%, Aldrich) and sodium nitrate salt ($NaNO_3$) were added. This mixture was stirred to uniformity, cast into glass containers and placed in an oven to dry and cure at 80°C until weight loss ceased (typically 3 days). Cured samples were cylindrical and typically weighed 15 g, had diameters of about 2 cm, and lengths of about 3.5 cm. All samples were prepared using equal parts, by weight, of the PSB latex and epoxy resin. A variety of waste forms were prepared having nominal salt loadings in the range of 8-40 wt% of $NaNO_3$ in the cured samples. The amount of water in the uncured waste forms was also varied in the range of 0.5 to 1.5 wt ratio of water to polymer component (epoxy plus PSB). Table I lists the samples and their compositions.

Table I. Sample Characteristics

Sample Name	Waste Loading[a]	Water Content[b]	Interior Salt Content[a] By EDS	Interior Salt Content[a] By SE[c]
8L	8.0	0.5	-	-
15L	15	0.5	9.2	9.9
20L	20	0.5	9.2	
21H	21	1.5	-	12.4
27H	27	1.5	9.8	12.8
40I	40	0.75	-	9.4

[a]wt% $NaNO_3$, [b](wt. H_2O)/(wt. Polymer), [c]Salt Extraction (SE)

Sample microstructures were characterized using carbon-coated sections using a JEOL 5900 LV SEM with a built-in Everhardt-Thronley secondary electron detector and a Robinson series VI scintillation-based backscattered electron BSE detector. Compositional data was obtained using an EDAX EDS detector with Genesis v1.0 software with integrated digital imaging and mapping capabilities.

The salt content of portions of the sample was also measured by salt extraction. Thin sections were immersed in water at 20°C to leach out the salt. The thickness of the thin sections was kept intentionally thin (2.25 to 5.30 mm) to minimize the time required to characterize the behavior of species having low diffusivities. The conductivity of the leachate was measured using an ion probe and correlated to salt content. Extraction was complete when variations in the conductivity of the leachate were no longer detected.

RESULTS

Each of the samples listed in Table I appear to be homogeneous on a macroscopic scale when visually inspected. Despite this visual homogeneity, SEM analysis revealed that these samples are inhomogeneous at the microscopic scale. Additionally, each sample contains a crust of salt on the external surfaces of the monolith. The characteristics of this crustal layer are shown in Figure 1, in which dark regions are polymer and light regions are salt.

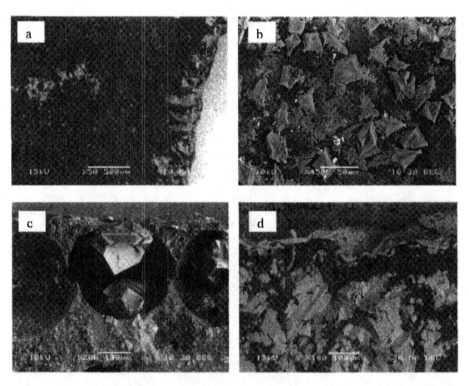

Figure 1. Backscattered SEM images of surface crust features: a) cross section containing the external surface (to the right) shows a crustal region several hundreds of microns in thickness; b) dendritic crystals growing on external surface; c) large, well-formed salt crystals in near-surface voids; d) continuous layer of salt on external surface with large salt regions just below surface.

As seen in Figure 1, the surface crust may contain a combination of dendritic crystals which have grown on the exterior of the monolith, large salt crystals in voids or bubbles just below the surface, a continuous layer or coating of salt on the exterior surface, and/or large salt regions just below the surface of the monolith.

To characterize further the distribution of salt in the waste forms, the quantity of salt in various portions of a monolith of sample 40I was determined by sectioning and salt extraction. Prior to sectioning, the surface salt that was not incorporated into the sample was removed. Table II shows the average composition in wt% $NaNO_3$ for the different sections identified in Figure 2. The top, crustal section of the sample contains the highest concentration of salt at 31.3 wt%. The disk taken from the top but below the crustal disk contains 23.6 % salt. Two samples were taken from the middle of the monolith: one was a disk containing the perimeter of the monolith (Section 3), while the other was just the core of the monolith (Section 4). The middle disk has a composition of 17.2 % salt while the core contains 9.4 % salt. These results confirm that the salt distribution in these waste forms is inhomogeneous, with higher salt concentrations near the external surfaces and lower concentrations in the middle of the waste form.

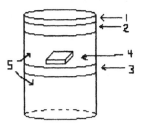

Figure 2. Schematic showing the sections of sample 40I leached to determine average salt content.

Table II. Salt Distribution in Sample 40I

Section (See Figure 2)	Location Description	Salt Content (wt% $NaNO_3$)
1	Crust	31.3
2	Top disk	23.6
3	Middle disk	17.2
4	Middle core	9.4
5	Rest of monolith	18.8

SEM/EDS was used to characterize the manner in which the salt was incorporated into the inner portions of the waste forms. As seen in Figure 3, the microstructure of the interior or central portion of sample 15L consists of corpuscles or sacs of polymer that are mostly filled with salt. These sacs are imbedded in a continuous matrix phase. EDS analysis over large areas (> 100 μm × 100 μm) reveals that the interior of this sample contains about 9.2 wt% salt. A similar result is obtained when the size of the scanned area is reduced so as to sample only the matrix phase between polymer sacs, indicating that the matrix phase contains small, finely dispersed salt crystals. An initial salt content of 9.9 wt% was measured for this sample by salt extraction (see Table I). This value compares favorably with the value obtained by salt extraction when one considers that the salt extraction results were obtained using a larger portion of the interior, a portion that contains sample volumes nearer to the salt-rich surfaces of the waste form.

Figure 3. SEM backscattered image of the interior of sample 15L showing corpuscles or sacs of polymer containing salt imbedded a continuous matrix.

Figure 4 shows the corpuscles or sacs observed in sample 8L. Here the salt crystals inside the sacs are relatively large and possess a well-defined morphology. The interior characteristics of other samples of this study contained similar features, namely corpuscles or sacs of salt imbedded in a polymer matrix containing small, well-dispersed salt crystals.

Samples 20L and 27H were prepared in an attempt to increase the salt content in the interior of the waste form. Sample 20L contains sacs of salt similar to those observed in previous samples. However, it also contains a family of pores with sizes in excess of 100 μm (Figure 5). EDS analysis of the interior portions of this sample show that increasing the nominal loading of this waste form does not lead to a substantial increase in salt content in the interior of the sample. The average composition in the interior of sample 20L was 9.2 wt% salt as measured by EDS. Even increasing the target salt loading to 27 wt% did not significantly increase the concentration of salt measured in the center of the waste form monolith. EDS measurements at the center of sample 27H show a composition of 9.8 wt% $NaNO_3$ while salt extraction indicates a composition of 12.8 wt%. These values

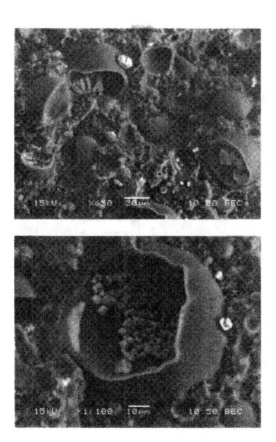

Figure 4. SEM backscattered images of the interior of sample 8L. Compared to 15L, the polymer sacs contain fewer, larger, and better defined salt crystals.

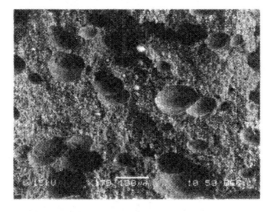

Figure 5. SEM backscattered image of the interior of sample 20L showing large pores.

are not too different from the value of 9.4 wt% measured in the interior or core region of sample 40I as determined by salt extraction (see Table I).

The microstructural characteristics of fully extracted samples are shown in Figure 6. In both samples 15L and 27H, the polymer corpuscles or sacs appear to be devoid of salt after extraction. EDS analysis indicates that almost all of the salt leaches out of the thin sections used for extraction studies, typically all but about 1 wt% or less. This indicates that the fine salt crystals distributed in the matrix phase surrounding the sacs leach out as well. These results confirm the use of salt extraction studies for these samples to characterize salt content in thin sections.

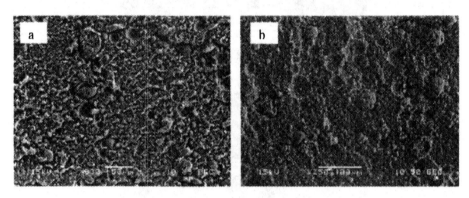

Figure 6. SEM backscattered images of leached samples. Polymer sacs appear devoid of salt and EDS indicates that the matrix phase contains little or no salt, as well: a) sample 15L, b) sample 27H.

DISCUSSION AND CONCLUSION

The key to making this material a viable waste form is believed to be the occurrence of a phase inversion during the curing process. Substantial evaporation must occur prior to curing so that the water phase becomes discontinuous and the polymer phase becomes continuous, thus encapsulating the water-soluble salt in the polymeric matrix. Since the water evaporates from the exterior surfaces of the waste form, it migrates from the interior and carriers the highly soluble sodium nitrate with it. This migration produces the heterogeneous distribution of salt in the waste from as detected by EDS and measured in the different sections of sample 40I by salt extraction (see Table II). Evaporation of the concentrated salt solution that migrates towards the surfaces of the waste form causes the formation of the dendritic crystals, cubic crystals, and large regions of embedded salt on or near the surfaces of the waste form as shown in Figure 1.

While some salt migrates from the interior of the waste form to the surface, the EDS and ES results from Table I confirm that at least 10 wt% of salt is encapsulated by the waste form. The micrographs of Figures 3-5 show that the salt is encapsulated as crystals in sacs or corpuscles of polymer embedded in a polymer matrix phase that also contains small salt crystals. These corpuscles may be generated during the mixing process, as the epoxy-PSB emulsion is forming. They become trapped in the polymer phase after the phase inversion occurs. The results of Figure 6 show that the salt is extracted from both the corpuscles and matrix during the leaching of thin sections for long times. However, results described in a companion paper indicate that the rate of leaching in these samples is sufficiently low as to qualify them for use as hosts for low-level radioactive wastes contaminated with toxic metals[10]. This is particularly true since the leaching rates obtained in the companion study characterize the diffusion of the fastest, most mobile species in the waste form, the salt.

An ideal waste form encapsulates a large amount of waste and prevents the waste from leaching out of the matrix. Attempts to increase the waste loading in the interior of the samples were not successful using the procedures of the current flowsheet. As seen in Tables II an IV, samples 20L, 21H, and 27H contain between 10-12 wt% salt in their interior, despite their higher nominal salt loading. In sample 20L, it is likely that large pores form (see Figure 5) because the low liquid content of the sample produces an emulsion that is rather stiff prior to curing. This prevents the emulsion from shrinking as water evaporates during curing, leading to the formation of internal porosity. It is likely that the amount of salt encapsulated by the matrix is strongly influenced by the parameters that control the emulsification, phase inversion, and curing processes. A similar conclusion is reached in the companion study regarding the leaching behavior. More work should be done to determine the influence of the cross-linking agent, surfactant, and curing temperature on the phase inversion and the final waste form properties.

This work has shown that the water-based emulsion process creates waste forms that successfully encapsulate at least 10 wt% highly-soluble salt in the sample interior and higher concentrations at the periphery of the waste form. This analysis indicates that higher salt contents might be achieved if the migration of the solutions towards the sample surface during curing can be controlled. This control will likely be achieved through adjustment and optimization of the parameters influencing the phase inversion process.

ACKNOWLEDGEMENTS

This work was supported by the DOE Office of Science and the National Science Foundation *Faculty and Student Teams Program* and by the Process Science and Engineering Division within the Environmental Technology Directorate at Pacific Northwest National Laboratory. The assistance of Rod Quinn, Loni Peurrung, Bradley R. Johnson, Michael Schweiger, Jim Davis, and Royace Aikins is gratefully acknowledged. This work was completed at the Applied Process Engineering Laboratories (APEL) at Pacific Northwest National Laboratory.

REFERENCES

[1] V. Maio, R.K. Biyani, R. Spence, R.G. Loomis, G.L Smith and A. Wagh, "Testing of Low-Temperature Stabilization Alternatives for Salt Containing Mixed Wastes-Approach and Results to Date"; pp. 514-521 in *Proceedings of the International Topical Meeting on Decommissioning and Decontamination and on Nuclear and Hazardous Waste Management, SPECTRUM '98*, American Nuclear Society, ISBN: 0-89448-635-7, 1998.

[2] A. Quach, G. Xia, R. Evans, A.E. Sáez, B.J.J. Zelinski, H. Smith, G.L .Smith, D.P. Birnie III and W.P. Ela, "Leach Resistance of Encapsulated Salts in Polymeric Waste Forms Fabricated Using an Aqueous-Based Route," *Ceramic Transactions*, this volume.

[3] J.R. Conner, *Chemical Fixation and Solidification of Hazardous Wastes*, New York, Van Nostrand Reinhold, 1990.

[4] T.M. Krishnamoorthy, S.N. Joshi, G.R. Doshi and R.N. Nair, "Desorption Kinetics of Radionuclides Fixed in Cement Matrix," *Nuclear Technology*, **104**, 351 (1993).

[5] M. Leist, R.J. Casey and D. Caridi, "The Management of Arsenic Wastes: Problems and Prospects," *J. Hazardous Materials*, **B76**, 125 (2000).

[6] C. Sanchez, G.J. de Soler-Illia, F. Ribot, T. Lalot, C.R. Mayer and V. Cabuil, "Designed Hybrid Organic-Inorganic Nanocomposites from Functional Nanobuilding Blocks," *Chem. Mater.*, **13**, 3061-3083 (2001).

[7] B. Boury and R.J.P. Corriu, "Nanostructured hybrid organic-iorganic solids from molecules to materials," *Chem. Org. Sil. Comp.*, **3**, 565-640 (2001).

[8] B.J.J. Zelinski, J. Young, K. Davidson, A. Aruchamy, D.R. Uhlmann, T. Suratwala, G.L. Smith and H.D. Smith, "Sol-Gel Stabilization of Radioactive Salt Wastes," in *Proceedings of the 18th International Congress on Glass*, Edited by M.K. Choudhary, N.T. Huff, and C.H. Drummond III, The American Ceramic Society, Westerville, Ohio, 1998.

[9] L. Liang, H. Smith, R. Russell, G. Smith and B.J.J. Zelinski, "Aqueous Based Polymeric Materials for Waste Form Applications," *Ceramic Transactions*, **132**, 359-368 (2002).

[10] H. Smith, G.L. Smith, G. Xia and B.J.J. Zelinski, "Morphology and Composition of Simulant Waste Loaded Polymer Composite – Phase Inversion, Encapsulation, and Durability," *Ceramic Transaction* (2003).

LEACH RESISTANCE OF ENCAPSULATED SALTS IN POLYMERIC WASTE FORMS FABRICATED USING AN AQUEOUS-BASED ROUTE

Anh Quach,[1] Gordon Xia,[2] Rachel Evans,[3] A. Eduardo Sáez,[1] Brian J. Zelinski,[3] Harry Smith,[2] Gary Smith,[2] Dunbar P. Birnie III[3] and Wendell P. Ela[1]

[1]Department of Chemical and Environmental Engineering
University of Arizona
Tucson, AZ 85721

[2]Pacific Northwest National Laboratory
902 Battelle Blvd.
Richland, WA 99352

[3]Department of Materials Science & Engineering
University of Arizona
Tucson, AZ 85721

ABSTRACT
The leaching behavior of a highly-soluble salt (sodium nitrate) encapsulated in novel polymeric waste forms was investigated. The waste forms are solid monoliths composed of cured blends of polystyrene-butadiene and an epoxy resin. The synthesis and encapsulation process are accomplished simultaneously via an aqueous-based route recently developed. Leaching of the salt was carried out by exposing sections of the waste forms to large volumes of well-stirred water. The measured time dependence of the leaching process is described quantitatively by a model based on the diffusion of the salt through the waste form. The results obtained suggest that diffusion occurs through limited but significant continuous porosity. Preliminary experiments indicate that this porosity may be reduced to improve leach resistance by the use of a post-curing heat treatment.

INTRODUCTION

Many industrial processes create wastes containing high concentrations of salts mixed with toxic metals such as cadmium, lead, and arsenic. These processes include semiconductor manufacturing mining/mineral processing, agricultural desalination, glass manufacturing, and pulp and paper mills, among others. Water treatment utilities use ion removal processes such as packed-bed adsorption, membrane filtration and ion exchange to trap toxic metals and other contaminants in iron or aluminum hydroxide containing sludges or residuals that often contain large quantities of salt. Another source of salt waste includes the 200 million kilograms of chloride, sulfate, and nitrate salts, contaminated with toxic metals and low-level radioactive elements, which are distributed among the various Department of Energy sites around the nation.[1] Attempts to dispose of this low-level radioactive waste generate an additional 5 million kilograms of salt wastes per year.

Disposal of solid wastes containing soluble salts is problematic because the salt component is vulnerable to leaching and high loadings tend to destabilize most matrices currently used for waste encapsulation. Encapsulation is necessary because the contaminating toxic metals must be prevented from entering the environment in an uncontrolled manner.

In a companion manuscript[2] we present details on a new technology to develop polymeric waste forms to encapsulate toxic wastes using an aqueous synthesis route. The synthesis method and the microstructure of the waste forms are characterized in detail. In this work we explore the leaching behavior of a highly soluble salt (sodium nitrate) that has been encapsulated in polymeric waste forms fabricated via an aqueous synthesis route from a polystyrene-butadiene latex and a commercial epoxy resin. Experiments are conducted to measure the amount of salt leached from the sample into pure water as a function of time, and a diffusion model is used to analyze the experimental data.

MATERIALS AND METHODS

To fabricate the samples, polystyrene-butadiene (PSB) latex (BASF, Styronal ND 656), epoxy resin (Buehler, Epo-Kwick Resin), and the surfactant sorbitan monooleate (Span-80) were mixed for 30 minutes to create an emulsion. To this, the cross-linking agent diethylenetriamine (DETA, 99%, Aldrich) and sodium nitrate salt ($NaNO_3$) were added. This mixture was stirred to uniformity, cast into glass containers and placed in an oven to dry and cure (usually at 80°C) until weight loss (water evaporation) ceased (typically 3 days). Cured samples were cylindrical and typically weighed 15 g, had diameters of about 2 cm, and lengths of about 3.5 cm. All samples were prepared using equal parts, by weight, of the PSB latex and epoxy resin. A variety of waste forms were prepared having nominal salt loadings in the range of 8-40 wt% of $NaNO_3$ in the cured samples. The amount of water in the uncured waste forms was also varied in the range of 0.5 to 1.5 wt ratio of water to polymer component (epoxy plus PSB).

Sample microstructures were characterized using carbon-coated sections and a

JEOL 5900 LV SEM with a built-in Everhardt-Thronley (ET) secondary electron (SE) detector and a Robinson series VI scintillation-based backscattered electron BSE detector. Compositional data was obtained using an EDAX EDS detector with Genesis v1.0 software with integrated digital imaging and mapping capabilities.

To characterize the leaching behavior of the waste forms, thin, flat sections were extracted from the middle of samples and immersed in a known excess volume of well-stirred water at 20°C to leach out the salt. The thickness of the thin sections was kept intentionally thin (2.25 to 5.30 mm) to minimize the time required to characterize the behavior of species having low diffusivities. Measured variation in the electrical conductivity of the liquid was used to determine the amount of salt leached into the solution as a function of time. To determine the total initial salt content, samples were leached until variations in the conductivity of the leachate were no longer detected.

THEORY

Two models were developed to interpret leaching data. Both models consider that the rate of leaching is controlled by the diffusion of salt in the sample. The first model conceptualizes the sample as a slab of finite thickness, whereas the second model, applicable for relatively short times after leaching commences, considers the sample as a solid of infinite thickness.

Slab Model

The solid sample is considered to be an infinitely long and wide slab with a thickness 2L. Under these conditions, the diffusion process can be considered one-dimensional, and concentration profiles will be established only in a direction (x) perpendicular to the slab's face. The concentration of salt at any point in the sample (c, mass of salt per unit sample volume) is governed by the differential equation

$$\frac{\partial c}{\partial t} = D \frac{\partial^2 c}{\partial x^2} \tag{1}$$

where D is the diffusivity of the salt in the sample. Since the volume of water into which the salt is leached is large and well stirred, the boundary conditions for this equation are

$$c=0, \quad x=0,2L \tag{2}$$

and the initial condition is

$$c=c_0, \quad t=0 \tag{3}$$

The problem can be solved analytically by separation of variables. From the solution, the average concentration of salt in the sample can be calculated from

$$\bar{c} = \frac{1}{2L} \int_0^{2L} c\,dx \tag{4}$$

which yields

$$\frac{\bar{c}}{c_0} = \frac{8}{\pi^2} \sum_{n=0}^{\infty} \frac{1}{(2n+1)^2} \exp\left[-\frac{(2n+1)^2 \pi^2 Dt}{(2L)^2}\right] \tag{5}$$

All samples used in the leaching experiment had exposed surface salt formed when the sample was cut from the original waste form. This exposed salt dissolved in the first few hours of the leaching experiment. The percent of salt retained in the sample, S, at a given time does not include the initial fraction of salt exposed on the surface, characterized by the percent P_c:

$$S = \frac{\bar{c}}{c_0}(100 - P_c) \tag{6}$$

The diffusivity (D) and percent exposed salt (P_c) were fit to experimentally-determined curves of S vs. t, obtained from the measurement of the concentration of salt in the leaching liquid.

Semi-Infinite Solid Model

For times short enough that the concentration profile inside the solid has not developed all the way to the center of the slab, the sample may be considered a semi-infinite solid. In this case, the diffusion process is governed by equation (1), but the boundary conditions are

$$c=0, \quad x=0 \tag{7}$$

$$c=c_0, \quad x \to \infty \tag{8}$$

The initial condition is given by equation (3). The solution is

$$c = c_0 \operatorname{erf}\left(\frac{x}{\sqrt{4Dt}}\right) \tag{9}$$

The flux of salt from the sample into the liquid can be calculated from this profile, and a mass balance of salt in the liquid leads to the following expression for the leached salt concentration in the liquid (c_f) as a function of time:

$$c_f = c_{f0} + \frac{2A}{V} c_0 \sqrt{\frac{Dt}{\pi}} \tag{10}$$

where c_{f0} is the concentration of salt in the liquid after the exposed surface salt has dissolved, A is the surface area of leached sample and V the volume of liquid. This equation can be fitted to experimental data on salt concentrations in the liquid at short times after the dissolution of the exposed surface salt, using c_{f0} and D as adjustable parameters.

The diffusivity of salt in the waste form can be used to calculate the leachability index, L_i, defined as:

$$L_i = \log(\beta / D) \qquad (11)$$

where $\beta = 1 \, cm^2 / s$. This equation is an appropriate expression for the leachability index as defined by the American National Standards Institute (ANSI/ANS-16.1-1986).

RESULTS

All the data gathered in the leaching experiments were analyzed by means of the slab model presented above. Data on percent salt retention (S) vs. time were adequately fitted by equations (5) and (6), using P_c and D as adjustable parameters. Figure 1 shows an example of leaching data as well as the fit obtained by the model. The parameter P_c, which for this particular sample is relatively high (35%), has no practical importance in terms of the characteristics of the waste form, since it represents solid salt that is exposed when the leached sample is cut out from the waste form. On the other hand, the diffusivity is independent of sample preparation, and is a characteristic of the waste form structure.

To further verify that the leaching process can be described by the diffusion mechanism proposed in this work, we also applied the semi-infinite model to selected data. To apply this model, we plot the salt concentration in the leaching liquid, c_f, vs. time. According to equation (10), the behavior should follow a linear relation, with intercept equal to the concentration of dissolved exposed surface salt, c_{f0}, and slope proportional to $\sqrt{D}$. Figure 2 presents this analysis for the short time region of the results in Figure 1. This representation clearly shows the relatively rapid dissolution of the exposed surface salt at the beginning of the process, followed by a period in which the semi-infinite solid model represents the data with accuracy, and the departure from the semi-infinite solid model at longer times, which indicates the penetration of the concentration profiles to the middle of the sample.

All the samples analyzed exhibited diffusivities in the order 10^{-8}-10^{-7} cm^2/s. This order of magnitude is consistent with the diffusion of salt through a network of limited number and relatively small, interconnected pores that are filled with water. The diffusivity is then an effective diffusivity, which is affected by the sample microstructure.

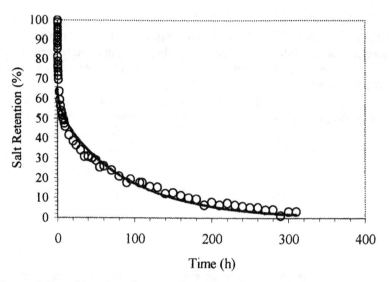

Figure 1. Leaching data for a section of a waste form. The disk was 2.53 mm thick with a diameter of about 2 cm. The sample has a 15% salt content. The solid line represents a fit of the data with the slab model. The initial fast drop in salt content corresponds to the rapid dissolution of exposed surface salt present on the sample's external surface. Fitting parameters: Pc=35%, D=2.0x10^{-8} cm^2/s.

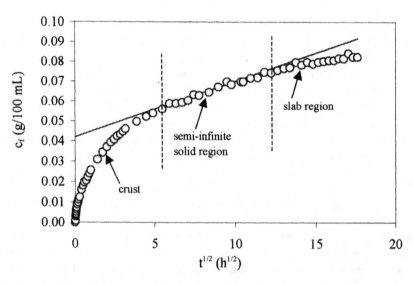

Figure 2. Application of the semi-infinite solid model to the data of Figure 1 at short times. The straight line has a slope that corresponds to the diffusivity calculated from the slab model.

The amount of salt encapsulated in the waste form affects the effective diffusivity. Table I shows results for samples with various salt contents. The effective diffusivity increases with salt content, which suggests an increase in the void fraction of the waste form. The measured diffusivities correspond to leachability indices greater than 6, which is considered satisfactory for an encapsulation process.[3] The low leachability index is maintained, despite the relatively high salt content, which is beyond what is typically obtained in conventional grout encapsulation processes.[4,5]

Table I. Diffusivity and leachability index of sodium nitrate from waste forms with different salt contents.

Nominal salt content (wt%)	Diffusivity (cm^2/s)	Leachability index
15	2.5×10^{-8}	7.6
21	30×10^{-8}	6.5
27	75×10^{-8}	6.1

Based on the hypothesis that open porosity is the main feature of the microstructure that affects leaching behavior, we decided to perform experiments by exposing samples to temperatures higher than the usual curing temperature (80°C). The objective was to assess if further curing or softening and consolidation of the polymeric matrix would induce a collapse of the pore structure and improve even more the leaching resistance of the waste form. Preliminary results indicate that this is the case. As seen in Figure 3, annealing samples at 120°C and 150°C for 2 days significantly reduces the leaching rates of a waste form with a nominal salt loading of 8 wt%. In addition, at the end of the leaching experiments, SEM images of sample sections from this waste form clearly show salt remaining in the polymeric matrix, as illustrated in Figure 4. Waste forms with the higher nominal salt loading of 27 wt% were also subjected to a post-curing heat treatment to quantify this effect. The diffusivities and leachability indices for these samples are presented in Table II. Treated waste forms have appreciably lower effective diffusivities than the original waste form. The mechanism behind the improved leaching resistance of treated samples needs to be further investigated, but our results show that there is the potential to make this encapsulation process even more efficient by manipulating the microstructure of the resulting waste form.

CONCLUSION

The leaching behavior of soluble salts from polymeric waste forms synthesized by a novel aqueous-based route indicates that this encapsulation process has the potential to be more effective than conventional technologies. The leaching experiments conducted in this work represent a stringent test of the encapsulation capability of the waste forms, since the salt used has a high solubility, and external mass transfer limitations have been removed.

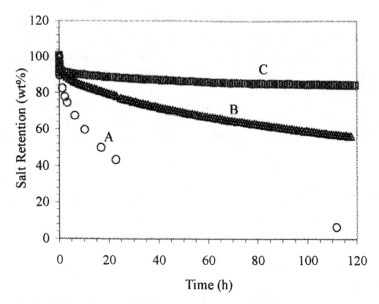

Figure 3. Leaching curves of salt retention percentages vs. time for waste forms (nominal loading of 8 wt% $NaNO_3$) heat-treated under different conditions: A) cured at 80°C; B) cured at 80°C then reheated at 120°C for 2 days; C) cured at 80°C then reheated at 150°C for 2 days.

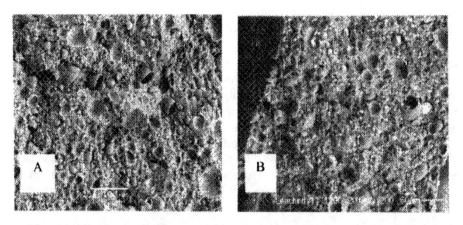

Figure 4. Backscattered images of leached samples of a waste form (nominal loading of 8 wt% $NaNO_3$) cured at 80°C: A) without reheating; B) reheated at 120°C for 2 days. The salt particles in sacs were almost leached out in sample A, while a considerable amount of salt particles still remained in the sacs in sample B.

Table II. Effects of temperature of 2 day post-curing heat treatment on leaching behavior for waste forms (nominal loading of 27 wt% NaNO₃)

Initial curing temp. (°C)	Post-curing temp. (°C)	Diffusivity (cm²/s)	Leachability index
80	-	7.5×10^{-7}	6.1
80	120	1.0×10^{-7}	7.0
80	150	1.1×10^{-8}	8.0

ACKNOWLEDGMENTS

This work was supported by the DOE Office of Science and the National Science Foundation *Faculty and Students Team Program* and by the Process Science and Engineering Division within the Environmental Technology Direactorate at Pacific Northwest National Laboratory. The assistance of Rod Quinn, Loni Peurrung, Bradley R. Johnson, Michael Schweiger, Jim Davis, and Royace Aikins is gratefully acknowledged. This work was completed at the Applied Process Engineering Laboratory (APEL) at Pacific Northwest National Laboratory.

REFERENCES

[1]Maio, V., R.K. Biyani, R. Spence, G. Loomis, G. Smith and A. Wagh, "Testing of Low-Temperature Stabilization Alternatives for Salt Containing Mixed Wastes-Approach and Results to Date;" pp. 514-521 in *Proceedings of the International Topical Meeting on Decommissioning and Decontamination and on Nuclear and Hazardous Waste Management SPECTRUM '98*, American Nuclear Society, La Grange Park, Illinois, 1998.

[2]Evans, R., A. Quach, G. Xia, B.J.J. Zelinksi, W. Ela, D.P. Birnie III, A.E. Sáez, H. Smith and G. Smith, "Microstructure of Emulsion-Based Polymeric Waste Forms for Encapsulating Low-Level, Radioactive and Toxic Metal Wastes," previous paper, 2003.

[3]USNRC, *Stabilization/Solidification of CERCLA and RCRA Wastes-Technical Position on Waste Form*, Revision 1, January 1991.

[4]Krishnamoorthy, T.M., S.N. Joshi, G.R. Doshi and R.N. Nair, "Desorption Kinetics of Radionuclides Fixed in Cement Matrix," *Nuclear Technology*, **104**, 351 (1993).

[5]Leist M., R.J. Casey and D. Caridi D., "The Management of Arsenic Wastes: Problems and Prospects," *J. Hazardous Materials*, **B76**, 125 (2000).

THERMAL PROCESSING OPTIMIZATION FOR SIMULATED HANFORD WASTE GLASS (AZ 101)

A. Giordana, W.G. Ramsey, T.F. Meaker, B. Kauffman, M. McCarthy, K. Guilbeau
Diagnostic Instrumentation and Analysis Laboratory at Mississippi State University
205 Research Boulevard
Starkville, Mississippi 39759, USA

J.D. Smith, F.S. Miller, T. Sanders, E.W. Bohannan
University of Missouri—Rolla
1870 Miner Circle
Rolla, MO 65409

J.Powell, M.Reich, J. Jordan, L. Ventre, R.E.Barletta, A.A. Ramsey, G.Maise, B. Manowitz, M. Steinberg, F. Salzano,
Radioactive Isolation Consortium, LLC
708 East Broad Street
Falls Church, VA 22046

ABSTRACT

This paper presents the results of a Time-Temperature-Transformation (TTT) study of the effect of different heat treatments on the final form and durability of the wasteform produced by the vitrification of Hanford High Level Waste (HLW). A borosilicate glass formulation containing 60 weight % of a Hanford Tank AZ 101 simulant approximated the radioactive waste glass. The TTT diagram was generated by analyzing thirty-six wasteforms produced by melting the frit-waste mixture at 1450°C, and then heat-treating the melts at specified temperatures for a predetermined length of time. The temperatures studied ranged from 500°C to 1200°C, and the length of the heat treatments ranged from 0.75 hour to 768 hours (32 days). The wasteforms produced were analyzed by Toxicity Characteristic Leaching Procedure (TCLP), X-ray Diffraction (XRD), and Scanning Electron Microscopy (SEM) – Energy Dispersive Spectrometry (EDS). Rietveld analysis was performed on some of the XRD results. Crystalline forms detected include transition metal (predominantly Fe) spinel, zircon, zirconia, and iron oxide.

These crystalline forms are characteristic to specific heat treatment conditions. The types or concentrations of crystals have little to no impact on the Product Consistency Test (PCT) performance of the wasteform. The results show a strong relationship between heat treatment conditions (time and temperature) and TCLP test response for cadmium only. Samples heat-treated at higher temperatures or longer periods of time show a considerably higher cadmium response (by TCLP) than samples heat-treated at lower temperatures or for shorter times.

INTRODUCTION

This paper presents the results of a study conducted under Department of Energy (DOE) contract DE-AC26-00NT40801 for Additional Tests of an Advanced Vitrification System.[*] The work was performed by the Diagnostic Instrumentation and Analysis Laboratory (DIAL) at Mississippi State University (MSU) and the Radioactive Isolation Consortium (RIC), LLC, together with the University of Missouri – Rolla (UMR).

The work is focused on the in-can melter alternative being explored by the Office of River Protection (ORP) and DOE as part of a mission to accelerate the disposal of Hanford HLW. The target result is a borosilicate glass wasteform compliant with the current revision (Rev. 2) of the Waste Acceptance Product Specifications (WAPS) for Vitrified High-Level Waste Forms.[1] The wasteform must be a high waste loading glass having the properties of chemical durability, thermal stability, and mechanical stability.

This study was conducted with an upper temperature limit of 1450°C for the glass melting operation. The melter temperature, higher than that used in previous studies[2, 3], was chosen to increase the percentage of incorporated waste.

A non-radioactive simulant of the contents of the AZ-101 Hanford Tank, produced by NOAH Technologies from a DOE recipe, was used as a substitute for the actual radioactive waste. The simulant and the study that resulted in this waste glass formulation , containing 60% waste loading by weight, are discussed elsewhere.[4]

---

[*] DISCLAIMER--This report was prepared as an account of work sponsored by an agency of the United States Government. Neither the United States Government nor any agency thereof, nor any of their employees, makes any warranty, express or implied, or assumes any legal liability or responsibility for the accuracy, completeness, or usefulness of any information, apparatus, product, or process disclosed, or represents that its use would not infringe privately owned rights. Reference herein to any specific commercial product, process, or service by trade name, trademark, manufacturer, or otherwise does not necessarily constitute or imply its endorsement, recommendation, or favoring by the United States Government or any agency thereof. The views and opinions of authors expressed herein do not necessarily state or reflect those of the United States Government or any agency thereof.

The oxide weight composition of the glass used in this study, as measured at DIAL, is shown in Table I below.

Table I– Oxide weight composition of the glass used in the present study.

Oxide	Wt % (normalized to 100 g)
Al2O3	16.87
As2O3	0.07
B2O3	6.11
BaO	0.08
CaO	0.73
CdO	0.69
CeO2	0.15
Cr2O3	0.16
Cs2O	0.17
CuO	0.05
Fe2O3	20.66
K2O	0.37
La2O3	0.60
MgO	0.16
MnO	5.60
Na2O	1.04
NiO	1.14
P2O5	0.20
PbO	0.32
Sb2O5	0.30
SeO2	0.00
SiO2	29.94
SO3	0.37
SrO	4.52
TeO2	0.14
TiO2	0.16
ZnO	1.92
ZrO2	7.49

EXPERIMENTAL DETAILS

A 36-point experiment was used to generate the TTT diagram. Identical batches of the glass formulation were placed in 250 cm^3 alumina crucibles and melted at 1450°C. The crucibles were then placed in a pre-heated furnace and isothermal heat tratments were performed at temperatures lower than the liquidus temperature of the glass. The samples were then cooled down in the heat treatment furnace. The temperatures of the isothermal heat treatments were 500,

640, 780, 920, 1060, and 1200°C. The lengths of the heat treatments were 0.75, 3, 12, 48, 192, and 768 hours.

Specimens randomly selected from different areas of the wasteform were mounted in epoxy, then ground and polished to a 1 μm or better finish. The specimens were then characterized using a scanning electron microscope (SEM) equipped with a light element energy dispersive x-ray spectroscopy (EDS) detector. As a standardization method, a sample of cast AZ-101 waste glass (previously characterized by dissolution followed by spectroscopic analysis) was mounted alongside the samples examined, allowing comparative EDS spectra to be gathered under identical conditions.

ANALYSIS OF CRYSTALLINE PHASES

Fig. 1 shows a backscattered electron image at 500x of a wasteform treated at 1200°C for 192 hours. Phase A was identified by EDS as the glass base, phase B as iron spinel, phase C as zircon, and phase D as zirconia. These identifications are consistent with the XRD analysis.

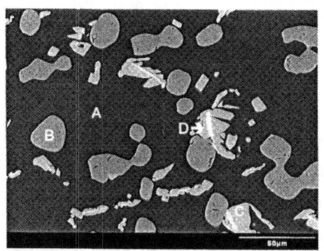

Fig. 1 - Backscattered electron image at 500x of a wasteform treated at 1200°C for 192 hours.

Qualitative XRD analyses were performed to determine the crystalline phases and were complemented by Rietveld analysis of the spectra. A TTT diagram was constructed, using the percent crystallinity determinations resulting from the Rietveld analyses, confirmed by the SEM-EDS analysis of the crystals compositions. The Rietveld analysis provided a relative uncertainty of about 10% for the zircon and zirconia weight percentage. The uncertainty in the spinel phase is considered somewhat higher because of the variability of the spinel composition across the glass matrix.

A known quantity of NIST traceable alumina was added to all samples prior to XRD analysis. A check of the accuracy of the Rietveld analysis was provided by analyzing mixtures of 1) a sample of crystal-free, water quenched waste glass and 2) a NIST traceable material (such as pure zirconia) having a crystal structure similar to a crystal phase identified by qualitative XRD. The concentration of the NIST traceable material was in the range of expected crystal concentrations in the wasteform studied.

The Rietveld analysis detected six crystalline phases, consistent with the EDS identification: 1) tetragonal ZrO_2, 2) monoclinic ZrO2, 3) $ZrSiO_2$, 4) Fe/Al spinel (modeled as Fe_3O_4 with Al substituted; EDS provided an estimate of about 33% Al cations in the magnetite matrix), 5) spinel ($MnAl_2O_4$ best pattern match), and 6) a phase having $FeMnO_3$ as best pattern match.

TTT DIAGRAMS

The TTT diagram shown below in Fig. 2 was constructed based on the Rietveld results and illustrates the crystalline forms observed in the wasteforms produced with the different heat treatments. The colored lines indicate approximately in which conditions the different crystalline forms were observed.

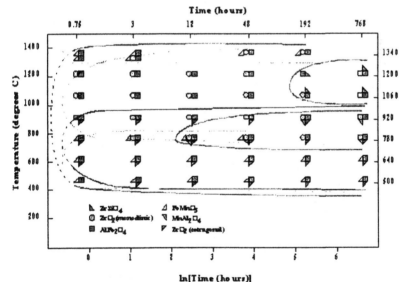

Fig. 2 – Crystalline phases in function of heat treatment as determined by Rietveld analysis

Fig. 2 makes it apparent that the crystallinity of the wasteforms is dominated by spinel and zirconium oxide(zirconia).

It is noted that spinel and zirconia occur early in the treatment. The spinel kinetics are in agreement with previous studies[5]. The zirconia displays rapid formation at temperatures below 1200°C.

The tetragonal form of zirconia is predominant in the heat treatments below 700°C, while the monoclinic form is predominant above 920°C. It is possible that the tetragonal zirconia is a metastable phase that originated during cooldown from 1450°C to the heat treatment temperature.

Formation of zircon ($ZrSiO_4$) occurs at temperatures above 950°C for the 192 hours treatments. It is likely that the zircon develops slowly from the existing zirconia phase and the surrounding glassy matrix.

The TTT diagrams in Fig. 3 and Fig 4 below were generated using a methodology similar to one developed at the Pacific Northwest National Laboratory[6]. The curves have been generated by extrapolation, assuming no crystallinity (i.e. below detection level) at the very beginning of the heat treatment.

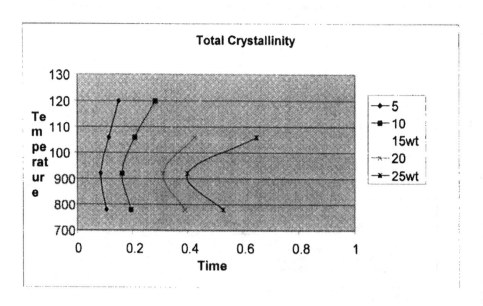

Fig. 3 - Diagram of the amount of total crystallinity in the wasteforms generated by the TTT study. The iso curves are interpolations aimed to facilitate the reading of the plot

The TTT diagram in Fig. 3 provides an indication of the amount of total crystallinity, comprising all the phases detected in the wasteform at a given time. The iso lines correspond to 5%, 10%, 15%, 20%, and 25% total crystallinity.

Referring to Figure 3, it is noted that crystal formation occurs very early in the thermal treatment at every temperature. The total crystallinity is greatest at 920°C. Extrapolations show that, at this temperature, the crystalline content of the glass exceeds 10% after only 10 minutes, and 25% after 30 minutes.

The "nose" of the curves occurs at 920 °C, suggesting that treatment at this temperature may be especially conducive to crystallization.

Spinel Phase Only

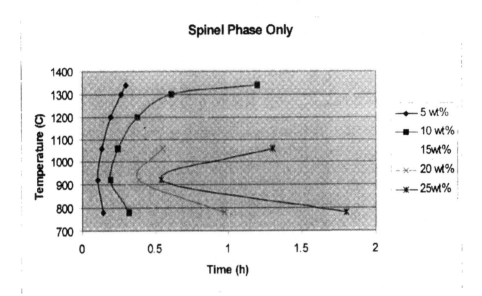

Fig. 4 – Diagram of the amount of total spinel phase in the wasteforms generated by the TTT study. The iso curves are interpolations aimed to facilitate the reading of the plot

The TTT diagram in Fig. 4 provides an indication of the amount of total spinel phase present in the wasteform at a given time. As in the previous figure, the iso lines correspond to 5%, 10%, 15%, 20%, and 25% total crystallinity.

Referring to Figure 4, it is noted that spinel formation occurs very early in the thermal treatment at every temperature, with the "nose" of the curves also evident at 920°C. At higher temperatures, the curvature of the iso lines suggests that spinel phase formation seems to occur later than the other phases.

WASTEFORM DURABILITY
PCT[7] and TCLP[8] analysis of the wasteforms generated by the different thermal treatments showed that the release of the WAPS-reportable elements was not significantly affected by the different processes, as exemplified by the boron release shown in Fig. 5 below.

PCT Results as a Function of Heat Treatment Temperature & Time: B

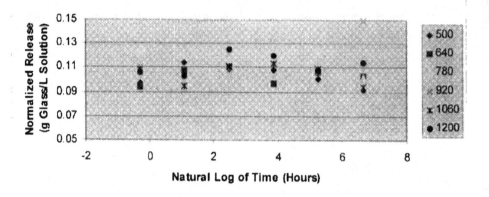

Fig. 5 – Normalized release of Boron for the different isothermal treatments as function of time.

The only leachate that appeared to vary significantly with the different isothermal processes was cadmium. The amount of cadmium released, as measured by TCLP, is shown in Fig. 6 below. While Cd release for the higher temperature, longer times was above the 0.11 mg/l Universal Treatment Standard (UTS) limit, it was well within the 1.0 mg/l Cd release criteria required by the WAPS.[1,9]

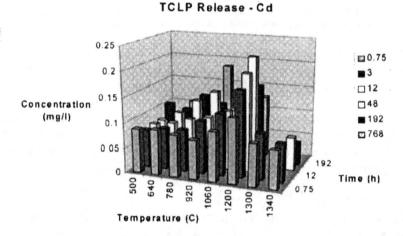

Fig. 6 – Cadmium release in function of the different isothermal treatments, as measured by TCLP.

CONCLUSIONS

We have performed a 36-point study of a borosilicate glass incorporating 60% by weight of waste simulant. The study investigated the effect of different isothermal heat treatments performed after high-temperature melting.

Our results show that the wasteforms produced in our experiment have a significantly higher inherent crystallinity than wasteforms produced at lower temperatures and incorporating lower levels of HLW simulant.[2,3] This is due to the high waste loading of our glass. The samples produced during this project contain on the order of 5+ volume percentage of high-Fe spinel and a lower concentration of zirconia as compared with about 0.1 volume percentage of high Fe spinel crystals at the glass-crucible interface in the lower temperature, lower waste content wasteforms. [2,3]

A TTT diagram was produced indicating the heat treatment time/temperature relationship with the various crystalline phases. Spinel was observed in all samples. Zircon formation was limited to high temperature and long-term heat treatments.

The heat-treated samples contained crystal contents as high as 30+ weight percent, but the release of WAPS-reportable elements was not found to change appreciably as a function of crystallinity. Only the cadmium release appeared to be somewhat affected by the heat treatment. All of the wasteforms produced were WAPS-compliant.

The study was performed as part of the development of Advanced Vitrification System (AVS), a variable temperature, in-can, inductively heated melter system currently being developed for high-level waste processing.

REFERENCES

[1] Waste Acceptance Product Specification for Vitrified High-Level Waste Forms, US Department of Energy, Office of Environmental Management, EM-WAPS Rev. 02, Washington, DC, December 1996.

[2] W.K. Kot, H. Gan, and I.L. Pegg, *Physical and Rheological Properties of Waste Simulants and Melter Feeds for RPP-WTP HLW Vitrification*, Final Report, VSL-00R2520-1, Rev. 0, Prepared for GTS Duratek, Inc. and BFNL, Inc., by the Vitreous Stale Laboratory, the Catholic University of America, Washington, DC 20064, October 31, 2000.

[3] W. K. Kot and I.L. Pegg, *Glass Formulation and Testing with RPP-WTP HLW Simulants*, Final Report, VSL-00R2540-2, Rev. 0, Prepared for GTS Duratek, Inc.

and BFNL, Inc., by the Vitreous Stale Laboratory, the Catholic University of America, Washington, DC 20064, February 16, 2001, provided to DIAL by the DOE.

[4] W.G. Ramsey et al., *Time-Temperature Transformation Study of Simulated Hanford Tank Waste (AZ-101) and Optimization of Glass Formulation for Processing Such Waste*, to appear in the proceedings of the Waste Management 2003 Conference, held February 23-27, 2003, in Tucson, AZ.

[5] See for example J. Alton, T.J. Plaisted, and P. Hrma, J. Non-Cryst. Sol. 311 (2002) 24-35.

[6] P.Izak, P. Hrma, B.W. Arey, T.J. Plaisted, J. Non-Cryst. Sol. 289 (2001) 17-29.

[7] American Society for Testing and Materials (ASTM) Designation C 1285-97.

[8] Toxicity Characteristic Leaching Procedure, Test Method 1311, in "Test Methods for Evaluating Solid Waste, Physical Chemical Methods," EPA Publication SW-846.

[9] 40CFR 261.24

AGING BEHAVIOR OF A SODALITE BASED CERAMIC WASTE FORM

Jan-Fong Jue, Steven M. Frank, Thomas P. O'Holleran, Tanya L. Barber, Stephen G. Johnson, and K. Michael Goff,
Argonne National Laboratory
P. O. Box 2528
Idaho Falls, ID 83403-2528

Wharton Sinkler
UOP
25 E. Algonquin Road
Des Plaines, IL 60017

ABSTRACT

A sodalite based ceramic waste form is one of the end products of a pyro-process developed by Argonne National Laboratory to treat metallic spent nuclear fuel from the EBR-II reactor for permanent storage. The sodalite based ceramic waste form contains a small quantity of actinides as well as fission products. In order to predict long-term degradation behavior of the sodalite based ceramic waste form in a repository, a short-lived actinide, ^{238}Pu (half-life is 87.7 years), was loaded into the ceramic waste form to accelerate the alpha-damaging process. The ^{238}Pu-doped ceramic waste form was fabricated by uniaxial hot pressing. The consolidated sodalite based ceramic waste form was subjected to aging for four years. The change in microstructure due to radiation damage was characterized periodically during the aging process using electron microscopy. Alpha-decay damage did not amorphize sodalite and actinide-bearing phases after four years of aging. No microcracks were found in the aged samples. Occasionally, bubbles and voids were found. These bubbles and voids are interpreted as pre-existing defects. However, some contribution to these bubbles and voids from helium gas cannot be ruled out.

INTRODUCTION

Argonne National Laboratory has developed an electrochemical method to treat spent metallic nuclear fuels for permanent storage. This pyro-process is detailed elsewhere [1,2]. The end products of this process are two waste forms.

One is a metallic waste form that contains actinide-bearing phases in a metallic matrix. The other one is a sodalite based ceramic waste form that contains sodalite $[Na_8(AlSiO_4)_6Cl_2]$ and glass as the major components while halite, fission products, and actinide-bearing phases are minor constituents. The radiation generated by the decay of actinides and fission products can cause damage to the waste forms so that the integrity of the waste forms in a repository is deteriorated with time. Radiation effects on waste forms have been reviewed by Ewing et al [3].

Radiation damage from alpha decays of the actinide containing phases has been reported in different ceramic systems [3-9]. High-energy alpha particles and heavy recoil nuclei from alpha decays generate defects in the waste form. Alpha particles, which have a range of 10 to 20μm, can displace atoms and eventually transform to helium gas. The recoiling nuclei have a shorter range (10-20nm) but can displace more atoms within their range. It was estimated that the recoiling nuclei account for about 90% of the displacement energy due to the alpha decay process [3]. Phase transformation, amorphization, phase separation, swelling, and gas bubble formation are possible forms of damage from alpha decay [3-11]. If the damage is extensive, cracking of the waste forms may happen which increases the exposed area to possible corrosive agents such as groundwater. Allen and coworkers use ion irradiation (He^+, Pb^+, and Kr^{++}) to simulate the radiation effects of alpha particles and recoiling nuclei on a surrogate ceramic waste form [12]. Amorphization due to ion beam bombardment was observed. In this work, an actinide (plutonium - 238) was added to a surrogate ceramic waste form. Due to the short half-life of ^{238}Pu, the ^{238}Pu-doped ceramic waste form aged for four years will accumulate alpha decay events corresponding to what the real ceramic waste form would receive over a period of ten thousand years. X-ray diffraction (XRD) and scanning electron microscopy (SEM) have been used to characterize the radiation effects of alpha decay on the phase stability and microstructure of the ceramic waste forms. XRD data from the ^{238}Pu-doped ceramic waste form samples revealed a unit cell volume expansion in plutonium oxide in the early stages of aging. Further aging only changed the unit cell volume of plutonium oxide slightly. This is consistent with the behavior of self-damaged plutonium oxide reported in the literature [5]. SEM reveals no surface cracking in the ^{238}Pu-doped ceramic waste form samples after aging for four years. This paper focused on the characterization of the aging behavior of the ^{238}Pu-doped ceramic waste form using transmission electron microscopy (TEM).

EXPERIMENTAL PROCEDURE
(a) Waste Form Preparation
A ^{238}Pu-containing salt is blended with zeolite 4A powder $[Na_{12}(SiO_2)_{12}(AlO_2)_{12}.xH_2O]$ at 500°C for 34 hours. The zeolite powder has a particle size of about 4μm. The chemical composition of the salt is given in Table

1. The weight ratio between the salt and zeolite powder (1 to 6.35) is chosen so that there is an average of 3.8 chlorine ions per zeolite unit cell. The salt-occluded zeolite powder was then mixed with glass frit in a one-to-three weight ratio. The chemical composition of glass is given in Table 2. Uniaxial hot pressing was conducted at 750°C for 4 hours. The maximum applied pressure was ~35 MPa. For comparison, a ceramic waste form sample without ^{238}Pu was also fabricated using the same fabrication procedure. Aging was conducted at room temperature for 4 years.

Table 1 - *Composition of salt*

Salt compound	Weight percent (wt%)
LiCl	26.250
KCl	23.279
^{238}PuCl$_3$	35.474
NaCl	2.484
^{239}PuCl$_3$	8.046
NdCl$_3$	1.1386
CsCl	0.7335
CeCl$_3$	0.6818
SrCl$_2$	0.2951
BaCl$_2$	0.3501
LaCl$_3$	0.3567
PrCl$_3$	0.3367
YCl$_3$	0.2050
RbCl	0.0967
SmCl$_3$	0.2000
KBr	0.0067
KI	0.0450

Table 2 - *Composition of glass*

Compound	Weight percent (wt%)
SiO$_2$	66.5
B$_2$O$_3$	19.1
Al$_2$O$_3$	6.8
Na$_2$O	7.1
K$_2$O	0.5

(b) TEM Characterization

Transmission electron microscopy (TEM) samples were prepared using a modified dimpling-ion milling method. The ceramic waste form samples with and without [238]Pu were broken into small pieces. The small segments were then bonded inside a copper ring (3mm in diameter, ~1mm in thickness) using the Gatan G-1 epoxy. After curing the mixture at 120°C for 15 minutes, the disc-shape sample was ground using silicon carbide abrasive papers to about 100 μm in thickness. A Gatan Dimple Grinder and cubic boron nitride pastes were used to form a dimple at the center of the disc. The center of the dimple had a thickness of less than 30 μm. A Gatan Precision Ion Polishing System (PIPS) was used to further thin down the dimpled sample until perforation. A 50 Å conductive carbon coating was applied on the ion-milled sample using a Gatan Ion Beam Coater (IBC). TEM characterization was conducted using a JEOL 2010 transmission electron microscope operated at 200 keV. Energy-dispersive X-ray spectra were obtained using an Oxford Instruments EDS system.

RESULTS AND DISCUSSION

(a) Typical TEM Microstructure

After hot pressing, [238]Pu-doped ceramic waste form samples have a density of 2.42 g/cm³. XRD results show that the salt-loaded zeolite was converted into sodalite during hot pressing. A typical TEM microstructure of a hot pressed [238]Pu-doped ceramic waste form sample is shown in Figure 1. In this figure, there is a sodalite region about 3 μm across with an average grain size of about 0.4 μm. Different sodalite regions are separated by a darker appearing glass phase. Actinide-bearing phases are black particles mainly found in the glass area, while some small actinide-bearing particles can been seen in the sodalite area as well. Several light-colored round inclusions in the glass area were identified to be halite using selected area electron diffraction (SAD) and energy-dispersive X-ray spectroscopy (EDS). The microstructure of the sample without plutonium is similar except lacking actinide-bearing particles.

(b) Radiation Damage

There are three kinds of radiation damage that are specifically looked for in this TEM study:

(i) Amorphization of sodalite and actinide-bearing phases due to alpha decay: If amorphization is caused by alpha particles, which have a range of 10 - 20 μm, it should be found throughout the sample. Amorphous areas having a chemical composition similar to sodalite, or exhibiting a high plutonium concentration, should be abundant in the examined sample. If the damage is coming from recoiling particles, no wide spread amorphization of sodalite is expected due to the short range of these particles (10-20nm). Amorphized regions will be limited to actinide-bearing particles and sodalite grains adjacent to them.

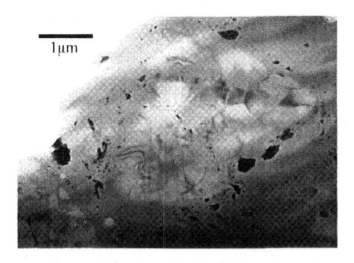

Figure 1. A bright-field transmission electron micrograph showing the typical microstructure of the [238]Pu-doped ceramic waste form.

(ii) Formation of bubbles and voids: Gas bubbles can come from several possible sources, such as (1) oxygen and chlorine from electrolysis [13-14], (2) helium from alpha particles [3], (3) inner gas trapped during ceramic waste form fabrication, and (4) argon from TEM sample preparation (i.e. ion milling). The origins of voids may be (1) evaporation of materials under a focused electron beam [12], (2) bubbles after releasing trapped gas, (3) segregation of vacancies, or (4) pre-existing defects after sample fabrication (less than theoretical density). To distinguish bubbles from voids under TEM is not straightforward. No effort was given to distinguish these two defects. Segregation of bubbles or voids on grain boundaries, interfacial boundaries, or existing defects (such as pre-existing bubbles and voids) can take place if diffusion is fast enough. However, as mentioned earlier, if the source of bubbles and voids is alpha radiation, the bubbles and voids should exist throughout the sample.

(iii) Micro-cracking along grain boundaries or interfacial boundaries: Segregation of defects such as bubbles and voids can form micro-cracks. Cracking of the ceramic waste form may also be the results of the volume change of one or more phases. Volume changes can come from (1) phase separation or crystallization of glass, (2) structural damage or amorphization of sodalite, (3) structural damage or amorphization of actinide-bearing phases. Interfacial boundaries, especially around the actinide-bearing phases, are the key areas to be characterized. Extensive cracking can increase the open surface area to possible corrosive media or weaken the bonding between the actinide-bearing phases and

the host phases (mainly glass). A higher release rate of actinides to the environment may be the outcome.

Figure 2 is a comparison of general TEM microstructure between a ceramic waste form sample without ^{238}Pu (no radiation damage) and a ^{238}Pu-doped ceramic waste form sample after aging at room temperature for four years. As can be seen in Figure 2, there is no notable difference between these two samples.

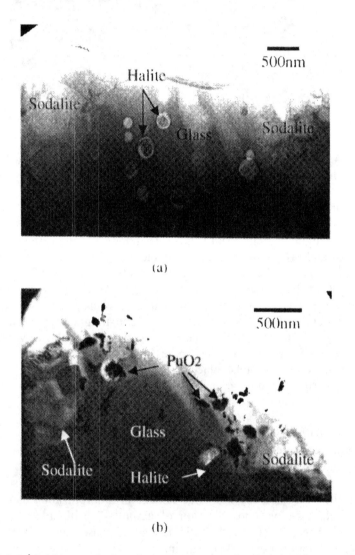

(a)

(b)

Figure 2. Bright-field transmission electron micrographs showing the microstructure of (a) a ceramic waste form sample without ^{238}Pu, (b) a ^{238}Pu-doped ceramic waste form sample aged for four years.

Selected area electron diffraction data from the aged sample show that sodalite and plutonium-bearing phases remain crystalline. It is concluded that radiation damage due to alpha decay did not cause amorphization of crystalline phases in the ^{238}Pu-doped ceramic waste form up to the accumulated alpha decay events of this study.

Occasionally, in the aged ^{238}Pu-doped ceramic waste form samples, bubbles and voids were found in the glass and sodalite regions as shown Figure 3(a). These features were commonly found in ceramic waste form samples without ^{238}Pu. Figure 3(b) is an example. As mentioned earlier, the radiation damages due to alpha particles should generate bubbles and voids throughout the sample. If the kinetics is feasible, bubbles and voids could segregate along grain boundaries and interfacial boundaries. These were not observed in the ^{238}Pu-doped ceramic waste form samples after aging for four years. Thus, these bubbles and voids in the ^{238}Pu-doped ceramic waste form samples after aging are interpreted as pre-existing defects. However, the possibility of helium gas diffusing to these pre-existing bubble and voids can not be ruled out. A significant contribution of helium gas to the existing defects may change their size and density. No statistically significant differences in the bubble size (20-200nm) and density were observed between the samples with and without alpha decay damage.

Higher magnification TEM bright field images reveal no microcracks on the interfaces between large actinide-bearing particles and glass. An example is given in Figure 4. This indicates that the small unit cell expansion of plutonium oxide observed by XRD in the early stages of aging did not create enough mechanical stress to degrade the bonding between the actinide-bearing phases and the host phase.

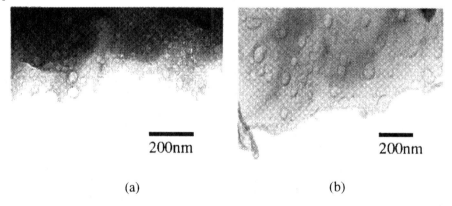

200nm   200nm

(a)   (b)

Figure 3. (a) A bright-field transmission electron micrograph showing bubbles in the ^{238}Pu-doped ceramic waste form sample. (b) A bright-field transmission electron micrograph showing bubbles in a ceramic waste form sample without ^{238}Pu.

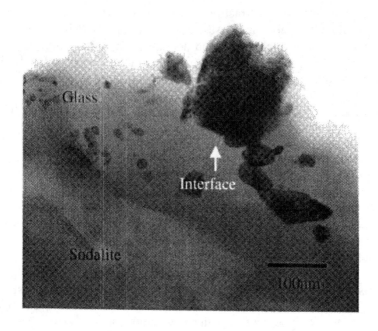

Figure 4. A bright-field transmission electron micrograph showing a large plutonium-bearing particle in the glass phase. No microcracking was observed.

CONCLUSIONS

Alpha-decay damage did not amorphize sodalite and actinide-bearing phases after four years of study. The accumulated alpha-decay events correspond to what real ceramic waste forms will receive in approximately ten thousand years.

No microcracks were found in the actinide-bearing ceramic waste form samples aged for four years.

Occasionally, bubbles and voids were found in the aged [238]Pu-doped ceramic waste form samples. Bubbles and voids with similar size and density were also found in ceramic waste form samples without actinides. These bubbles and voids are interpreted as pre-existing defects. However, some contribution to these bubbles and voids from helium gas cannot be ruled out.

ACKNOWLEDGMENTS

This work was supported by the Department of Energy, Nuclear Energy Research and Development Program, under contract No. W-31-109-ENG-38. Special thanks to Dr. D. E. Janney, Dr. J. I. Cole, and Mr. M. Surchik.

REFERENCES

[1]M. F. Simpson, K. M. Goff, S. G. Johnson, T. J. Bateman, T. J. Battisti, K. L. Toews, S. M. Frank, T. L. Moschetti, T. P. O'Holleran, and W. Sinkler, "A Description of the Ceramic Waste Form Production Process from the Demonstration Phase of the Electrometallurgical Treatment of EBR-II Spent Fuel," *Nuclear Technology*, **134** 263-277 (2001).

[2]W. Sinkler, D. W. Esh, T. P. O'Holleran, S. M. Frank, T. L. Moschetti, K. M. Goff, S. G. and Johnson, "TEM Investigation of a Ceramic Waste Form for Immobilization of Process Salts Generated during Electrometallurgical Treatment of Spent Nuclear Fuel," *Ceramic Transactions*, **107** 233-240 (2000).

[3]R. C. Ewing, W. J. Weber, and F. W. Jr. Clinard, "Radiation Effects in Nuclear Waste Forms for High-Level Radioactive Waste," *Progress in Nuclear Energy*, **29** 63-127 (1995).

[4]W. J. Weber, "Ingrowth of Lattice Defects in Alpha irradiated UO_2 Single Crystals," *Journal of Nuclear Materials*, **98** 206-215 (1981).

[5]T. D. Chikalla and R. P. Turcotte, "Self-Radiation Damage Ingrowth in $^{238}PuO_2$," *Radiation Effects*, **19** 93-98 (1973).

[6]Y. Inagaki, H. Furuya, and K. Idemitsu, "Microstructure of Simulated High-Level Waste Glass Doped with Short-Lived Actinides, ^{238}Pu and ^{244}Cm," *Materials Research Society Symposium Proceedings*, **257** 199-206 (1992).

[7]W. J. Weber, "Alpha-Decay-Induced Amorphization in Complex Silicate Structure," *Journal of the American Ceramic Society*, **76** 1729-1738 (1994).

[8]W. J., Weber, "Radiation-Induced Defects and Amorphization in Zircon," *Journal of Materials Research*, **5** 2687-2697 (1990).

[9]S. M. Frank, T. L. Barber, T. Disanto, K. M. Goff, S. G. Johnson, J-F. Jue, M. Noy, T. P. O'Holleran, and W. Sinkler, "Alpha-Decay Radiation Damage Study of a Glass-Bonded Sodalite Ceramic Waste Form," *Materials Research Society Symposium Proceedings*, **713** 487-494 (2002).

[10]S. Sato, H. Furuya, T. Kozaka, Y. Inagaki, and T. Tamai, "Volumetric Change of Simulated Radioactive Waste Glasses Irradiated by the $^{10}B(n, a)^7Li$ Reaction as Simulation of Actinide Irradiation," *Journal of Nuclear Materials*, **152** 265-269 (1988).

[11]M. S. El-Genk, and J. Tournier, "Estimates of Helium Gas Release in $^{238}PuO_2$ Fuel Particles for Radioisotope Heat Sources and Heater Units," *Journal of Nuclear Materials*, **280** 1-17 (2000).

[12]B. G. Storey, and T. R. Allen, "Radiation Damage of a Glass-Bonded Zeolite Waste Form Using Ion Irradiation," *Materials Research Society Symposium Proceedings*, **481** 413-418 (1998).

[13]D. G. Howitt, H. W. Chan, J. F. DeNatale, and J. P. Heuer, "Mechanism of the Radiolytically Induced Decomposition of Soda-Silicate Glasses," *Journal of the American Ceramic Society*, **74** 1145-1147 (1991).

[14]E. Johnson, J. Ferrer, and L.T. Chadderton, "Radiolytic Radiation Damage of Sodalite," *Physica Status Solidi*, **(a)49** 585-591 (1978).

PROCESS FOR SELECTIVE REMOVAL AND CONCENTRATION OF ACTINIDES AND HEAVY METALS FROM WATER

B.P. Kiran and Allen W. Apblett
Department of Chemistry
Oklahoma State University
Stillwater, OK, 74078.

ABSTRACT

The ability of molybdenum hydrogen bronze, HMo_2O_6, to absorb heavy metals and radionuclides from water was investigated. It was found that it could remove substantial amounts of metal ions from water and was selective for metal with large radii. The products from uranium, thorium, lead, and neodymium uptake were discovered to be insoluble molybdate phases. A cyclic process was developed whereby HMo_2O_6 adsorbed uranium from aqueous solution and then the uranium and molybdenum trioxide were separated by treatment with aqueous ammonia. Solid ammonium uranate was isolated by filtration and the aqueous ammonium molybdate was converted back to HMo_2O_6.

INTRODUCTION

Heavy metals and actinides are a common contaminant of ground water and can arise from natural and anthropogenic sources. Remediation of such contaminated waters is problematic due to: (i) the presence of heavy metals in typically very low concentrations (100-500 μg/L); (ii) the ground water itself being found up to depths of several hundred meters below the surface; and (iii) the coexistence of alkali and alkaline-earth metals in much higher concentrations (30-300 mg/L), further complicating the remediation processes. In spite of all these difficulties, the remediation of ground water contaminated with metals is essential because of their high toxicity to humans and other living organisms [1, 2]. Since heavy metals are cumulative poisons or exhibit biological effects in small doses, even low concentrations of heavy metals can be problematic.

Commonly used above-ground water treatment processes do not provide an adequate solution to heavy metal remediation. Processes converting soluble metal salts to corresponding insoluble hydroxides proved to be ineffective because the hydroxides still have a small but finite solubility. Calculated equilibrium concentrations for lead, cadmium and mercury [3], over the entire pH range, highly exceed the permitted values [4]. Another approach to above ground remediation, electrochemical reduction, takes advantage of the relative ease of reduction of heavy metal as compared to lighter, non-toxic metals. However, the very high cost of this method makes it an unattractive alternative for ground-water remediation [5].

Reactive permeable barriers are among the most promising methods for the remediation of heavy-metal contaminated groundwater. Such barriers have been designed and proven effective for a variety of heavy metals. Zero-valent iron is one of the more successful materials for this purpose. For example, iron reduces the mobile, soluble ions such as CrO_4^- to the non-toxic, insoluble and therefore immobile form Cr^{3+}. Permeable reactive barriers constructed using iron suffer from the drawback that they are nonspecific and react with a wide variety of dissolved species, leading to a period of life that is shorter than what would be expected from stoichiometric considerations [6-,7,8]. In addition to iron, zeolites [9,10]; materials from biological sources such as seaweed, algae, and bacterial biomass [11,12]; humic acids and peat [13,14]; and activated carbon [15] have been tested as potential materials for construction of reactive barriers. The low cost of zeolites, high uptake selectivities [16] and their stabilities make them an attractive material for this application, but their low uptake capacities, slow reaction kinetics and low hydraulic conductivity work against them [17]. Materials from biological sources have low uptake capacities and selectivities [11,12]. Humic acid and peat have high selectivities towards heavy metals but have poor uptake capacities and also cannot be used in the presence of calcium and magnesium [13,14]. Activated carbon also has poor uptake capacities and selectivities for the uptake of heavy metals. The objective of the research described reported herein was to explore the effectiveness of molybdenum hydrogen bronze for the uptake of heavy metals.

Molybdenum hydrogen bronze, HMo_2O_6, is a promising reagent for environmental remediation. It has a number of unique properties that suggest it could perform better than other reductants for treatment of contaminated waters and the construction of permeable reactive containment barriers. For example, when reductions of inorganic or organic pollutants are performed in a column-type reactor, the color change from royal blue to white would greatly facilitate monitoring of the column's remaining reductive capacity. Unlike other reductants that can be employed in the presence of water and oxygen (such as iron), molybdenum blue has an open layered structure that allows the entire reductive capacity to be used and enhances the rate of reaction by providing a tremendously increased area for the reaction to take place. Since both reduced and oxidized forms of the oxide materials have layered structures through which reactants and products can intercalate, passivation due to build up of oxidized product on the surface does not occur. This is in significant contrast to iron that can form a crust of rust that arrests further reaction of the iron particles with contaminant species. Finally, molybdenum blue is easily recycled after use in redox reactions since regeneration only requires treatment with hot butanol in the presence of a trace of HCl or with zinc/HCl. In fact, the regeneration process with butanol only produces butaraldehyde as a by-product and, in actual industrial production, this could be captured and sold as a commodity chemical.

EXPERIMENTAL

All reagents were commercial products (ACS Reagent grade or higher) and were used without further purification. Bulk pyrolyses at various temperatures were performed in ambient air in a digitally-controlled muffle furnace using approximately 2 g samples, a ramp of 10°C/min and a hold time of 4 hr. X-ray powder diffraction (XRD) patterns were recorded on a Bruker AXS D-8 Advance X-ray powder diffractometer using copper K_α radiation. Crystalline phases were identified using a search/match program and the PDF-2 database of the International Centre for Diffraction Data [18]. Scanning Electron Microscopy (SEM) photographs were recorded using a JEOL Scanning Electron Microscope. Colorimetry was performed on a Spectronic 200 digital spectrophotometer using 1 cm cylindrical cuvettes.

Measurement of the Uptake of Metals by Molybdenum Bronze

Molybdenum blue was tested for the ability to remove Pb^{2+}, Th^{4+}, UO_2^{2+} and Nd^{3+} from aqueous solution. HMo_2O_6 (1.0 g) was reacted with 100 ml of individual (approximately 0.1M) solutions of Pb^{2+}, Th^{4+}, UO_2^{2+} and Nd^{3+}. In all cases, nitrate salts were used with the exception of uranyl where both a nitrate and an acetate salt were tested. After stirring magnetically for a sufficiently long time for complete reaction, as indicated by complete disappearance of the blue color, the mixtures were separated by filtration through a 20 µm nylon membrane filter. The solid products were washed copiously with distilled water and then were dried in a vacuum desiccator. They were subsequently characterized by infrared spectroscopy, thermal gravimetric analysis, and X-ray powder diffraction. The uranium and neodymium concentrations in the treated solutions were analyzed using UV/Visible spectroscopy (λ= 415 nm and 521 nm, respectively). Solutions were treated with nitric acid before analysis to ensure no speciation of metal ions would interfere with the measurement. Lead was determined gravimetrically as lead chromate [19]. Quantitation of thorium was performed colorimetrically using the blue complex (λ= 575 nm) formed between thorium and carminic acid [20].

Selectivity Determination

The selectivity of molybdenum blue for actinides was tested by competition experiments with calcium. Thus, the reactions between uranyl nitrate and molybdenum blue were repeated in the presence of 1.0, 2.0, and 5.0 molar equivalents of calcium nitrate per mole of uranyl ion and the uptake of uranium was determined by UV/Visible spectroscopy. The selectivity for heavy metals was tested by carrying out reactions between lead nitrate and molybdenum blue in presence of 1 molar equivalent of calcium nitrate per mole of lead ions and the uptake of lead was determined gravimetrically as described earlier.

Recovery of Uranium and Molybdenum Trioxide

Uranium was recovered from the iriginite phase by treatment with a strong base. Thus, 1.1 gram of the uranyl molybdate complex was stirred overnight with 100 ml 15% solution of ammonium hydroxide. The reaction mixture was separated by filtration through a 20 µm nylon membrane filter. The solid product was washed copiously with distilled water and then dried in a vacuum desiccator. The product was subsequently characterized by infrared spectroscopy, thermal gravimetric analysis, and X-ray powder diffraction. The filtrate was evaporated and the solid obtained was analyzed by infrared spectroscopy, thermal gravimetic analysis and X-ray powder diffraction.

Determination of Rate of Uranium Uptake

The rate of uranium uptake was determined by carrying out the following experiment. A 200 ml (0.1M) solution of uranium was stirred with 2 g of molybdenum bronze. Aliquots of 5 ml of the reaction mixture were withdrawn at regular intervals and the pH was measured. Uranium was quantified by the procedure described above. After completion of the reaction, the reaction mixture was filtered as described above. The residue and the filtrate were analyzed separately.

RESULTS AND DISCUSSION

Molybdenum bronze was tested for its ability to remove Th^{4+} (as a model for plutonium(IV)}, UO_2^{2+} (of interest in its own right and as a model for PuO_2^{2+}), and Nd^{3+} (as a surrogate for the later transuranics, radioactive lanthanides, and Pu^{3+}) from aqueous solution. Also, the uptake of lead as a model heavy metal was investigated. The experiments that were performed were designed to determine the capacity of the blue reagents for the various metals and to identify the mechanism of metal uptake. Molybdenum bronze was reacted with an aqueous solution of each of the metals. The stoichiometry was adjusted so that there was at least a one-fold excess of contaminant metal ions {on the basis of one molar equivalent of metal ion per Mo(V) site}.

The experimental conditions and results for the molybdenum bronze/metal ion reactions are listed in Table I while the results of the analyses and binding capacity calculations are given in Table II. Molybdenum blue has a remarkable capacity for absorption of actinides and heavy metals. It absorbed 122% by weight of uranium, 37% by weight of thorium, 61.6% by weight of neodymium, and 110% by weight of lead. The substitution of acetate ions for nitrate ions had a small, negative effect on the uptake of uranium. These extremely high capacities bode well for the eventual application of molybdenum blue in environmental remediation.

Table 1. Experimental Conditions for Metal Uptake Reactions

Metal Solution (0.1M)	Weight of Molybdenum Blue (g)	Weight of Solid Product (g)	Color of Solid Product
Uranium acetate	1.04	2.15	Yellow
Uranium nitrate	1.05	2.32	Yellow
Thorium	1.00	1.40	White
Neodymium	1.10	1.48	Grey
Lead	1.04	2.74	White

At the outset of research, it was thought that adsorption of metals by molybdenum blue would occur either by ion-exchange or redox reactions. The latter possibility is negated because the colors of the final products are those of the contaminant metal ions in their oxidized states. The disappearance of the blue color of Mo(V) indicated that oxidation of the molybdenum bronze had occurred. If the metal ions are not the oxidizing agents, the only other possibilities are nitrate ions or oxygen. Since uranyl acetate also forms a yellow product like uranyl nitrate, oxygen appears to be the oxidizing agent.

Another possible mechanism for the uptake of metals is a simple ion- exchange reaction (Equation 1) involving the hydrogen ions of the molybdenum blue.

$$H_xMoO_3 + (x/n)A^{n+} \longrightarrow A_{x/n}MoO_3 + xH^+ \quad\quad (1)$$

The uptake of the metals in terms of milliequivalents per gram of molybdenum blue were 4.27 for neodymium, 5.14 for uranium, 5.29 for lead, and 1.59 for thorium. Thus, the moles of metal that can be absorbed by molybdenum blue varies with the charge of the metal ion as would be expected for an ion-exchange mechanism. Within the group of doubly-charged metal ions, the moles of metal absorbed are almost equivalent. In this case, the uptake of metals may be expressed as approximately 1.5 moles per mole of HMo_2O_6 and is therefore larger in magnitude than the number of Mo(V) centers or exchangeable protons. This result indicates that the molybdenum(VI) centers in molybdenum bronze also play a role in metal binding and a simple ion-exchange mechanism does not occur. The uptake of neodymium (1.24 moles per mole of molybdenum bronze) and that of thorium (1.1 moles per mole of molybdenum bronze) also exceed the capacity suggested by Equation 1.

Table II. Results of Metal Uptake Experiments

Metal Solution	Initial Conc.	Final Conc.	Uptake (mmol)	Metal Capacity (mmol/g)	Metal Capacity (weight %)
Uranium Acetate	0.1 M	0.053 M	4.7	4.5	108%
Uranium Nitrate	0.1 M	0.046 M	5.4	5.1	122%
Thorium Nitrate	0.1 M	0.084 M	1.6	1.6	37.0%
Neodymium Nitrate	0.1 M	0.047 M	5.3	4.8	69.5%
Lead Nitrate	0.1 M	0.045 M	5.5	5.3	110%

A major concern for the application of molybdenum bronze in the field is its selectivity for actinides and heavy metals as opposed to benign cations normally found in natural waters. Therefore, the selectivity of molybdenum bronze for uranyl ion over calcium ions was determined. The results are displayed in Table III and demonstrate that molybdenum bronze is highly selective for uranium. Even a five-fold higher concentration of calcium ions over uranyl ions had little effect on the absorption of uranium.

Table III. Results of Competition Experiments Between Calcium and Uranium

Uranium: Calcium Ratio	Initial Uranium Concentration	Final Uranium Concentration	Weight Percent of Uranium Absorbed
1:0	0.1M	0.056M	122%
1:1	0.1M	0.055M	121%
1:2	0.1M	0.049M	100%
1:5	0.1M	0.052M	111%

Selectivity of molybdenum bronze for lead ion over calcium ions was also determined. The results are displayed in Table IV and demonstrate that molybdenum bronze is also highly selective for lead. with a five-fold higher concentration of calcium ions versus lead ions having little effect on the absorption of lead.

Table IV. Results of Competition Experiments Between Calcium and Lead

Lead: Calcium Ratio	Initial Lead Concentration	Final Lead Concentration	Weight Percent of Lead Absorbed
1:0	0.1M	0.045M	110%
1:1	0.1M	0.045M	110%
1:5	0.1M	0.050M	111%

X-ray powder diffraction analysis of the solid product from lead uptake by HMo_2O_6 revealed that it consisted mainly of $PbMoO_4$ (wulfenite). The interaction between HMo_2O_6 and Pb^{2+} is so strong that the molybdenum oxide layers are destroyed to yield a normal ortho-molybdate salt. The other metals, uranium and thorium also show similar behavior. The product of uranium uptake with molybdenum bronze was uranium molybdate ($UMo_2O_{12}H_6$, iriginite) while that of thorium uptake was $Th(MoO_4)_2$. Neodymium, however, behaved differently and formed an amorphous phase. However when heated to 800°C crystallization to neodymium molybdate, $Nd_2(MoO_4)_3$ occurred. This result suggests that the neodymium metal ions intercalate between the layers of HMo_2O_6 (staging) and react to give what might be phases that consist of negatively-charged slabs of MoO_6 octahedra with the contaminant ions residing between the layers, which then converts to the normal molybdate phase on heating. These results suggest that the mechanism of metal uptake by the molybdenum bronze is complete destruction of its layered structure and formation of a normal molybdate salt of the metal in which Mo is present as MoO_4 tetrahedra or MoO_6 octahedra.

Figure 1 shows the SEM photographs of the solid products obtained from the reaction between molybdenum bronze and various metals.

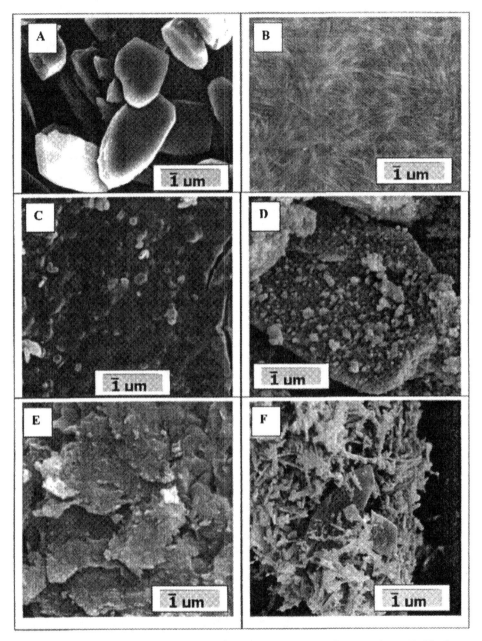

Figure 1. SEM pictures of (A) Molybdenum Bronze. (B) Uranium Nitrate Product. (C) Thorium Nitrate Product. (D) Lead Nitrate Product. (E) Uranium Acetate Product. and (F) Neodymium Product. (SEM= 5000 × magnification)

All the reacted solids show a change from the original molybdenum bronze's morphology and particle size {Figure 1(A)}.The SEM image of the product of reaction between molybdenum bronze and uranyl nitrate, shown in Figure 1(B), shows the product to consist of long fibers, a complete change of morphology from that of the original molybdenum bronze. The formation of fibers could occur by formation of reactive sites on the original molybdenum bronze surface followed by outward anisotropic growth. Since the structure of the product iriginite consists of uranium and molybdenum oxide/hydroxide chains, such a growth pattern would not be surprising. In the case of uranium acetate, however, the structure is one of thin flakes and not fibers {Figure 1(E)}. It is well documented that in hydrothermal synthesis, the nature of the anion exerts a strong influence on the morphology of the generated ceramic products. In particular, strongly-coordinating and chelating ligands such as acetate usually generate different morphologies as compared to non-coordinating anions such as nitrate. The difference in the morphologies of the uranyl nitrate and acetate products observed in this investigation might then suggest a second mechanistic possibility in which the molybdenum bronze particles are completely dissolved in a dissolution/precipitation process that generates new particles with different morphologies. Complete morphological rearrangement is also seen in the case of thorium nitrate {Figure 1(C)}, in which the product consists of very long relatively-flat glassy particles with embedded smaller particles. The overall appearance is one of a partially melted solid. The product of lead nitrate reaction with the molybdenum bronze shows an almost identical SEM image {Figure 1(D)} as that of the original molybdenum bronze, except that the surface is not as smooth. The neodymium product {Figure 1(F)}consists of larger chunks similar to the starting molybdenum bronze's morphology along with small needle-like particles. This results suggests that intercalation of neodymium into molybdenum blue does occur leaving apparently intact particles behind. The presence of small needle-like phases may be due to delamination/ reaggregation of molybdate layers or might be caused by dissolution/reprecipitation reactions. Whatever the mechanism, both morphologies of the product are isomorphous to X-rays.

Cyclic Process for Uranium Uptake

The reaction of an excess of an excess of molybdenum hydrogen bronze (2.0 g) with 200 ml of a 0.1 M solution of uranyl nitrate was found to follow pseudo-first order kinetics with respect to uranium concentration. The observed rate constant was found to be 0.0107 hr^{-1} for this particular set of conditions.

The uranyl molybdate product (irignite) obtained on the reaction of molybdenum bronze with uranyl nitrate was treated with a 15% solution of ammonium hydroxide. The reaction was stirred overnight and the reaction mixture was separated by filtration. The X-ray powder diffraction pattern of the residue corresponded to ammonium uranate {$(NH_4)_2U_3(OH)_2O_9.2H_2O$}, which has applications in the nuclear power industry. The ammonium uranate can be further converted to UO_3 upon heating to 600˚C. Evaporation of the filtrate produced ammonium molybdate , {$(NH_4)_2(Mo_2O_7)$}, that was identified by XRD. Molybdenum trioxide (MoO_3) could be recovered on heating the ammonium molybdate product to 242°C as determined by thermal gravimetric analysis.

The results of the uranium recovery experiment suggest a complete process cycle in which uranium is selectively absorbed by the molybdenum bronze forming a uranyl molybdate (iriginite) phase. Uranium, in the form of ammonium uranate, can be recovered from this molybate phase by treatment with ammonium hydroxide. This ammonium uranate can be collected and sold to the nuclear power industry or can be heated to 600°C to produce UO_3. MoO_3 can be recovered by evaporation followed by heating of the filtrate. Hence a complete cycle can be developed in which the only reagents used are ammonium hydroxide and butanol along with heat; by- products produced are butanal, ammonia and water and the main reaction product is ammonium uranate. Potentially the ammonia could be recovered and reused and the butanol could be used industrially.

CONCLUSIONS

In conclusion, it has been demonstrated that molybdenum blue has an extremely high capacity for absorption of contaminant metals. Considerable information has been collected concerning the mechanism of metal absorption and the results obtained so far suggest that the formation of normal molybdate salts of the metal predominate. However, in the case of neodymium, it is believed that intercalation of the metal ions between the layers of HMo_2O_6 followed by reaction to yield solids in which the metal ions are trapped as counterions to the freshly-generated molybdate sites, precedes the formation of normal molybdate salts. These reactions are highly selective for heavy metals or metals that are chemically-soft or that have a large radii from water, and suggest considerable promise for application in environmental remediation and as reactive barriers for the prevention of the spread of contaminant plumes.

REFERENCES

[1] B. A. Choudhary and R. K. Chandra, *Prog. Food Nutr. Sci.*, **11**, 55 (1987).

[2] G. S. Shukla and R. L. Singhal, *Can. J. Physiol. Pharmacol.*, **62**,1015 (1984).

[3] C. Papelis; K. F. Hayes and J. O. Leckie, *HYDRAQL: a program for the computation of chemical equilibrium composition of aqueous batch systems including surface-complexation modeling of ion adsorption at the oxide/solution interface*, Stanford University, Stanford, CA, **1988**

[4] United States Environmental Protection Agency, Office of Water. http://www.epa.gov/safewater

[5] C. W. Fetter, *Contaminant Hydrogeology*, Chapter 9. Macmillan Publishing Company, New York, 1993.

[6] D. W. Blowes; C. J. Ptacek; S. G. Benner; C. W. T. McRae; T. A. Bennett and R. W. Puls, *J. Contam. Hydrol.*, **45**, 123 (2000).

[7] S. G. Benner; D. W. Blowes; W. D. Gould, R. B. Herbert and C. J. Ptacek, *Environ. Sci. Technol.*, **33**, 2793 (1999).

[8] D. W. Blowes, C. J. Ptacek and J. L. Jambor, *Environ. Sci. Technol.*, **31**, 3348 (1997).

[9] A. I. Bortun, S. A. Khainakov; V. V. Strelko; and I. A. Farbun, *Ion Exchange Developments and Applications*: Special Publication- Royal Society of Chemistry, **182**, 305, 1996.

[10] F. Sebesta; J. John; A. Motl; and E. W. Hooper, *Ion Exchange Developments and Applications*: Special Publication- Royal Society of Chemistry, **182**, 305, 1996.

[11] M. M. Figueria; B. Volesky; V. S. T. Ciminelli; and F. A. Roddick, *Water Res.*, **34**, 196 (2000).

[12] D. Kratochvil; and B. Volesky, *Water Res.*, **32**, 2760 (1998).

[13] L. M. Yates; and R. V. Wandruszka, *Environ. Sci. Technol.*, 33(12), 2076 (1999).

[14] P. A. Brown; S. A. Gill; and S. J. Allen, *Water Res.*, **34**, 3907 (2000).

[15] A. Gierak, *Adsorpt. Sci. Technol.*, **4**, 47 (1996).

[16] A. I. Bortun; L. N. Burton; and A. Clearfield, *Solvent Extr. Ion Exch.*, **15**, 909 (1997).

[17] Z. Li; H. K. Jones; R. Bowman; and R. Helferich, *Environ. Sci. Technol.*, **33**, 4326 (1999).

[18] "Powder Diffraction File (PDF-2)" (International Centre for Diffraction Data, Newtown Square, PA).

[19]. A. I. Vogel, G. H. Jeffery, J. Bassett, J. Mendham, and R. C. Denney, *Vogel's Textbook of Quantitative Analysis*, Longman Scientific and Technical: Burnt Mill, Harlow Essex, UK, pp. 458-459, 1989.

[20] F. D. Snell, C. T. Snell, and C.A. Snell, "Thorium by Carminic Acid" in *Colorimetric Methods of Analysis*, Vol. IIA, D. Van Nostrand Co.: Princeton, N.J., pp. 518-519, 1959.

KEYWORD AND AUTHOR INDEX

Frit 320, 111

Gan, H., 279
gas emissions, 167
Gee, J.T., 217
geologic repository, 227
Giordana, A., 351
glass composition, 289, 297
glass compositions, 261
glass tank furnaces, 159, 249
Goff, K.M., 361
Goles, R., 239
Gombert, D., 239
Guerrero, H.N., 121, 131
Guilbeau, K., 351

Hanchar, J.M., 31
Hand, R.J., 101
hazardous wastes, 227, 297
Headrick, W.L., 167
heat distribution control, 143
heat transfer, 69
heavy metals, 371
Heckendorn, F.M., 217
Herman, C.C., 79, 239
high-level nuclear waste, 79, 101,
179, 197, 239, 279
Hollenberg, G., 41
Houston, H.M., 179
Hrma, P., 69, 93, 261
Hyatt, N.C., 101
hyperfine interaction, 31

immobilization, 227, 239, 309
Imrich, K.J., 217
in-container vitrification, 261
iron, 11, 289
iron phosphate glass, 309
Iverson, D.C., 217

Jaeger, H., 31
Jantzen, C.M., 79, 319
Jenkins, C.F., 217
Johnson, S.G., 361
Jordan, J., 351
Jue, J.F., 361

Kauffman, B., 351
Kelly, S., 227
Keyvan, S., 167
Kim, C.W., 309
Kim, D.-S., 69, 261, 297, 309
Kiran, B.P., 371
Koopman, D.C., 79
Kot, W.K., 279
Kurosky, R., 41

Ladirat, C., 197
land disposal restrictions, 227
leach resistance, 341
leaching tests, 11, 249, 331
Lee, W.E., 101
Lessor, D.L., 143
Leturcq, G., 11
Li, H., 21
lithium aluminate, 41
Lofaj, F., 289
Lorier, T.H., 269
low-activity waste, 93, 227, 249,
261, 309, 331
lumped parameter analysis, 121

magnesium, 289
Maise, G., 351
Manowitz, B., 351
Matyáš, J., 69, 261
McCarthy, M., 351
McGrier, P.S., 269
Meaker, T.F., 351
melt rate, 111, 121, 131, 269
melter materials, 217
melter operation, 57

melter removal, 217
melter shutdown, 179
microstructure, 331, 361
microwave system, 159
Miller, D.H., 111
Miller, F.S., 351
mineral phases, 319
mini-melter tests, 111
Misercola, A.J., 179
molybdenum hydrogen bronze, 371
Mooers, C.T.F., 289
Moore, R.E., 159, 167
Morgan, S., 101
Mougnard, P., 197
Muller, I.S., 149, 289
Musick, C.A., 227

Naseri-Neshat, H., 131
Neidt, T., 309

O'Holleran, T.P., 361
oxidation, 79

Palmer, R.A., 179
Peeler, D.K., 239, 309
Pegg, I.L., 149, 279
Pickett, J.B., 79
pilot scale glass furnace, 167
Pletzke, K., 31
poly(acrylic acid), 149
polymeric materials, 331, 341
porosity, 341
powder synthesis, 41
power requirements, 121, 239
Powell, J., 351
Prod'homme, A., 197
pyro-process, 361

Quach, A., 331, 341

Rabiller, H., 11
radioactive wastes, 331
radionuclides, 319, 371
radionuclides separation, 11

Ramsey, A.A., 351
Ramsey, W.G., 351
REDOX model, 79
reflectance spectroscopy, 3
refractory walls, 159, 217
regulations, 227
Reich, M., 351
remote visual inspection, 217
rheological properties, 149
rhodium oxide, 279
Richardson, J., 239
Ricklefs, J.S., 93
Romero, C., 167
Rose, P.B., 101
Rossow, R.A., 167

Sáez, A.E., 331, 341
Salzano, F., 351
Sanders, T., 351
Scales, C.R., 101
scanning electron microscopy, 11,
69, 309, 341, 351
Schumacher, R.F., 249
Schweiger, M.J., 261
short range order, 31
silicon controlled rectifiers, 143
simulants, 261
Sinkler, W., 361
sludge batch 2, 111, 269
Smith, D.E., 261
Smith, G.L., 331, 341
Smith, H.D., 331, 341
Smith, J.D., 351
Smith, M.E., 111
sodalite based ceramic waste
form, 361
sodium nitrate, 341
solubility, 3, 21
spinel, 279
spray drying process, 41
Stefanovsky, S., 239
Steinberg, M., 351

sulfate, 249
sulfate segregation, 93
sulfur, 309
surface modification, 149
surfactants, 149
Swanberg, D.J., 227

temperature, 279, 351
thermal analysis, 69
thermal processing optimization, 351
Thomson, S.J., 3
titanate ceramics, 11, 21
Tonn, D., 41
toxic metal wastes, 331
toxicity characteristic leaching procedure, 297, 351

underground tanks, 227
uranium, 3

Vance, E.R., 3, 21
vapor hydration test, 289
Varghese, B., 159
Velez, M., 159, 167

Ventre, L., 351
Vienna, J.D., 93, 239, 261, 297, 309
vitrification, 69, 79, 149, 179, 197, 249, 351

waste glass melters modeling, 121, 131
waste glass melting, 57, 93
waste-loading tests, 269

Xia, G., 331, 341
X-ray diffraction, 11, 69, 351

Zamecnik, J.R., 79
Zelinski, B.J., 331, 341
Zhang, Z., 3
Zhao, H., 149
Zhu, D., 309
zircon, 31
Zoughi, R., 159